Atlas of Metal-Ligand Equilibria in Aqueous Solution

J. KRAGTEN

Chemical Department,
Natuurkundig Laboratorium,
Universiteit van Amsterdam

Translation Editor: Dr. Mary Masson, University of Aberdeen

ELLIS HORWOOD LIMITED
Publisher · Chichester

Halsted Press: a division of
JOHN WILEY & SONS
New York · London · Sydney · Toronto

The publisher's colophon is reproduced from James Gillison's drawing of the ancient Market Cross, Chichester

First published in 1978 by

ELLIS HORWOOD LIMITED
Coll House, Westergate, Chichester, Sussex, England
this being the 50th book published by Ellis Horwood in his 50th year of publishing

Distributors:

Australia, New Zealand, South-east Asia:
JOHN WILEY & SONS AUSTRALASIA PTY LIMITED
1-7 Waterloo Road, North Ryde, N.S.W., Australia

Canada:
JOHN WILEY & SONS CANADA LIMITED
22 Worcester Road, Rexdale, Ontario, Canada

Europe, Africa:
JOHN WILEY & SONS LIMITED
Baffins Lane, Chichester, Sussex, England

North and South America and the rest of the world:
HALSTED PRESS, a division of
JOHN WILEY & SONS INC.
605 Third Avenue, New York, N.Y. 10016, U.S.A.

© 1978 J. Kragten/Ellis Horwood Ltd., Publishers

Library of Congress Cataloging in Publication Data

Kragten, J., Atlas of metal-ligand equilibria in aqueous solution.
1. Ligands — Charts, diagrams, etc.
2. Coordination compounds — Charts, diagrams, etc.
3. Chemical equilibrium — Charts, diagrams, etc.
I. Title.
QD474.K7   541'.2242   77-12168
ISBN 85312-084 6 (Ellis Horwood Ltd., Publishers)
ISBN 0-470-99309-X (Halsted)

Set in 11pt Press Roman by Coll House Press
Printed Offset Litho in Great Britain by Biddles of Guildford, Surrey.

All rights reserved. No part of this publication may be reproduced, stored in a retrieval system or transmitted, in any form or by any means, electronic, mechanical photocopying recording or otherwise, without prior permission.

QB
474
.K7
1978

To
Coby
Edith
Annelies

92801

Put it before them briefly so they will read it, clearly so they will appreciate it, picturesquely so they will remember it and, above all, accurately so they will be guided by its light.

JOSEPH PULITZER

Table of Contents

Foreword 11
Author's Preface 13

Chapter 1 15
  1.1 The Plots: Theory 16
    1.1.1 Hydroxides 16
      Hydroxide precipitation 19
      Polynuclear hydroxide complex formation 21
      Areas of predominance of mononuclear complexes 22
    1.1.2 Other ligands 22
      The side-reaction coefficient 22
      Hydroxide precipitation 24
      Polynuclear hydroxide formation 25
    1.1.3 Air-saturated solutions 25
      The influence of CO_2 25
      The influence of O_2 27
  1.2 The Plots: Some Practical Points 29
    1.2.1 Significance of the lines in the plots 29
    1.2.2 Superposability of the plots 29
    1.2.3 The amount of ligand in solution 30
    1.2.4 Data sources and selection criteria 31
    1.2.5 Constants and experimental conditions 32
    1.2.6 Interpretation 33
  1.3 Some Practical Applications 34
    1.3.1 Conditional constants 34
    1.3.2 Manipulations of solutions 34
      Dilution 34
      Neutralization 36
  1.4 Arrangement of the plots 39

Table of Contents

Chapter 2
    Silver(I)...40
Chapter 3
    Aluminium(III)...52
Chapter 4
    Barium(II) – including the carbonate system.............70
Chapter 5
    Beryllium(II)..80
Chapter 6
    Bismuth(III)...94
Chapter 7
    Calcium(II) – including the carbonate system...........104
Chapter 8
    Cadmium(II) – including the carbonate system...........114
Chapter 9
    Cerium(III)...160
Chapter 10
    Cobalt(II) – including the carbonate system...........172
Chapter 11
    Chromium(III)...214
Chapter 12
    Copper(II)..222
Chapter 13
    Dysprosium(III)...246
Chapter 14
    Erbium(III)...258
Chapter 15
    Europium(III)...272
Chapter 16
    Iron(II) – including the carbonate system..............284
Chapter 17
    Iron(III)...322
Chapter 18
    Gallium(III)..344
Chapter 19
    Gadolinium(III)...358
Chapter 20
    Hafnium(IV)...372
Chapter 21
    Mercury(II)...382
Chapter 22
    Indium(III)...400
Chapter 23
    Lanthanum(III) – including the carbonate system........416

Table of Contents

Chapter 24
    Magnesium(II)..436
Chapter 25
    Manganese(II) – including the carbonate system.........................452
Chapter 26
    Neodymium(III)..496
Chapter 27
    Nickel(II)..508
Chapter 28
    Lead(II)..534
Chapter 29
    Palladium(II)...576
Chapter 30
    Scandium(III)...586
Chapter 31
    Samarium(III)...600
Chapter 32
    Tin(II)...612
Chapter 33
    Tin(IV)...622
Chapter 34
    Strontium(II) – including the carbonate system........................626
Chapter 35
    Terbium(III)..636
Chapter 36
    Thorium(IV)...648
Chapter 37
    Titanyl,Ti(IV)..660
Chapter 38
    Thallium(III)...670
Chapter 39
    Uranium(IV)...682
Chapter 40
    Uranyl,U(VI)..688
Chapter 41
    Vanadium(III)...700
Chapter 42
    Vanadyl,V(IV)...704
Chapter 43
    Yttrium(III)..720
Chapter 44
    Ytterbium(III)..730
Chapter 45
    Zinc(II)..742

Chapter 46
 Zirconium(IV)..766
Appendix
 Stability Constants for Proton–Ligand Complexes 799
 Tabular Index...780

Foreword

Present-day theoretical treatments of metal–ligand equilibria owe much to Ringbom's *Complexation in Analytical Chemistry*. Since this book was published in 1963, many authors have found it useful to follow Professor Ringbom in his use of 'conditional constants' and 'side-reaction coefficients'. However, until now, in order to work with conditional constants it has been necessary to consult tables or carry out tedious calculations. I believe, therefore, that Dr. Kragten has done a great service for analytical chemists in producing this collection of plots of side-reaction coefficients for 45 metals in combination with 29 ligands, which give the necessary information at a glance. The accompanying plots of pM' against pH, which illustrate directly the areas of predominance of the various species and indicate the conditions under which hydroxides will precipitate and polynuclear hydroxo-complexes will form, also provide valuable information for the analyst. It is a great pleasure to be associated with the publication of this most useful addition to the analytical literature.

M. R. MASSON
University of Aberdeen

2 August 1977

Author's Preface

There is increasing interest at the present time in making critical compilations of the equilibrium constants for metal–ligand complexes. Without such a compilation, the chemist is faced with a mass of confusing and possibly imprecise data. Original papers often contain insufficient information about the experimental set-up, the temperature, the ionic strength, the age and exact nature of any solid phase, etc., and the reader may not be able to criticize the data adequately. Thus, critical compilations satisfy an important need in analytical and coordination chemistry.

Nevertheless, only limited use is made of these critical compilations. The complexity of the equilibrium systems means that it is not easy to obtain — directly from the constants — a clear insight into the behaviour of the metal ions under the various experimental conditions possible. Usually, an intermediate graphical representation is used as an aid to understanding, but the extensive calculations and plotting of graphs necessary for each metal–ligand combination are tedious and time-consuming. Moreover, if someone does perform the necessary calculations, he will restrict himself to his own particular problem, and this in turn means that the few results available in the literature are not generally useful. Thus we can see that a gap exists between the data in the literature and daily laboratory practice, and that it should be filled in such a way that the information is readily available in an easily accessible and generally applicable form. In compiling this Atlas, my aim has been to fill this gap and to help the chemist working in industrial and research laboratories to solve everyday problems.

Although the greatest care has been taken in selecting the constants to be used, it was impossible to check each one for reliability. Specialists in a particular field may well be able to detect errors in the plots relating to their own speciality, and I hope that anyone who does detect such an error will not hesitate to contact me, so that I can incorporate the necessary correction in any future edition.

I am grateful to the copyright-holders, John Wiley and Sons, and to the authors, Dr. C. F. Baes Jr. and Dr. R. E. Mesmer, for permission to make use of data in *The Hydrolysis of Cations*.

Author's Preface

I am greatly indebted to Mr Steven Arlman for his assistance in preparing the computer program, his stimulating input of ideas, his painstaking plot administration and his assistance in plot-scanning. Without him, the book could not have appeared for some years. I am grateful to the staff of the Physical Laboratory for giving me the opportunity to prepare this Atlas by allowing Mr Arlman and myself to have sufficient free time, and by supporting the work with a considerable proportion of the computer budget.

I thank Mr Joop van Loenen and the production staff of SARA (the Academic Computer Centre, Amsterdam) for their pleasant cooperation during the production of the computer plots, and for their patience in ensuring that the plots were of the high quality needed.

I wish to express my gratitidue to Mr A. Ph. Reynaert, whose practical experience, achieved by meticulous observation, was the inspiration for the Atlas, to Prof. Gerrit den Boef and Prof. Adam Hulanicki for their interest and encouragement, and to Mrs Mariet Mölders and Mrs Tineke Koster for their assistance in typing.

I am grateful to the Series Editors, Dr. R. A. Chalmers and Dr. Mary Masson, for their help in preparation of the manuscript, and my final thanks go to Felicity, Ellis and Clive Horwood for their enthusiastic cooperation, and for being such a friendly publishing family.

14 August 1977

J. KRAGTEN
Natuurkundig Laboratorium
Universiteit van Amsterdam

CHAPTER 1

Introduction

It is unthinkable that anyone should attempt to describe the properties of metal ions in aqueous solution without considering hydrolysis reactions. Most cations hydrolyse in aqueous solution, because they form strong bonds with oxygen atoms, and because, as a result of the self-ionization of water, hydroxide ions are always present at concentrations which can vary over the unusually wide range of $10-10^{-14}$ M. With some metals the bonds with oxygen are so strong that the species $M(OH)_2^{n+}$ do not dissociate further even in concentrated strong acids. Such species are sometimes written as MO^{n+}, particularly when their exact composition is not known. In the calculations, the species MO^{n+} was treated as if it were a metal ion of valency n.

When other ligands are present in solution, there will be competition between the ligands, L, and the products of hydrolysis. A general formula for the final product is $M_a O_b (OH)_c (H_2O)_d L_e H_f$. However, without forgetting the correct formula for the product, it is possible to express the reactions in simpler terms if hydroxide ions and hydrogen ions are considered to be the only ligands competing with L. (This is possible because O^{2-} and H_2O can be expressed in terms of OH^- and H^+.) In all cases of interest, this representation is adequate for our purposes.

We have studied the reactions of 45 metals with 29 ligands, and about 600 of the combinations turned out to be of real interest. Different methods of graphical presentation are possible, but since most workers are not interested in detailed information about the exact distribution of metal ions between the various possible metal–ligand and metal–hydroxide species, but want to be able to find out such basic things as the pH at which a hydroxide precipitates, the conditions for formation of polynuclear complexes, and the conditional constant for reaction of the metal with EDTA or DCTA, I decided to present two types of plots.

The concept of the side-reaction coefficient, first introduced by Schwarzenbach, was extensively used by Ringbom in his excellent monograph *Complexation in Analytical Chemistry* [1]. In the first type of plot, the side-reaction coefficient (α) for reaction of the metal ion with hydroxide and ligand is plotted against pH for various ligand concentrations.

The second type of plot is of pM' *vs* pH. Such plots illustrate the influence of metal–ligand complex formation on the regions where polynuclear hydroxo-complexes form and where hydroxide precipitates. The material is presented in such a way that the graphs for the separate metal–ligand combinations are superposable; the graph for a system containing more than one ligand can be composed from the graphs for the individual metal–ligand combinations.

In preparation of this Atlas, certain simplifying assumptions had to be made, and one of these was that the metal–ligand species are not polynuclear. In reality, binuclear and polynuclear species are formed in concentrated solutions with some metal–ligand combinations. If this is known to occur in solutions with pM' < 2, a warning, d, is given in the tabular index to indicate that the pM'–pH diagram may not be completely reliable when pM' < 2. (See, for example, DTPA and citrate.) If such species are formed in more dilute solutions, no plots are given, or the ligand is not considered at all (e.g. dimercaptopropanol, BAL).

Some metal–ligand complexes have limited solubility. If the metal–ligand complex precipitates only in concentrated solutions, a warning, p, is given in the tabular index. If precipitation occurs in more dilute solutions, no plots are given (e.g. oxalates, fluorides) or the ligand is not considered at all (phosphate).

Another possibility that had to be neglected was the formation of mixed-ligand complexes.

The greatest care was taken in selecting values for the constants from the literature, and the computer output (plots and print-out) was double-checked against the original literature. Nevertheless, it must be realized that there is some uncertainty in the positions of the lines on the plots, and that the accuracy of their positions depends on the accuracy of the data used. Only data which seemed obviously unreliable have been rejected; the relevant combinations are marked U in the tabular index. Where there was some doubt, the data were still used to produce plots, in view of the likelihood of unsuspected errors in the rest of the data. It was impracticable to check the accuracy of the constants (there were more than two thousand) for all the metal–ligand combinations. However, in many cases, large errors in some of the constants lead to only relatively minor inaccuracies in the positions of the lines.

1.1 THE PLOTS: THEORY

1.1.1 Hydroxides

Hydroxide equilibria form the basis of all the plots, and in this section the formation of the various hydroxide species will be discussed. The influence of other ligands will be considered later.

The reaction scheme for hydroxide formation can be set out as below; for argument's sake the metal ion is supposed to be tervalent, M^{3+}.

$$M^{3+} \rightleftharpoons \underbrace{MOH^{2+} \rightleftharpoons M(OH)_2^+} \rightleftharpoons M(OH)_3^0 \rightleftharpoons M(OH)_4^- \rightleftharpoons M(OH)_5^{2-} \cdots$$

$$\Updownarrow \qquad\qquad\qquad \Updownarrow$$

$$M_2(OH)_2^{4+} \qquad\qquad \{M(OH)_3 \cdot nH_2O\}_{solid} \qquad\qquad (1)$$

$$\qquad\qquad\qquad\qquad\qquad \text{or}$$

$$\Updownarrow \qquad\qquad\qquad \{M_2O_3 \cdot mH_2O\}_{solid}$$

$$M_p(OH)_q^{(3p-q)+}$$

The system has two independent variables; the concentrations of the various species depend on both the total concentration of M and the pH. If an additional constraint is set, only one degree of freedom remains, a relationship exists between the metal-ion concentration and pH, and a curve reflecting the particular constraint can be drawn on the pM′–pH plot.

The borderline of the precipitation region is found by determining the concentration at which $M(OH)_3^0$ just precipitates at various pH-values. By setting $[M(OH)_i] = [M(OH)_{i+1}]$ for different values of i, we get the borderlines of the regions of predominance of the various mononuclear species. We can derive the borderline of the region of predominance of polynuclear complexes by setting $x = 1$ in the expression $p \cdot [M_p(OH)_q] = x \cdot \sum_i [M(OH)_i]$. At this borderline, 50% of the metal will be present as polycomplexes, so it is not very useful to analytical chemists, who are more concerned to know where polycomplexation begins. A useful approximation to this is the borderline where only 1% of the metal is present as polycomplex. It is this 1% borderline that appears in the plots; it is found by setting $x = 0.01$.

According to Ringbom [1], a system of co-existing equilibria can be divided into a main reaction and the competitive side-reactions, covering all other reactions. The parameter [M′], the total concentration of the metal ion not involved in the main reaction, is much more useful than the the concentration [M^{3+}] of free metal ion, to which it is related by Ringbom's side-reaction coefficient $\alpha_{M(X)} = [M']/[M^{3+}]$. If OH$^-$ is the only side-reaction component, we have

$$\alpha_{M(OH)} = ([M] + [MOH] + [M(OH)_2] + \cdots + 2[M_2(OH)_2] + \cdots)/[M] \qquad (2)$$

(charges will be omitted for convenience and greater generality).

For convenience, we will restrict ourselves for the moment to the case where polycomplexation does not occur to a measurable extent, which always happens when [M′] is small enough.

For the reaction

$$M + iH_2O \rightleftharpoons M(OH)_i + iH \qquad (3)$$

the overall stability constant $*\beta_i$ is introduced in accordance with the IUPAC notation:

$$*\beta_i = \frac{[M(OH)_i][H]^i}{[M]} \qquad (4)$$

Substitution of this equation in Eq. (2) and omission of the polycomplex terms gives:

$$\alpha_{M(OH)} = 1 + 10^{(pH + \log *\beta_1)} + 10^{(2pH + \log *\beta_2)} + \cdots 10^{(m\,pH + \log *\beta_m)} \quad (5)$$

The side-reaction coefficient in (5), which will be called α_{mono}, is a function of only the pH, it implies that for a given value of pH, $[M]^{3+}$ can be determined if $[M]$ is known. In principle the concentrations of all species in reaction scheme (1) are known if there is no polycomplexation or precipitation.

The pH-range can be divided into distinct regions in each of which one of the hydroxide complexes predominates. Because of the form of equation (5), in each of the regions the corresponding term in (5) predominates, and accordingly $\alpha_{M(OH)}$ can be approximated by

$$\log \alpha_{M(OH)} \cong (i\,pH + \log *\beta_i) \quad (6)$$

$$(i = 0, 1, 2, \cdots m; \log *\beta_0 = 0)$$

Because of the linear relationship between $\log \alpha_{M(OH)}$ and pH in equation (6) a graph of $\log \alpha$ vs pH is a series of straight lines (see Fig. 1.1), the curve for $\log \alpha_{M(OH)}$ vs pH being the rounded-off line connecting the line-segments corresponding to the highest values of $\log \alpha$. Note that the difference between the exact curve and the straight lines is generally smaller than $\log 2$, and cannot be larger than $\log m$; in practice it is thus negligible in comparison with the usual uncertainties in $\log *\beta_i$. Equation (6) can be regarded as an acceptable approximation.

When one or more polynuclear complexes with the general formula $M_p(OH)_q$ are formed in solution, α_{mono} [Eq. (5)] has to be supplemented with the corresponding polycomplex terms; the system of straight lines outlined above can be extended with lines corresponding to these polycomplexes.

The stability constant of the complex can be defined as

$$*\beta_{qp} = \frac{[M_p(OH)_q][H]^q}{[M]^p} \quad (7)$$

(note the reversal of p and q in accordance with IUPAC notation). Introduc-

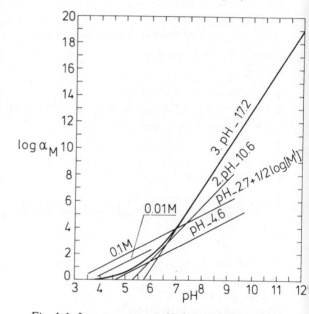

Fig. 1.1 Log $\alpha_{M(OH)}$ vs pH for scandium, calculated with $\log *\beta_1 = -4.6$, $\log *\beta_2 = -10.6$, $\log *\beta_3 = -17.2$, $\log *\beta_{22} = -5.8$, $\log *\beta_{43} = -14.0$ and $\log *\beta_{53} = -17.5$. The exact curve (without polycomplexation) and the straight-line approximations are drawn, as well as the straight lines for the polycomplex $Sc_2(OH)_2$ at total scandium concentrations of 0.01 and 0.1M. (From [2], by permission of the copyright holders, Pergamon Press).

tion of (7) into the general formula (2) for the side-reaction coefficient leads to the addition of the term

$$\sum \frac{p[M_p(OH)_q]}{[M]} = \sum p \frac{*\beta_{qp}[M]^{(p-1)}}{[H]^q} = \alpha_{poly} \quad (8)$$

to Eq. (5) and so to the total side-reaction coefficient:

$$\alpha_{M(OH)} = \alpha_{mono} + \alpha_{poly} \quad (9)$$

There are two possibilities. The first is that the mononuclear complexes predominate, in which case α_{poly} can be neglected and Eq. (6) can be used for α_{mono} as before. The second is that α_{mono} can be neglected in (9). If, as often happens, one of the polycomplexes — let us say p,q — predominates, replacement of [M] in Eq. (8) by $[M']/\alpha_{M(OH)}$ then gives

$$\alpha_{M(OH)} = \alpha_{poly} = p*\beta_{qp}[M']^{(p-1)}/[H]^q \alpha_{M(OH)}^{(p-1)} \quad (10)$$

Rewritten in logarithmic form this leads to

$$\log \alpha_{M(OH)} = \left(\frac{q}{p}\right)pH + \left\{\frac{\log *\beta_{qp} + \log p}{p}\right\} + \left(\frac{p-1}{p}\right)\log[M'] \quad (11)$$

This equation leads to a set of parallel straight lines in the plot of $\log \alpha$ vs pH, the positions of which depend on the total concentration of metal involved in the side-reactions ([M']). When more than one polynuclear complex is formed, every such complex gives rise to a set of straight lines corresponding to Eq. (11). It should be noted that Eq. (11) also holds for mononuclear complexes ($p = 1$) and thus can be regarded as the general formula for the straight lines in the log $\alpha_{M(OH)}$ vs pH plot. In Fig. 1.1, log α_{poly} is drawn for different scandium concentrations. The binuclear hydroxide $Sc_2(OH)_2^{4+}$ is formed predominantly only at scandium concentrations above 10^{-3}M.

Equation (11) will now be used to derive the borderlines for precipitation and 1% polycomplexation.

Hydroxide precipitation

For a metal ion M^{n+} a precipitate of general composition $M(OH)_n$ is formed when its limiting concentration is exceeded. This concentration may be expressed as

$$[M(OH)_n]_{max} = *K_{sn} \quad (12)$$

which can be combined with Eq. (4) to give the well-known formula for the solubility product:

$$\frac{[M^{n+}]_{max}}{[H]^n} = *K_{sn}/*\beta_n = *K_{s0} \quad (13)$$

[M^{n+}] can be eliminated from this equation by substituting [M']/$\alpha_{M(X)}$ for it. This leads to the general formula for the borderline of the precipitation region; and the same formula holds when a ligand is present.

$$p[M']_{max} = -\log {^*K_{s0}} + n\,pH - \log \alpha_{M(X)} \qquad (14)$$

Assuming that hydroxide formation is the only side-reaction taking place in solution, we can substitute Eq. (5) in Eq. (14). Knowing that $\alpha_{M(OH)}$ can be split into separate terms for distinct pH-regions we can do the same here and substitute Eq. (6) in Eq. (14).

However, if a polycomplex predominates, Eq. (11) should be substituted. This equation also represents the mononuclear complexation and so Eq. (15) can be regarded as the general formula.

$$p[M']_{max} = (np-q)pH - (p\log {^*K_{s0}} + \log {^*\beta_{qp}} + \log p) \qquad (15)$$

Again, we get a linear relationship, but now between $p[M']_{max}$ and pH, so a number of straight lines occur in the pM'–pH plot. The curve for $p[M']_{max}$ is the rounded-off line connecting the line-segments corresponding to the *lowest* pM'-values (note the negative sign of $\log \alpha$ in (12)]. A better approximation to the curve is possible; although still not exact, it leads to values deviating by less than 0.03 log units from the correct ones in all cases of practical interest.

If we look at the right-hand side of Eq. (6), we can see that each term is derived from a corresponding one in Eq. (5). Conversely, $\alpha_{M(OH)}$ can be found by summing powers of ten, as in Eq. (5), but with the exponents equal to the terms on the right-hand side of Eq. (6). This can also be done for the general case (polycomplexes present) in which case $\alpha_{M(OH)}$ is found by summing powers of ten, with the exponents equal to the terms on the right-hand side of Eq. (11).

The same principle can be applied to Eq. (15), to give

$$[M']_{max} = \sum_p \sum_q 10^{-(np-q)pH + (p\log {^*K_{s0}} + \log {^*\beta_{qp}} + \log p)} \qquad (16)$$

This equation was used to calculate the borderlines of the regions of hydroxide formation

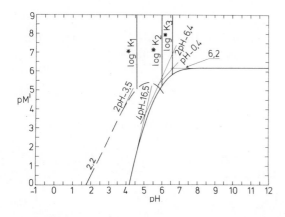

Fig. 1.2 The pM'–pH plot for scandium, based on log $^*K_{s0}$ = 11.0 and the log $^*\beta$ values given for Fig. 1.1. The precipitation borderline (———) and 1% borderline for the polycomplex $Sc_2(OH)_2$ (– – –) are given together with the straight-line approximations. The log $^*K_i$ lines are also drawn, indicating the predominance regions for mono-complexes $Sc(OH)_i$ (i = 0,1,2,3). (After [2], by permission of the copyright holders, Pergamon Press)

in all the pM'–pH diagrams. In Fig. 1.2 the exact precipitation borderline is drawn for Sc, together with the straight lines from Eq. (15). The borderline from Eq. (16) coincides completely with the exact line. If we take into account the uncertainty in the data used for the calculations, it is obvious that approximation by rounding-off the straight lines is adequate for many calculations done 'by hand', and that the results from Eq. (16) are suitable for all cases of practical interest.

Polynuclear hydroxide-complex formation

The concentration borderline for polycomplexation differs in principle from that for precipitation. The second is theoretically a sharp boundary, but the first is not, so we have to stipulate a concentration restriction.

The analytical chemist — along with other sorts of chemist — usually considers that polycomplexes are absent when the amount of metal involved in polycomplexation is less than 1% of the total. This limit is arbitrary, but no more so than the 50% limit for predominance. In analysis the idea is that, at the most, 1% of M will not be determined in quantitative analysis. In other fields a different choice may be made for other reasons, but since many chemists are interested in the 1% limit, this has been adopted as the additional constraint for calculation. It can be noted that both the 0.1% and the 50% limits differ by only a few tenths of a log unit from the 1% limit.

The 1% restriction can be expressed mathematically as

$$\frac{p[M_p(OH)_q]}{[M]} < \frac{10^{-2}[M']}{[M]} = 10^{-2} \alpha_{mono} \qquad (17)$$

Elimination of $M_p(OH)_q$ from (7) and (17), and replacement of [M] by $[M']/\alpha_{mono}$ leads to the general formula for the 1% borderline.

$$p[M']_{1\%} = \left\{ \frac{q\mathrm{pH} - p \log \alpha_{mono}}{(p-1)} \right\} + \left\{ \frac{\log {}^*\beta_{qp} + 2 + \log p}{(p-1)} \right\} \qquad (18)$$

After substitution of α_{mono} from Eq. (5) the exact curve $p[M']_{1\%}$ vs pH can be calculated in the pM'–pH plot. These two formulae were used for the calculation of the dashed 1% borderlines in the plots. Substitution of Eq. (6) again gives the straight line approximation of the exact curve. In Fig. 1.2 the line is mainly composed of two straight parts: $pM' = 2pH - 3.5$ and $pM' = -2pH + 17.7$.

When more than one complex has been reported in the literature, all the constants have been used in the calculations. For some metals the 1% borderline turned out to consist of various line-segments corresponding to the different polycomplexes. The composition of each of the complexes is written near the middle of the corresponding line-segment.

Equation (18) can also be used when ligands other than OH^- are present; α_{mono} has to be extended with the corresponding terms for these side-reactions. This situation will be discussed more fully later.

Areas of predominance of mononuclear complexes

The borderlines between the regions of predominance of the various complexes are found by equating successive terms in Eq. (5). Thus, the positions of the (vertical) borderlines are given by

$$\text{pH} = \text{p}^*K_i = \text{p}^*\beta_i - \text{p}^*\beta_{i-1} \qquad (19)$$

In a few cases, a value of log $^*K_i$ is smaller than the value of log $^*K_{i-1}$, and the log $^*K_i$ line therefore appears at lower pH than the log $^*K_{i-1}$ line. In such cases, the species $M(OH)_i$ is a minor one which never predominates, and the area of predominance of $M(OH)_{i-1}$ is bordered by the region of predominance of $M(OH)_{i+1}$. The true borderline in such a case is given by

$$\text{pH} = \tfrac{1}{2}(\text{p}^*\beta_i - \text{p}^*\beta_{i-2}) = \tfrac{1}{2}(\text{p}K_i + \text{p}K_{i-1}) \qquad (20)$$

The true border line has not been drawn in the few cases which occur, first, because the difference from the 'out-of-order' log $^*K_i$ lines is usually small, and, secondly, because we thought that the individual log $^*K_i$ lines would give more information than the 'mean line' given by Eq. (20). The 'out-of-order' sequence of the lines on the pM'–pH plot should give sufficient warning to the user.

1.1.2 Other Ligands

For any metal–ligand combination, the complexes formed have a wide range of possible compositions, even when the most general formula mentioned earlier is simplified to a combination of M, H, OH and L, and we restrict ourselves to soluble mononuclear metal–ligand complexes.

Fortunately most chemists are not interested in detailed information about the individual complexes, but want to have a quantity which can easily be used as a measure of the tendency for complex-formation. The side-reaction coefficient is very suitable for this purpose and this quantitity is presented in the upper plot of each page. The influence of the metal–ligand complexation on the solubility of the metal hydroxide and on hydroxide polycomplexation is given in the lower plot.

The side-reaction coefficient

This coefficient has already been defined as $\alpha_{M(X)} = [M']/[M^{n+}]$. In the previous section we restricted ourselves to hydroxides and this led to the definition of the side-reaction coefficient given in Eq. (2). When a ligand L is present with which M can form complexes such as ML, ML_2, ML_3, ... MHL, MHL_2, ... MH_2L, ... and MOHL, $M(OH)_2L$, this coefficient has to be extended with the terms corresponding to these species, because they are now included in the total concentration [M'], and so we get

$$\alpha_{M(L,OH)} = \alpha_{M(OH)} + \frac{[ML]}{[M]} + \frac{[ML_2]}{[M]} + \cdots\cdots + \frac{[MHL]}{[M]} + \frac{[MHL_2]}{[M]} + \cdots\cdots \qquad (21)$$

Combining each term with the corresponding formation constant gives an expression in which each term is a function of [L] and pH. However, L may be the anion of a weak acid, and this further complicates the equilibrium.

As the stability constant of the reaction $M + 2H_2L + HL \rightleftharpoons MH_5L_3$ is different from that of the reaction $M + H_4L + L + HL \rightleftharpoons MH_5L_3$ even though the product is the same, it will be obvious that — especially in connection with the data input of the computer — we need an unequivocal definition for the formation constant for such species as MH_5L_3. Suitable definitions are as follows.

$$\beta_{221} = \frac{[MH_5L_3]}{[M][H_2L][H_2L][HL]} \quad \text{and} \quad \beta_{410} = \frac{[MH_5L_3]}{[M][H_4L][HL][L]} \tag{22}$$

The number of subscript indices agrees with the number of concentrations $[H_iL]$ in the denominator and each index i itself corresponds to the number of protons (i) bound to L in $[H_iL]$. So in general

$$\frac{[MH_nL_m]}{[M][H_iL][H_jL][H_kL]\ldots} = \underbrace{\beta_{ijk\ldots}}_{m \text{ indices}} \tag{23}$$

Substitution of this equation in Eq. (21) for $\alpha_{M(X)}$ leads to terms of the general form

$$\beta_{ijk}[H_iL][H_jL][H_kL]\ldots \tag{24}$$

which depend on the pH and on [L']. For the stepwise dissociation of the acid H_rL yet another side-reaction coefficient can be introduced: $\alpha_{L(H)} = [L']/[L'^-]$ or $\alpha_{H_jL(H)} = [L']/[H_jL]$. In the case of L'^- we get

$$\alpha_{L(H)} = \frac{[L']}{[L'^-]} = \frac{[L] + [HL] + [H_2L] + \ldots\ldots [H_rL]}{[L]}$$

$$= 1 + 10^{(\log \beta_1 - \text{pH})} + 10^{(\log \beta_2 - 2\text{pH})} + \ldots\ldots + 10^{(\log \beta_r - r\text{pH})} \tag{25}$$

in which β_k refers to the reaction $L + kH \rightleftharpoons H_kL$. In the case of H_jL the side-reaction coefficient $\alpha_{H_jL(H)}$ can be determined from $\alpha_{L(H)}$ through

$$\frac{[H_jL]}{[L]} = \beta_j[H_j]^j = 10^{(\log \beta_j - j\text{pH})} \tag{26}$$

which leads to

$$\alpha_{H_jL(H)} = \alpha_{L(H)} \big/ 10^{(\log \beta_j - j\text{pH})} \tag{27}$$

Substitution of these side-reaction coefficients, which depend only on pH, in Eq. (23) gives for the terms in (21)

$$\frac{[MH_n L_m]}{[M]} = \frac{\beta_{ijk}\ldots}{\alpha_{H_iL(H)} \cdot \alpha_{H_jL(H)} \cdot \alpha_{H_kL(H)} \cdot \cdot} \cdot [L']^m = \beta'_{ijk}\ldots[L']^m \qquad (28)$$

in which $\beta_{ijk}\ldots$ can be regarded as a conditional constant which depends only on pH. Substitution in (21) finally gives:

$$\begin{aligned}
\alpha_{M(L,OH)} = \alpha_{M(OH)} &+ [L']\,(\beta'_0 + \beta'_1 + \beta'_2 + \ldots\ldots\ldots\ldots) \\
&+ [L']^2\,(\beta'_{00} + \beta'_{01} + \beta'_{11} + \ldots\ldots\ldots\ldots) \\
&+ [L']^3\,(\beta'_{000} + \beta'_{100} + \ldots\ldots\ldots\ldots\ldots) \\
&+ [L']^4\,(\beta'_{0000} + \ldots\ldots\ldots\ldots\ldots\ldots) \\
&+ \text{etc.} = \alpha_{M(OH)} + \alpha_{rest} \qquad (29)
\end{aligned}$$

Basic metal–ligand complexes are frequently formed in solution. The existence of basic complexes of ML_2, ML_3 etc. is not certain, and these complexes would contribute only marginally to the equilibrium, so their contribution to $\alpha_{M(L,OH)}$ has not been included in the computer calculations. Basic complexes of ML are considered, and for these

$$*\beta_{iOH} = \frac{[ML(OH)_i] \cdot [H]^i}{[ML]} \qquad (30)$$

is given, in addition to various $\beta_{ijk\ldots}$ values, alongside the log α plots in this Atlas. The terms $[ML(OH)_i]/[M]$ in Eq. (30), which have to be added to $\alpha_{(M(L,OH))}$, can be transformed into

$$\frac{[ML(OH)_i]}{[M]} = \beta_0 \cdot 10^{(\log *\beta_{iOH} + i\text{pH})} \cdot [L'] \qquad (31)$$

The values of $\alpha_{M(L,OH)}$ for the various metal–ligand combinations have been calculated from Eq. (29) and Eq. (31) has been added when necessary. For reasons given later, on p. 29, polynuclear hydroxide-complex terms have been omitted from Eq. 29); α_{mono} was substituted for $\alpha_{M(OH)}$.

Hydroxide precipitate formation

Transforming the general and mathematically exact formula for the borderline of precipitation – Eq. (14) – into

$$[M']_{max} = 10^{(\log *K_{s0}-n\text{pH})} \cdot \alpha_{M(L,OH)} \quad (32)$$

shows that $[M']$ has the same additive property as $\alpha_{M(L,OH)}$.

Substitution for $\alpha_{M(L,OH)}$ from Eq. (29) into Eq. (32) gives:

$$[M']_{max} = \left\{10^{(\log *K_{s0}-n\text{pH})} \cdot \alpha_{M(OH)}\right\} + \left\{10^{(\log *K_{s0}-n\text{pH})} \cdot \alpha_{rest}\right\} \quad (33)$$

The first term on the right-hand side of Eq. (33) can be replaced by Eq. (16). Thus only the second term, which depends on the calculated value of α_{rest}, has to be added to the values of $[M']_{max}$ previously calculated for the case where no L is present.

$[M']_{max}$ was found in this way for all the relevant pM'–pH plots in this Atlas.

Polynuclear hydroxide formation

The 1% borderline for polynuclear hydroxides in the absence of a ligand is calculated from Eq. (18), which was derived from Eq. (17) by replacing $[M']/[M]$ by α_{mono}. If ML complexes are formed, Eq. (17) has to be modified by substituting $(\alpha_{mono} + \alpha_{rest})$ for α_{mono}. Carrying out the same substitution in Eq. (18) gives the equation used to calculate the positions of the 1% borderlines in the presence of ligand L.

1.1.3 Air-saturated Solutions

In the introduction it was said that when a ligand other than OH^- forms insoluble compounds with a metal, it would not be considered in this Atlas, because the aim is to help the chemist mainly interested in metals in solution, and less interested in precipitation. Ligands such as phosphate and sulphide are not included. However, in two special cases an exception must be made: under normal working conditions (i.e. unless particular precautions are taken) CO_2 and O_2 are always present in aqueous solutions, and they may react with the metals to form insoluble compounds.

The influence of CO_2

The mean partial pressure of CO_2 in air is $10^{-3.52}$ atmosphere. For aqueous solutions of ionic strength = 0.1 at 25°C the equilibrium constant for the reaction $CO_2(\text{gas}) + H_2O \rightleftharpoons H_2CO_3$ is $K = 10^{-1.46}$ (mole/atm). From this, and the formation constants of H_2CO_3, $\log K_1 = 10.35$ and $\log K_2 = 6.35$, the following equilibrium concentrations in air-saturated solutions can be derived.

$$[H_2CO_3] = 10^{-5.0} M \quad (34)$$

$$[HCO_3^-] = 10^{(\text{pH}-11.35)} M \quad (35)$$

$$[CO_3^{2-}] = 10^{(2\text{pH}-21.7)} M \quad (36)$$

For a bivalent metal forming an insoluble carbonate the solubility product

$K_{s0}^{II} = [M^{2+}][CO_3^{2-}]$ can be combined with the side-reaction coefficient $\alpha = [M']/[M^{2+}]$. It leads to

$$p[M']_{max} = -\log K_{s0}^{II} + \log [CO_3^{2-}] - \log \alpha_{M(X)} \tag{37}$$

Combining (37) with (36) gives

$$p[M']_{max} = -(21.7 + \log K_{s0}^{II}) + 2pH - \log \alpha_{M(X)} \tag{38}$$

For a tervalent metal the solubility product $[M^{3+}]^2 [CO_3^{2-}]^3 = K_{s0}^{III}$ leads analogously to

$$p[M']_{max} = -(32.55 + \tfrac{1}{2}\log K_{s0}^{III}) + 3pH - \log \alpha_{M(X)} \tag{39}$$

In the case of a quadrivalent metal the solubility product $[M^{4+}][CO_3^{2-}]^2 = K_{s0}^{IV}$ leads to

$$p[M']_{max} = -(43.4 + \log K_{s0}^{IV}) + 4pH - \log \alpha_{M(X)} \tag{40}$$

Equations (38), (39) and (40) are all similar to the equation [(14)] for hydroxide precipitation and they can be written in the general form

$$p[M']_{max} = -\log K_{carb} + npH - \log \alpha_{M(X)} \tag{41}$$

With most metal ions CO_3^{2-} and HCO_3^- do not form soluble complexes, so they do not contribute to α_M. This implies that we can directly compare log $*K_{s0}$ with log K_{carb}. If log $K_{carb} <$ log $*K_{s0}$ we can conclude that a carbonate will precipitate in preference to a hydroxide in air-saturated solutions.

The difference between log $*K_{s0}$ and log K_{carb} is extremely large for Ca, Ba and Sr. In fact, because of the ease with which the carbonates precipitate, it is very difficult to study the hydroxide precipitation, and the values of $*K_{s0}$ for Ba and Sr are known only roughly. A slightly better value has been obtained for calcium, but it is only approximate. Thus, in the chapters on Ca, Ba and Sr, the only situation considered is that where the solid phase is the metal carbonate. For the species Cd(II), Co(II), La(III), Pb(II) and Mn(II), accurate values are available for both log $*K_{s0}$ and log K_{carb}. The values differ significantly enough to justify separate sets of plots of pM'–pH for air-saturated solutions. The plots of log α against pH, which remain unchanged, are repeated for convenience and uniformity.

For the metals Dy, Gd, Mg, Nd, Sm, Y and Yb the difference between log $*K_{s0}$ and log K_{carb} is small compared with the uncertainty in the values, and it is probable that mixtures of carbonate and hydroxide will precipitate. Carbonate precipitation then has no influence on the position of the precipitation borderline.

Copper(II) forms a hydroxide which is less soluble than copper carbonate and there is no tendency for a carbonate to form in acidic or weakly basic air-saturated solutions. At high pH, however, Cu has a strong tendency to form the soluble carbonate complex $Cu(CO_3)_2^{2-}$, which causes an increase in the solubility of the hydroxide precipitate.

Addition of the term

$$\frac{[Cu(CO_3)_2^{2-}]}{[Cu^{2+}]} = \beta[CO_3]^2 = 10^{(4pH - 43.4 + \log \beta)} \tag{42}$$

to the side-reaction coefficient $\alpha_{M(OH)}$ leads to an expression containing two terms with the same dependence on pH [i.e. Eq. (42), and $10^{(4pH + \log *\beta_4)}$ which refers to $Cu(OH)_4^{2-}$], and because $Cu(CO_3)_2^{2-}$ is predominant, Eq. (42) is the larger. Thus, the influence of the formation of $Cu(CO_3)_2^{2-}$ on the equilibrium can be determined by substituting the value $\log \beta - 43.4$ [from Eq. (42)] for $\log *\beta_4$ in the expression for $\alpha_{M(OH)}$. The resulting plot is given at the end of the chapter on copper.

It should be noted that all the plots for metal–carbonate–ligand systems refer to solutions which are in equilibrium with the CO_2 in the air. If a metal ion solution is deaerated, the concentration of H_2CO_3 decreases, and $[HCO_3^-]$ and $[CO_3^{2-}]$ will decrease proportionally. If the CO_2 concentration is decreased to 1% of its original value, $\log K_{carb}$ will decrease by 2, 3 or 4 log units, depending on the valency of the metal ion, and the borderline of precipitation will shift downwards to the same extent. Thus it is possible to gain some idea of the effectiveness of and need for deaeration.

The influence of O_2

When a metal can occur in different oxidation states, it may be oxidized to a state which can precipitate as the hydroxide or oxide because that is less soluble. Such reactions do not always proceed measurably. The formation of CeO_2 is an example; Ce(IV) solutions can be kept for months in a supersaturated state. Nevertheless one should be aware of the possibility of such reactions. For example, it is well known that manganous ions in alkaline solution react rapidly with O_2 to form a precipitate of MnO_2, but it is less widely known that the reaction also proceeds in weakly acidic media. Unless some action is taken (e.g. addition of ascorbic acid) this phenomenon interferes severely in complexometric microtitrations.

For the reaction

$$Mn^{2+} + 2H_2O \rightleftharpoons MnO_2 \text{ (solid)} + 4H^+ + 2e \tag{43}$$

the redox potential (in mV) is

$$E_1 = 1228 - 118.2 \text{ pH} - 29.5 \log [Mn^{2+}] \tag{44}$$

The redox potential (in mV) for O_2 in solution is

$$E_2 = 1227 - 59.1 \text{ pH} \tag{45}$$

If equilibrium is reached, the equilibrium concentration of Mn^{2+} ions is found by equating E_1 and E_2, which gives

$$p[Mn^{2+}]_{max} = 0 + 2pH \qquad (46)$$

and thus

$$p[Mn']_{max} = 0 + 2pH - \log \alpha_{Mn(OH)} \qquad (47)$$

In acidic solutions $\log \alpha_{Mn(OH)}$ may be made equal to zero, and it becomes evident that, at equilibrium, manganous ions are extremely insoluble; the precipitation borderline goes through the origin of the pM'–pH system.

Under equilibrium conditions where oxidation is possible the reaction $Mn^{2+} + H_2 \rightleftharpoons Mn(OH)_2 + H^{2+}$ can occur only when the redox potential is less than

$$E = 462 - 59.1 pH \qquad (48)$$

The case of Mn(II) is extreme, but other metals show the same phenomena. Oxidation of Pb^{2+} by O_2 can be split into two reactions,

$$Pb^{2+} + H_2O \rightleftharpoons PbO_2 \text{(solid)} + 4H^+ + 2e \qquad (49)$$

with potential

$$E = 1449 - 118.2 pH - 29.5 \log [Pb^{2+}] \qquad (50)$$

and the reaction

$$2H_2O \rightleftharpoons O_2 + 4H^+ + 4e^- \qquad (51)$$

for which, in air at atmospheric pressure:

$$E = 1228 - 59.1 pH \qquad (52)$$

If equilibrium is reached, the potentials in (50) and (52) can be equated to give:

$$p[Pb^{2+}]_{max} = -7.5 + 2pH \qquad (53)$$

and thus

$$p[Pb']_{max} = -7.5 + 2pH - \log \alpha \qquad (54)$$

Comparison with Eq. (14) shows that $\log K_{ox} = 7.5$ is smaller than both $\log {}^*K_{s0}$ for hydroxide precipitation and $\log K_{carb}$ for carbonate precipitation (see Pb chapter). As a result, Pb(II) may disappear from solution by precipitation as PbO_2, and this can occur in practice when reaction (49) is catalysed. Other metals behave similarly; the most important are Co(II) [apparent $\log {}^*K_{s0} = 8.8$], Cr(III) (in theory), Ni(II) [apparent $\log {}^*K_{s0} = 8.5$], Fe(II), V(III) and U(IV).

1.2 THE PLOTS: SOME PRACTICAL POINTS

1.2.1 The Significance of the Lines in the Plots

In all the plots of log α vs pH, the line represents α_{mono} (i.e. $\alpha_{M(OH)}$ without the polycomplex terms). The reasons for this are elucidated with scandium as example.

It is obvious from Fig. 1.1 that polycomplexes are significant only at metal concentrations above 10^{-3} M. If we look at the corresponding region of the pM'-pH plot (i.e. pM' <3), we can see that at pH values above about 4.5 scandium will precipitate, and that above pH 5, the concentration remaining is below 10^{-3} M. Thus, because of precipitation, the polycomplex lines for scandium, as drawn in Fig. 1.1, have no meaning above pH 4.5.

In principle, the same holds for all cases of polynuclear hydroxide complexation. Even when polycomplexation is extensive, the lines end up close to the pH-axis, and so are not very informative. Because the line-segments are more likely to be confusing than helpful, and since the 1% borderlines in the pM'-pH plots give much more information about the behaviour of the metal, the contribution of the polycomplexes has always been omitted in the calculation of the value of log α to be used for the plot of log α vs pH.

In addition to the line for log $\alpha_{M(OH)}$, each plot includes lines for log $\alpha_{M(L,OH)}$ at various values of [L'], the total free concentration of L. The difference between log $\alpha_{M(L,OH)}$ and log $\alpha_{M(OH)}$ is a measure of the dominance of the metal–ligand reaction over the metal–hydroxide reaction.

The start of the region of predominance on the low-pH side can be easily read from the plot and needs no comment. The dotted lines in the pM'-pH plot indicate the pH limit above which more than 1% of the metal–ligand complex has been hydrolysed. This limit is found from the condition log $\alpha_{M(L,OH)}$ − log $\alpha_{M(OH)}$ = 2; it cannot always be read from the log α plot because of the limited extent of the axes.

The solid lines in the pM'-pH plots represent the borderlines of the precipitation region for the various values of [L'] tabulated alongside the log α-pH plot. It should be noted that the concentrations do not always lead to lines in the plots.

The dashed lines are the 1% borderlines for polyhydroxides. In some cases where complexes of differing compositions are formed, the 1% borderline may consist of several line-segments. Indices are given near each line-segment. If the polycomplex region is penetrated through the line labelled s,t, more than 1% of M is transformed into the species $M_s(OH)_t$. However, deeper penetration will result in a mixture of all the polycomplexes.

To sum up, the pM'-pH plots are used in the following way. For a given initial total metal ion concentration pM', a horizontal line from the appropriate point on the ordinate will cut the polycomplexation borderline at the pH value at which the polycomplex concentration will correspond to conversion of 1% of the total metal ion concentration, and will cut the precipitation borderline at the pH value at which precipitation begins.

For mononuclear hydroxo complexes the leading plots show the predominance regions for each species, e.g. MOH will be the predominant species in solution (in the absence of other ligands) between the pH values indicated by the vertical lines for log $^*K_1$ and log $^*K_2$ (except for the 'out of order' values, see p. 22).

1.2.2 Superposability of the Plots

When more than one ligand which can form complexes with M is present in solution, the expression for the side-reaction coefficient, $\alpha_{M(L,OH)}$, must be extended.

In the theoretical section it was shown that, when a single hydroxide complex is predominant, the hydrolysis can be explained adequately in terms of that complex, whether it is mononuclear or polynuclear. It was also shown that a more accurate description of the system can be obtained by drawing the rounded-off line connecting the various line-segments corresponding to the various predominant species.

This is justified by the approximation $\log(a + b) = \log a$, allowable when $b < 0.1a$, and by the fact that the rounding-off procedure adds some tenths of a log unit to $\log a$, so that even when $a = b$ the error is decreased to a value less than the usual uncertainty in the position of the line.

The same sort of rounding-off procedure can be applied to the plots of $\log \alpha$ vs pH. When ML-complexes are dominant, the $\alpha_{M(L,OH)}$ line is applicable rather than the $\alpha_{M(OH)}$ one, and *vice versa*. When the ML-complexes take over dominance from the MOH-complexes, the $\alpha_{M(L,OH)}$ and $\alpha_{M(OH)}$ lines can be rounded off where they cross.

This procedure can be applied more generally. In the presence of a second ligand, Y, which forms complexes MY which are stronger than ML, the value of α will be determined mainly by the MY equilibrium. If the separate side-reaction coefficients for the two ligands are known, it is possible to draw the line corresponding to $\alpha_{overall}$ by drawing a rounded-off line connecting the line-segments corresponding to the highest values of $\log \alpha$. Thus, the $\log \alpha$ vs pH plots are superposable.

In a similar way, the line-segments corresponding to the *lowest* pM'-values on two or more pM'–pH plots may be connected in order to find overall borderlines for poly-complexation and precipitation (cf. scandium, Fig. 1.2). Thus, the pM'–pH plots are also superposable.

This superposability is important, because it means that the separate metal–ligand plots can be used in more general cases when more than one ligand is present in the solution. The superposition is valid, provided that no stable mixed-ligand complexes (i.e. complexes of the form MLY) are formed. Although such ternary complexes are known in some cases (Hg–EDTA–NH$_3$, Bi–EDTA–Cl, Cu–amines), their contribution to α and their influence on the pM'–pH plot is usually not very significant, although they may sometimes be the explanation for unexpected phenomena.

Some ligands, such as citrate and tartrate, are known to form ternary hetero-nuclear complexes when certain pairs of metals are present in solution. This phenomenon has also been neglected here.

1.2.3 The Amount of Ligand in Solution

The concentrations of L given alongside the plots are values of $[L']$, the total amount of L not bound to the metal. The total analytical concentration C_L is found by adding to these $[L']$ values the concentration of L bound to the metal:

$$C_L = [L'] + \left\{ [ML] + 2[ML_2] + 3[ML_3] + \ldots \right\} \tag{55}$$

C_L can be related to the analytical concentration of the metal C_M, by means of the average coordination number $\bar{n}$, defined as

$$\bar{n} = ([ML] + 2[ML_2] + 3[ML_3] + \ldots)/C_M \tag{56}$$

which leads to

$$C_L = [L'] + \bar{n} C_M \tag{57}$$

The average coordination number is a 'by-product' of our calculations. Plots of $\bar{n}$ do not appear in the Atlas, because the calculated values are not very useful. Frequently, it is known from other sources that $\bar{n}$ approaches an integral value and if it does not, the calculated value of $\bar{n}$ is subject to large errors because of uncertainty in the values of the constants, and so is of limited value. In nearly all cases of practical interest it is satisfactory to guess $\bar{n}$ when its value is not known from elsewhere.

Equations (55), (56) and (57) show why it was possible to use $[L']$ instead of C_L to express the concentration of ligand for the plots. Use of $[L']$ made the calculations simpler, less expensive in computer time, and easier to control. In addition, it is easier to use the plots, because the concentration $[L']$ corresponding to the line can be related directly to the value of pM' on the axis.

1.2.4 Data Sources and Selection Criteria

In the collection of information we were grateful for and made extensive use of the data in *Stability Constants* [3,4] and in *Critical Stability Constants* [5].

When several workers were in close agreement on a particular value the average was selected. Values showing considerable scatter were eliminated. In cases where agreement was poor and few results were available for comparison, the original literature was consulted. In a few cases, the selection could be guided by comparison with complexes of the same metal with similar ligands, and with complexes of the same ligand with similar metals. It had to be remembered that unexpected anomalous behaviour occasionally occurs, however.

When an important constant missing from the set of data selected was present in another set, the missing constant was sometimes obtained by adjusting the value from the other set in order to achieve internal consistency.

Single values have been included only when the resulting plots did not conflict with the practical experience of the author. In some cases, values could be improved by recalculation, after correction of the values of other constants on which they are based but which were known less accurately at the time of the original publication.

Nevertheless some values may be in error. Such errors will, we hope, be detected when the work is repeated by other investigators, and when the plots in this Atlas are compared with practical experience. When there are differences between experimental observations and the plots presented here, the discrepancy can be utilized to make corrections in the constants used for the plots. In this way, a consistent set of constants having good correlation with experiment should finally remain.

In view of this, we did not hesitate to introduce some unverified but apparently reliable data, in the hope that the relevant plots might form starting points for new investigations leading finally to refined data and plots.

An important guideline in deciding whether or not to make use of inaccurately known data was the influence of the variance in the data on the position of the line in the plot. Sometimes, a large variance in a constant gives rise to only small changes in the measurable experimental quantities and hence in the lines. Conversely, the origin of the large variances then becomes obvious. In such cases, the doubtful data have been used with confidence.

We were very concerned to make a reliable selection of hydrolysis constants of the metals, since these constants are the basis of what we hope will grow into a consistent collection including most of the metals. The work was begun in 1975, and, after we had carefully compiled hydrolysis constants for most of the metals, we were very happy to have our attention drawn to the work of Baes and Mesmer [6], *The Hydrolysis of Cations*. It turned out that our selections agreed very well with theirs in most cases. Where there was disagreement, we decided to follow their line, especially since their expertise had led them to adopt better criteria for deciding which of the many possible polynuclear complexes reported were most likely to be the correct ones.

A special word must be said about the rare-earth elements. Only limited reliable information is available about hydrolysis constants, especially for the lighter elements. We decided, therefore, to determine the constants for Ce(III). We were so pleased with the apparent reliability of the results [7] that we decided to use the values in the Atlas. One factor affecting the decision was that our values fitted very well in the correlation equation given by Baes and Mesmer [6]. Thus, we thought it permissible to guess unknown data for some of the other rare earths with the aid of this equation.

1.2.5 Constants and Experimental Conditions

Analytical chemical reactions are usually carried out at room temperature in solutions of moderate ionic strength; similar experimental conditions are used in electrochemistry, water control, and many other fields.

Whenever they were available, values for constants determined at 20°C and at an ionic strength of 0.1 were selected; otherwise, values for 25°C and ionic strength 0.1 were taken. In some cases, values were adjusted for these conditions by interpolation etc. When values for other conditions have been employed, the fact is noted in the introduction to the relevant chapter.

It was sometimes difficult to choose between the widely scattered values for the solubility products of the metal hydroxides. Scatter arises because the true equilibria are obscured by such phenomena as supersaturation, dependence of solubility on particle size, slow reaction in dilute solutions, and recrystallization. Since most investigators are likely to be interested in equilibria involving fresh precipitates, it was decided that all the $*K_{s0}$ data selected should refer to freshly-formed precipitates. The situation for aged precipitates can be deduced from the plots by shifting the borderline of precipitation by the difference in the log $*K_{s0}$ values. Any special information about a particular metal is given in the introduction to the relevant chapter.

1.2.6 Interpretation

As already mentioned, most of the constants used for composing the plots refer to an ionic strength of 0.1. This sometimes gives rise to a conflicting situation in the pM′–pH plots, however, because lines drawn into regions near the axes refer to acid and metal concentrations $> 0.1M$. Also, lines have been plotted for ligand concentrations $\gg 10^{-2} M$ without correction to the constants.

In a few cases, the constants were adjusted for an ionic strength of 2, because the metals concerned are usually manipulated in strongly acidic media, and a conflicting situation then occurs when dilute solutions are considered.

We did not try to correct for this. In our opinion, it is rather academic to discuss such corrections when inaccuracies in the constants usually cause larger errors in the positions of the lines. Moreover, many other effects have been overlooked because of insufficient information. (Formation of weak ternary complexes, dimerization in concentrated solutions, and uncertainty in the identity of certain of the complexes are examples of this.)

The information on metal-ion behaviour given in the plots is intended to be sufficiently precise to enable experimental observations to be explained at least qualitatively, and usually quantitatively.

Other contradictions in the diagrams originate from specific physical properties of the reactants. For instance, the solubility of EDTA in acid medium is limited, but the line corresponding to 0.1M is extended to the pH axis. Again, if an acidic medium is used, HCN will escape as a gas, so a line for 0.1M has no meaning at pH values less than 5.

It should also be noted that some metal-ligand species have limited solubility, or form polynuclear metal–ligand complexes. A warning is given in the tabular index when this can be expected in solutions more concentrated than $10^{-2} M$.

Metal–ligand combinations do not appear in the Atlas when (1) a metal–ligand complex precipitates at concentrations below $10^{-2} M$, (2) the complex formed is so weak at a ligand concentration of 0.1M that log α is less than 1 at all pH values, and (3) polynuclear metal–ligand species occur in dilute solutions (pM′ > 2).

It should be borne in mind that the literature data may have been determined in a limited region of the pM′–pH plot, and that their use in our calculations means in effect that they have to be extrapolated to cover the whole plot area. In such cases, there is no guarantee that the plot will not deviate significantly from experiment, and it is likely that some regions of the plot will be less reliable than others.

The tabular index contains additional qualitative information about each metal–ligand combination, and we recommend that it should always be consulted.

1.3 SOME PRACTICAL APPLICATIONS

1.3.1 Conditional Constants

The ligands EDTA, DCTA and DTPA are used extensively in complexometry. In order to be able to select suitable experimental conditions for titration of a metal of concentration C_M, it is necessary to know the value of the so-called conditional constant

$$K' = \frac{[ML']}{[M'][L']} \tag{58}$$

for various experimental conditions. All the concentrations in (58) are total free concentrations, which means that the side-reaction coefficients for M, L and ML have all been introduced into the equation for the stability constant $K = [ML]/[M][L]$.

For a good titration, the quantity $Z \ (= C_M K')$ has to exceed some minimum value; Z is a measure of the sharpness of the end-point. Values for Z can be read directly from the plots of $\log \alpha$ vs pH. Since EDTA forms only 1:1 complexes, and knowing that ML predominates over MOH, and that $[ML'] = [ML] + [MHL] + [MOHL] + \ldots$, we can say from Eq. (29) that

$$\alpha_{M(L,OH)} = [L'] \left\{ \beta'_0 + \beta'_1 + \ldots \right\} = \frac{[ML']}{[M^{n+}]}$$

Substitution in (58) gives, after some rearrangement

$$\log Z = \log \alpha_{M(L,OH)} - \log \alpha_{M(OH)} + \log \frac{C_M}{[L']}$$

Thus, $\log Z$ is equal to the difference between the line for $\log \alpha_{M(OH)}$, and the line for $\log \alpha_{M(L,OH)}$ for which the ligand concentration $[L']$ is equal to the desired value of C_M.

Knowing this, the influence of other ligands on K' — and thus on Z — can be found from the difference between the line for $\log \alpha_{M(EDTA,OH)}$ on the M – EDTA plot, and the line for $\log \alpha_{M(L,OH)}$ on the plot for the other ligand.

1.3.2 Manipulation of Solutions*

A particular set of experimental conditions can be represented by a point in a pM'–pH diagram. When conditions change during a manipulation this point describes a locus. In the following some common manipulations will be considered and their influence on accuracy in analytical chemistry will be discussed.

Dilution

Suppose for example that 100 mg of scandium are dissolved in 1.5 ml of perchloric acid (1 + 1) which afterwards acts as an indifferent medium. The solution will be about

---
*Reprinted in part from [2], by permission of the copyright holders, Pergamon Press.

1 ml in volume and contain 2 mmole of Sc and 2 mmole of surplus perchloric acid. Thus to a first approximation p[Sc'] = −0.3 and pH = −0.3 (point P_0 in Fig. 1.3 corresponds to these values). When the solution is diluted homogeneously with water, [Sc'] and [H] change by the same factor; the same value has to be added to both p[Sc'] and pH. Point P shifts along a 45° line; a tenfold dilution shifts the experimental conditions to agree with point P_1 (Fig. 1.3). If the dissociation of water is taken into account, the dilution line will finally approach pH 7 asymptotically. The production of H$^+$-ions because of the hydrolysis of scandium is neglected; it causes only a slight shift of the line.

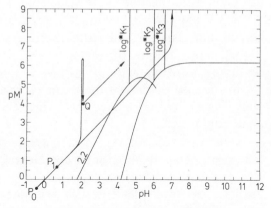

Fig. 1.3 The pSc'–pH diagram with some dilution lines. P_0 corresponds to an acid solution in which a piece of scandium has been dissolved, and P_1 to a 10-fold dilution of this solution.

In practice, dilution will not occur homogeneously. Considering the solution in this non-homogeneous state it follows that the dilution factors will range from zero to infinity. If we divide the solution into a large number of small parts, each of which is different but itself approximately homogeneous, the points P_i corresponding to these parts will lie on or near the dilution line and be spread over the whole line. In the case of Sc the region of polycomplexation is penetrated (Fig. 1.3); in the case of Bi the precipitation region is also entered (see page 95).

Generally, polycomplexes and precipitates form rapidly, which means that these compounds are formed even during a short penetration of the regions.

After homogenization of the solution the points P_i will telescope to one final point which theoretically may lie outside the polycomplex and precipitation regions. However, the compounds have a low reaction velocity and the restoration of equilibrium does not take place completely within a reasonably short time. This is the reason for less accurate results being obtained. Its importance depends upon the accuracy desired, the reaction velocities and the particular technique used for the analysis. For instance, in atomic-absorption spectrometry polynuclear complex formation will seldom decrease the absorption signal. In complexometry both phenomena usually interfere.

The question is in which way dilution interference can be prevented. If dilutions are made with 0.01M perchloric acid, the dilution line will asymptotically approach pH 2. If the dilution factor for Sc is 10^4, point Q in Fig. 1.3 is reached after homogenization. This medium can be diluted further with water without problems; the 45° line does not penetrate the polycomplex of precipitation region. From the diagrams for tin(IV) (see page 623) it follows that when a strongly alkaline solution is diluted, precipitation will also occur. P, lying near the horizontal axis at pH 14, shifts along a 135° line towards lower pH and will penetrate the precipitation region at about pH 11. Aluminium in alkaline solution is present as Al(OH)$_4^-$, so the slope of the tangent of the precipitation borderline is equal to −1. In this case dilution will not cause precipitation: the dilution locus of P has the same slope.

In the case of Bi (see page 95) P_0 lies nearly at the 1% polycomplex borderline. If such a Bi solution is diluted with an acid solution of pH 1, the (6,12)-polycomplex region is penetrated and about 10% of the bismuth will be converted into $Bi_6(OH)_{12}$. It turns out practically that the $Bi_6(OH)_{12}$ complex completely dissociates on further dilution at the pH 1 level. Whether this is a matter of kinetics or inaccuracy in the polycomplex equilibrium constant and so in the line position is not known at the moment, but a deeper penetration of the region to values above pH 3 always leads to inaccurate titration results with drifting end-points. The atomic-absorption results are also then less accurate.

In the case of tin(IV) the point P_0 near the origin lies behind the precipitation line, which agrees with the fact that when tin is dissolved in concentrated nitric acid precipitation occurs.

Dilution is one of the most common operations and will lead to specific problems for every metal. It also forms an integral part of many other manipulations but this is not always recognized. For instance when a tin(IV) solution is poured into a beaker that has just been rinsed and is still wet, some SnO_2 precipitates and an error of a few per cent will occur. Rinsing of a glass electrode after pH-adjustment, and adding a reagent solution which is not sufficiently acidic are other errors which can easily be committed unwittingly. When EDTA is added to a cold aluminium solution we deal almost solely with dilution during this manipulation, as EDTA hardly reacts with Al^{3+} at room temperature. Polynuclear hydroxide and colloidal solid $Al(OH)_3$ is formed, which – even on heating – reacts only slowly with EDTA, and a few per cent error can be expected. The remedy is acidification of the EDTA to pH 2.5 and homogeneous neutralization with hexamine by boiling.

In many cases it is possible to find a suitable pathway to avoid error, but as the different manipulations are closely related, let us discuss neutralization.

Neutralization

Suppose that a strongly acidic concentrated metal solution at pH_{P_0} is neutralized homogeneously with a dilute solution of a strong base of $pOH = pC_{base}$. The first additions act like a dilution, so approximately a 45° line is followed near P_0. When the solution has been diluted by the factor $10^{(pC_b - pH_{P_0})}$ its pH approaches pC_b and the pH will change with a jump to a pH-value near $(14 - pC_{base})$, just as in an acid-base titration. Further additions act only as dilution; the pH will asymptotically approach the value $(14 - pC_{base})$. Summarizing, we can say that the neutralization line consists of three parts: a 45° line through the starting point P_0, a horizontal part at the level $pM' = pM'_{P_0} + (pC_{base} - pH_{P_0})$ and a vertical part at $pH = pH_{base} = (14 - pC_{base})$.

When the basic solution has a concentration higher than that of the acid solution at pH_{P_0}, the 45° line will not occur and the horizontal line starts immediately at P_0. So far we have considered homogeneous equilibrium situations, and should now consider what happens in practice between addition and homogenization, when we deal with non-homogeneous situations. The neutralization reaction can be regarded as proceeding instantaneously, and because diffusion takes place very slowly, it can be assumed that the neutralization and dilution phenomena are always close to equilibrium in individual sections of the solution. So just as for dilution we can assume here that the corresponding points P for small parts of the solution are statistically spread along the neutralization line. Deviations from the line may be regarded as small. This means that the addition of a solution of sodium hydroxide will lead to a deep penetration of the precipitation region and that precipitate formation can be expected in most cases. Just as in dilution, the production of H^+-ions during the formation of soluble mono- or polynuclear hydroxides can be neglected in consideration of the position of the neutralization line; it will shift the line only a few tenths of a logarithmic unit.

The presence of anions of weak acids and bases has a large influence on the shape of the neutralization line. When, for instance, ammonia at a concentration C_{am} is added to a metal ion solution (pM'_{P_0}, pH_{P_0}) containing acetic acid at a concentration of C_{ac} and with P_0 near the origin, the following neutralization line will be approximately followed. When $pC_{am} > pH_{P_0}$, a 45° (dilution) line through P_0 will first be followed. At $pM' = pM'_{P_0} + (pC_{am} - pH_{P_0})$ a horizontal part starts. Near $pH = pK_{ac}$ (4.7) a vertical jump occurs as a consequence of the buffer action as the acetic acid is neutralized. The height of the jump depends on C_{ac} and C_{am} and can be approximated by $(pC_{am} - pC_{ac})$, i.e. the logarithm of the dilution factor and hence the amount by which pM' increases. A further small addition of ammonia causes a horizontal jump through the 'equivalence' point at $pH = 7.1$ until $pH \sim p*K_{am}$ (9.4), the ammonia 'half-titration' value. A further addition of ammonia largely dilutes the solution; the pH will slowly approach the value $pH = \frac{1}{2}(14 + p*K_{am} - pC_{am})$ for a pure ammonia solution.

This example demonstrates how a pathway can be constructed in the pM'–pH diagram. It also shows how pM'-jumps can be generated by the addition of weak acids and bases, jumps that can be used to create alternative pathways in order to prevent unwanted hydroxide reaction. An overshoot of pH during the neutralization of acids can be prevented by using buffer solutions with a pH 1 unit higher than that of the equimolar buffer; 80% of the buffer capacity then remains available for neutralization. When even this procedure leads to too-high pH-values, use can be made of an acidified hexamethylenetetramine (hexamine) solution which releases ammonia on heating, thus neutralizing the solution homogeneously and controllably. The addition of solids should be avoided. In the vicinity of solids high local concentrations will occur, and with salts of weak acids extreme pH-values may arise.

When, during a manipulation, ions which form a complex with the metal concerned are in solution or come into solution the manipulation locus has to be constructed in the relevant plot.

The Sn(IV)–Cl plots show that tin metal can be dissolved in concentrated hydrochloric acid without precipitation of SnO_2; point P_0 near the origin stays outside the SnO_2 region for chloride concentrations $> 2M$. If a solution with $[Cl] \sim 2M$ is diluted tenfold with 1M perchloric or nitric acid, the pH remains about 0, pM' becomes 1 and $[Cl]$ will be about 0.2M. The result is that P now lies in the precipitation region as the borderline has shifted faster than P. If the solution is diluted with 1M hydrochloric acid, $[Cl]$ gradually changes from 2M asymptotically towards 1M. Under these circumstances point P remains outside the precipitation region and no SnO_2 precipitates. Combining this with the diagram for Sn(IV)–EDTA, and knowing that EDTA can be used for a complexometric back-titration procedure, we can conclude that the EDTA solution to be added should be acidified to $pH = 0$ and have $[Cl] = 1M$. (The concentration of EDTA has to be $< 7 \times 10^{-3}$ M because of the limited solubility of H_4Y.) After addition of the EDTA, which quickly reacts with Sn at room temperature, we have to consider the EDTA lines in the pM′–pH plot.

The implication is that the strongly acidic Sn(IV)–EDTA solution should not be neutralized with a sodium hydroxide or ammonia solution. The best method is to add a solution of hexamine acidified to pH 5–6 (to prevent crossing the borderline during the addition) and then to heat the solution to raise the pH by hydrolysis of the hexamine. In this way very accurate tin determinations are possible.

From the pM′–pH diagram for Cu it is obvious that neither a precipitate nor a polycomplex will be formed when a piece of copper is dissolved in nitric acid and the solution is diluted with water up to point (4,4) (see pp. 223 and 245). When sodium hydroxide is added to adjust the pH to 5.5 for the complexometric titration with TAR as indicator, a 4–5% error is found.

Addition of dilute ammonia solution (10^{-2}M, p. 225) also gives a penetration of the precipitation region. In practice an error of about 1% is found. The addition of enough concentrated ammonia gives an error of only 0.3% because the neutralization line passes over the lines e,f,g corresponding to ammine-formation. However, it is found experimentally that a few slips can still occur. Very accurate results are obtained when the solution is neutralized with a solution of acetate or hexamine of pH 6. It may also be mentioned that when an ammoniacal solution of copper (10^{-2}M) is titrated at pH 9.5 with trien (triethylenetetramine) with an ion-selective electrode used for the end-point determination, the trien should also be made ammoniacal. If not, then if the Cu–NH$_3$ solution is diluted locally with titrant by a factor of 10 the borderline shifts 4 log units upwards while point P at (2,9.5) shifts only 1 unit, and P is passed by the borderline This explains the poor response of the electrode after a few determinations, because of the nearly irreversible hydroxide precipitation at the electrode surface.

The addition of EDTA to an aluminium solution has already been discussed. The pH can be slowly increased on heating if hexamine is added. When the Al–EDTA complex is completely formed and a final pH-adjustment is needed, in nearly all cases of practical interest ($C_{Al} < 10^{-2}$ M), the neutralization line lies above the precipitation line, so no precipitate will be formed. However, when the pH is raised so that the dotted pH limit is passed, more than 1% of the Al–EDTA will dissociate, forming Al(OH)$_4^-$, and this reverts only slowly to Al–EDTA when the pH is lowered. Thus, formation of the soluble tetrahydroxo-complex leads to irregular results in this case. Systematic errors can usually be expected with all metals when the dotted pH-limits are passed, because of low reversion velocities.

When Al(OH)$_4^-$ is formed as above, a decrease in pH can cause penetration of the precipitation region, because part of the system behaves as if EDTA were absent. It is obvious that the addition of a complexing ligand does not guarantee trouble-free manipulations.

The aim of this section has been to show how pM′–pH plots can be used to elucidate specific pitfalls which may exist in the common operations of analytical chemistry, and examples have been used to give some insight into manipulation errors and to show that a number of small sins can easily be committed and are mostly overlooked as an origin of errors.

1.4 ARRANGEMENT OF THE PLOTS

All the plots for one oxidation state of a metal are arranged in one chapter. The chapters are arranged alphabetically according to the chemical symbols of the metals; lower oxidation states come first.

The introduction of each chapter gives the hydrolysis constants used throughout the chapter, and attempts to give the reader some idea of the reliability which should be attributed to the plots for the particular metal.

The pM'–pH plot for the metal in the absence of complexing ligands other than OH^- ions appears on the next page. In this plot, the regions of predominance of the various mononuclear complexes are indicated by the log $*K_i$ lines. (Note that 'out-of-order' log $*K_i$ lines do not indicate the true predominance borderlines; see p. 22).

On the following pages, the plots for the various ligands appear. There is one page for each metal–ligand combination, the upper diagram being the plot of log α vs pH and the lower one the pM'–pH plot. The values of [L'] and the stability constants for the particular metal–ligand combination given alongside the plot of log α vs pH hold for both plots.

The formation constants of the proton–ligand complexes are tabulated at the end of the book (p. 799).

Within each chapter, the metal–ligand combinations are arranged alphabetically according to the name of the ligand.

REFERENCES

[1] A. Ringbom, *Complexation in Analytical Chemistry*, Interscience, New York, 1963.
[2] J. Kragten, *Talanta* 1977, **24**, 483.
[3] L. G. Sillén and A. E. Martell, *Stability Constants of Metal-Ion Complexes*, Special Publication No. 17, The Chemical Society, London, 1964.
[4] L. G. Sillén and A. E. Martell, *Stability Constants of Metal-Ion Complexes, Supplement No. 1*, Special Publication No. 25, The Chemical Society, London, 1971.
[5] R. M. Smith and E. E. Martell, *Critical Stability Constants*, Vols. 1 and 2, Plenum Press, New York, 1975.
[6] C. F. Baes and R. E. Mesmer, *The Hydrolysis of Cations*, Wiley, New York, 1976.
[7] J. Kragten and L. C. Decnop-Weever, *Talanta*, in the press.

CHAPTER 2

Silver (I)   Ag

Brown Ag_2O is precipitated on addition of alkali to silver ions in solution. The solid hydroxide, AgOH, is unstable and exists only transiently.

The following hydrolysis constants were selected for the calculation and construction of the plots in this chapter.

$$\log {}^*\beta_1 = -11.9 \qquad \log {}^*K_{s0} = 6.3$$
$$\log {}^*\beta_2 = -23.8$$

There is no reliable evidence about the formation of polynuclear silvery-hydroxide compounds.

In air-saturated aqueous solution, silver carbonate never precipitates; an apparent constant $\log K_{carb} = 10.6$ can be deduced from the solubility product of silver carbonate ($\log K_{s0} = -11.1$) and this apparent constant is much higher than the value of Ag_2O ($\log {}^*K_{s0} = 6.3$).

A decrease in reduction potential may lead to separation of metallic silver. The potential should, preferably, remain above the value $E = (799 - 59.1\ p[Ag^+])$ mV.

Silver(I)

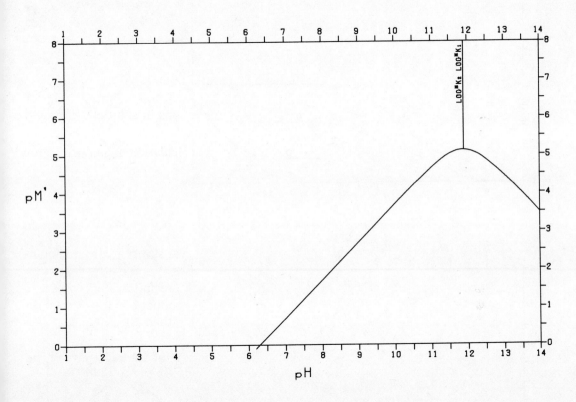

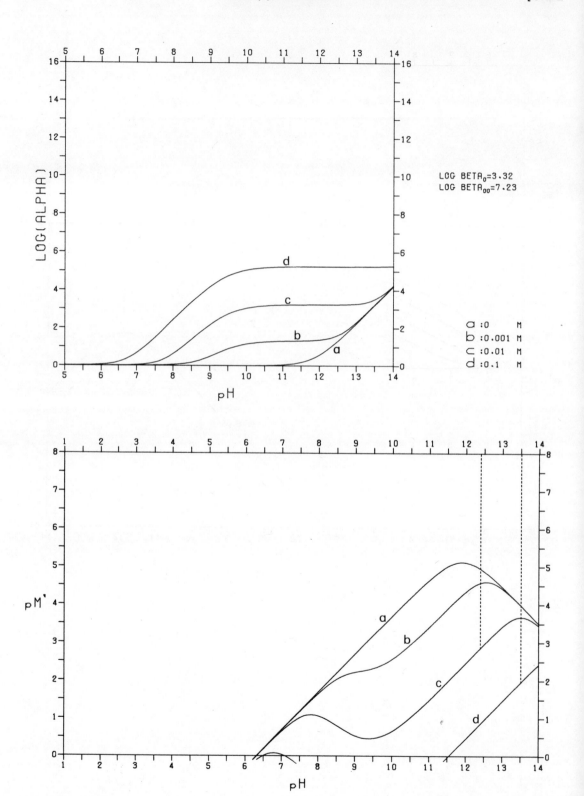

Silver(I)-DCTA

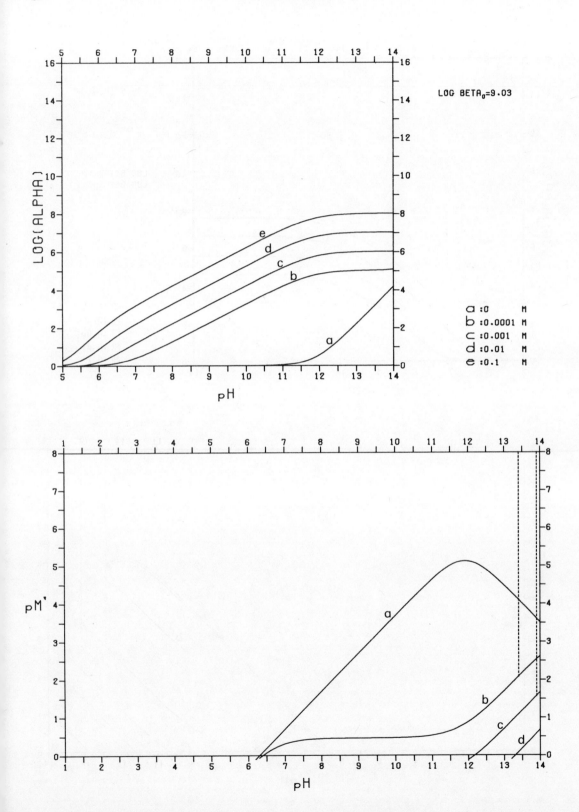

Silver(I)-DTPA

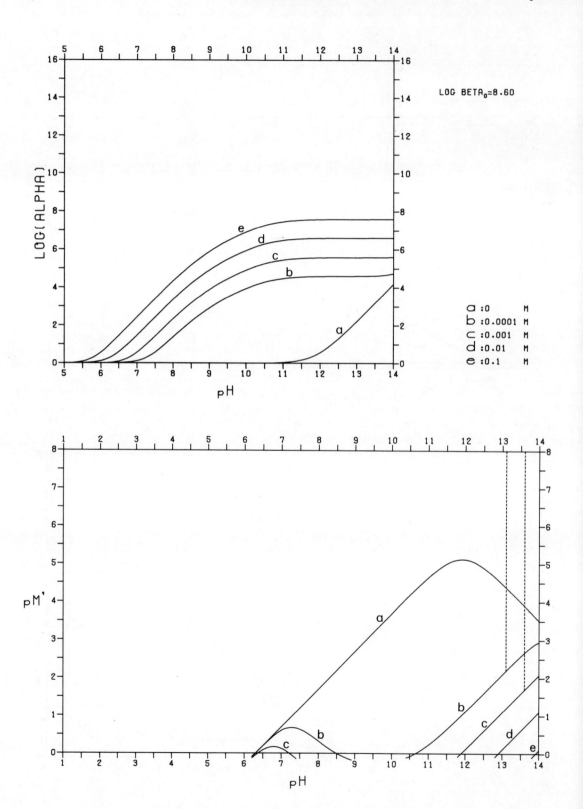

Silver(I)-EDTA

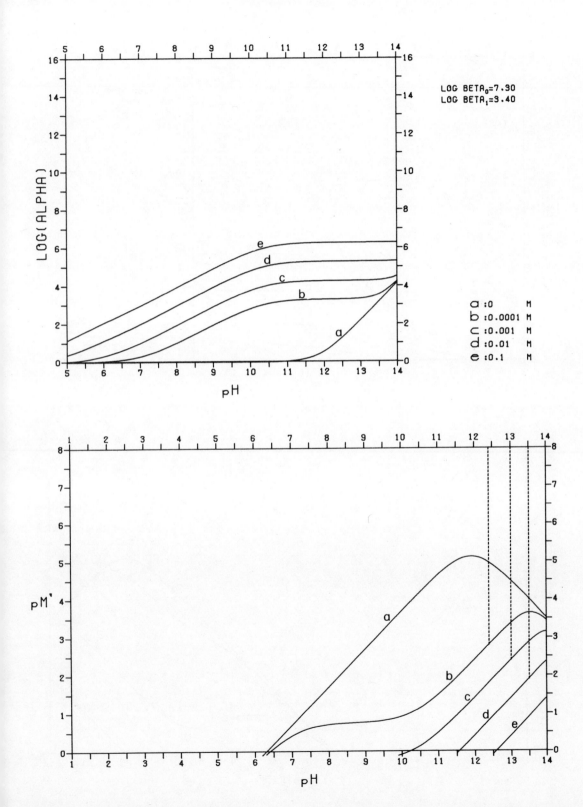

Silver(I)-Ethylenediamine

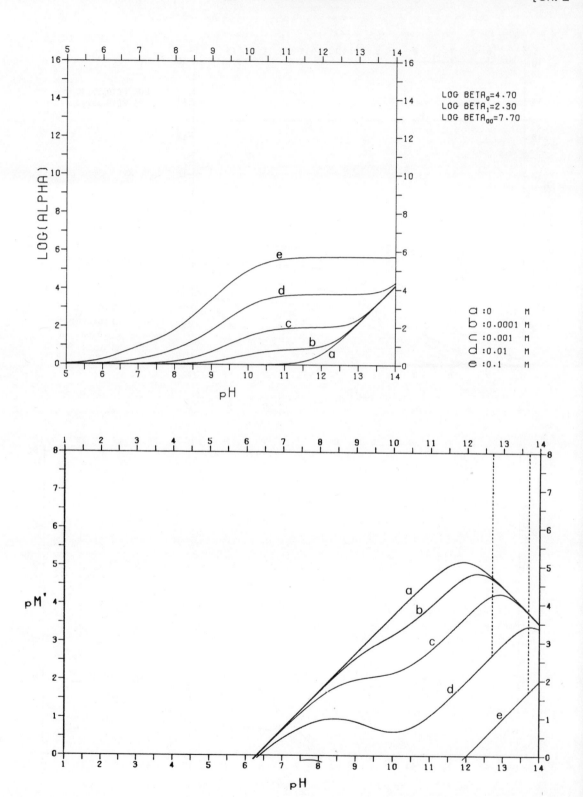

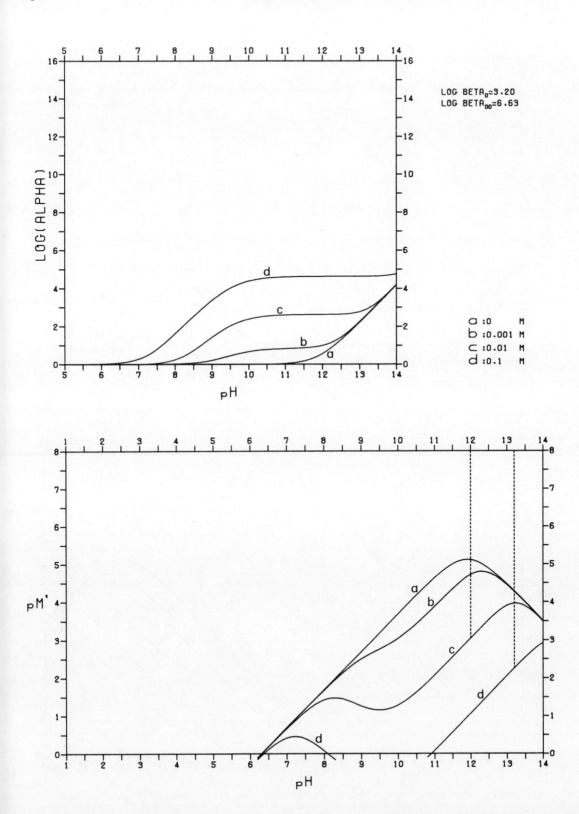

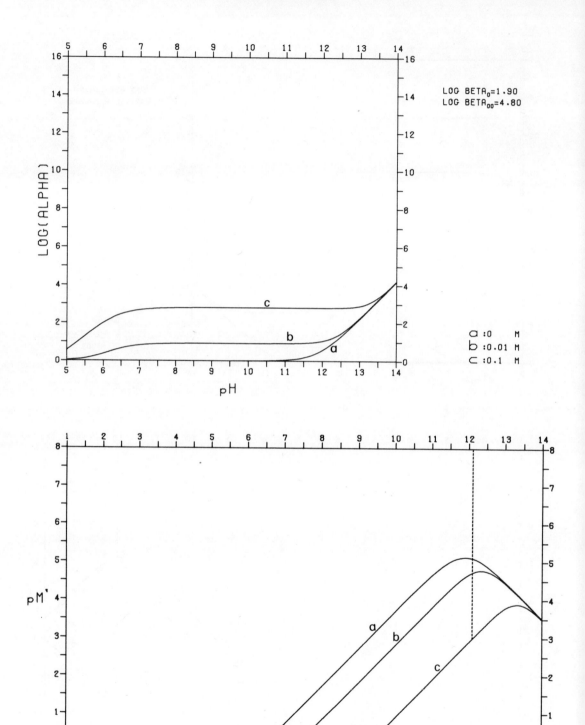

Silver(I)-1,10-Phenanthroline

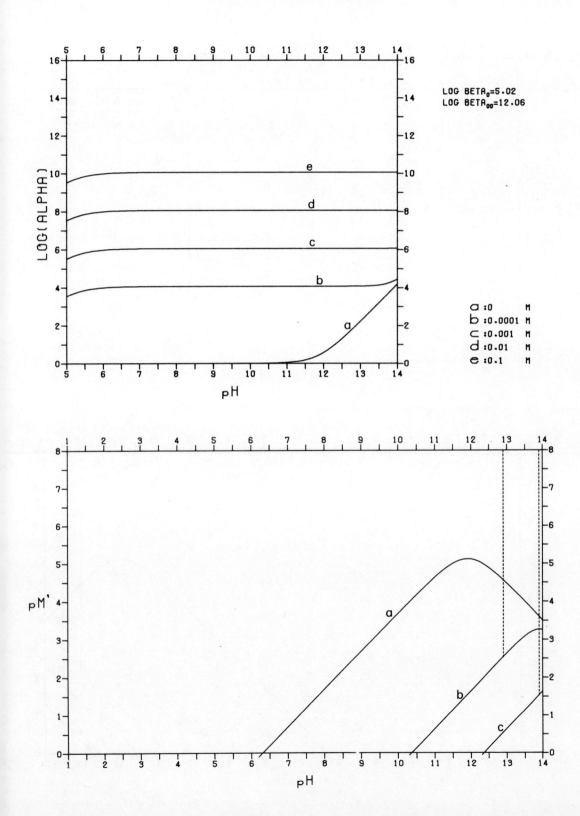

LOG BETA$_0$=5.02
LOG BETA$_{00}$=12.06

a : 0 M
b : 0.0001 M
c : 0.001 M
d : 0.01 M
e : 0.1 M

Silver(I)-Tetren [Ch. 2

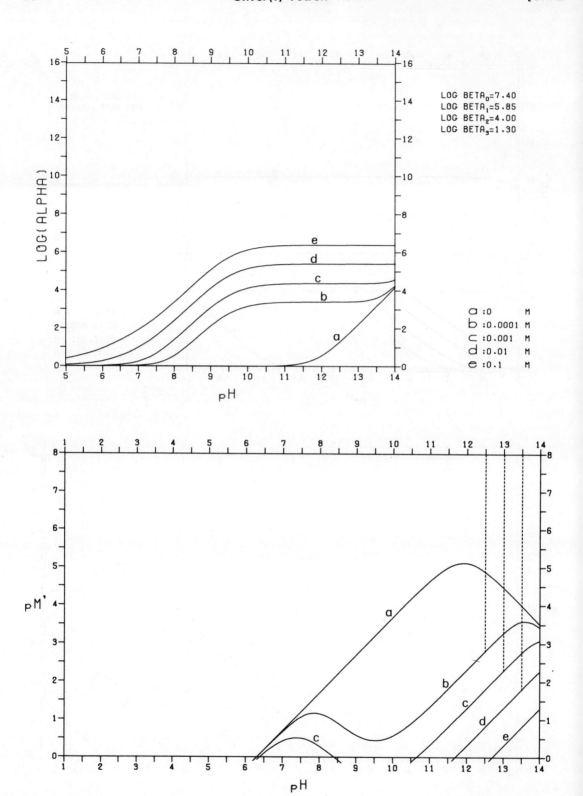

Silver(I)-Trien

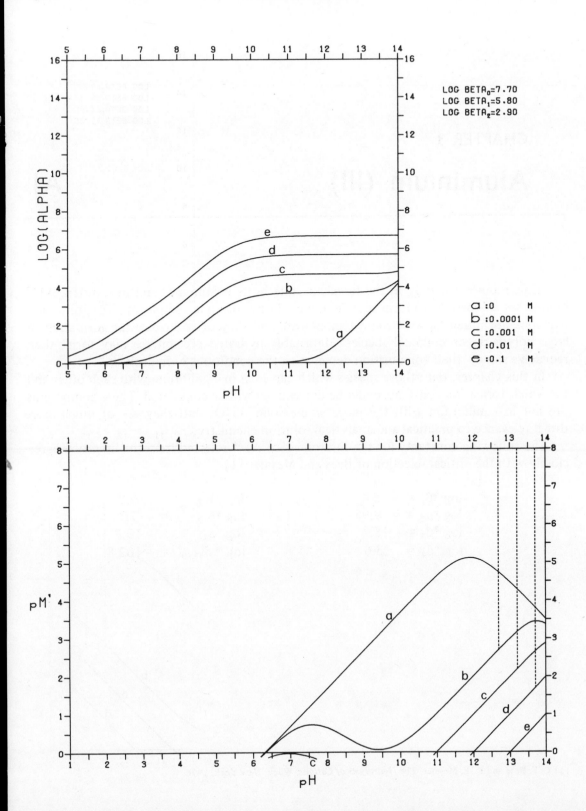

CHAPTER 3

Aluminium (III)   Al

Aluminium exists only in the tervalent state in its compounds and in solution. Al^{3+} hydrolyses extensively to form soluble polynuclear hydroxide complexes. The formation reactions are rather slow at room temperature and they often involve the formation of large polymeric or colloidal species. Metastable hydrated precipitates may form, then redissolve slowly; their composition depends on the conditions of hydrolysis.

In this chapter, only those species which can exist in equilibrium with each other and for which formation constants could be determined will be considered. These compounds are not in equilibrium with the more stable solid Al_2O_3, but they are of much more direct relevance to practical and analytical solution chemistry.

The following constants for the hydrolysis of Al^{3+} selected from the literature agree closely with the critical selection of Baes and Mesmer [1].

$$\log {}^*\beta_1 = -5.4$$
$$\log {}^*\beta_2 = -9.98$$
$$\log {}^*\beta_3 = -15.7$$
$$\log {}^*\beta_4 = -23.6$$

$$\log {}^*K_{s0} = 9.2$$
$$\log {}^*\beta_{22} = -7.7$$
$$\log {}^*\beta_{43} = -13.7$$
$$\log {}^*\beta_{32,13} = -102.8$$

[1] C. F. Baes and R. E. Mesmer, *The Hydrolysis of Cations*, Wiley, New York, 1976.

Aluminium(III)

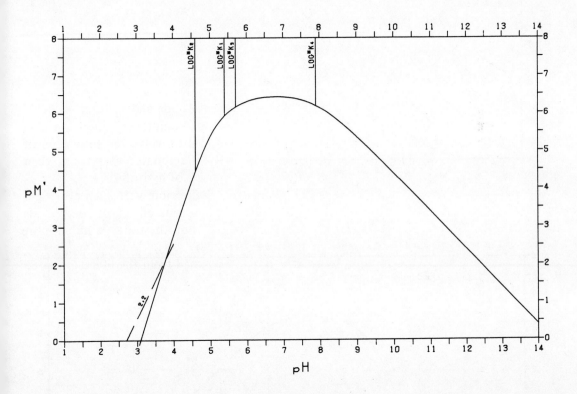

Aluminium(III)-Acetylacetone

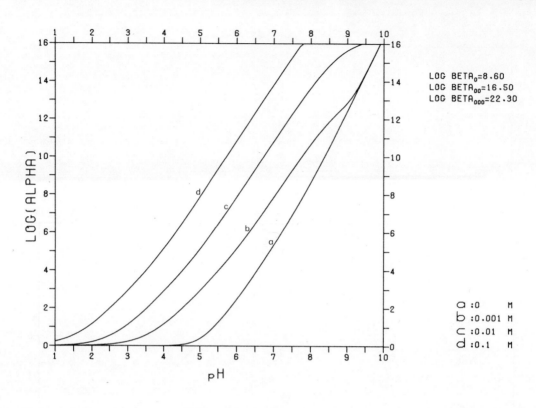

LOG BETA$_0$=8.60
LOG BETA$_{00}$=16.50
LOG BETA$_{000}$=22.30

a : 0   M
b : 0.001 M
c : 0.01   M
d : 0.1   M

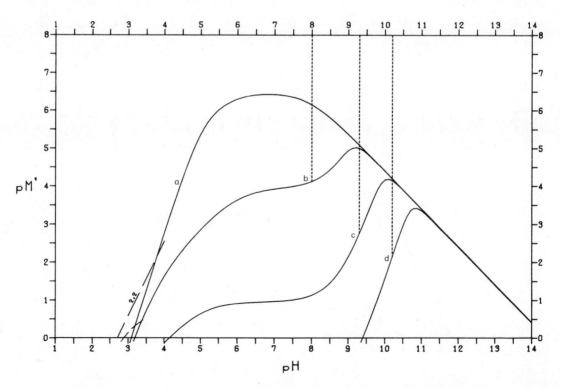

Aluminium(III)-Citrate

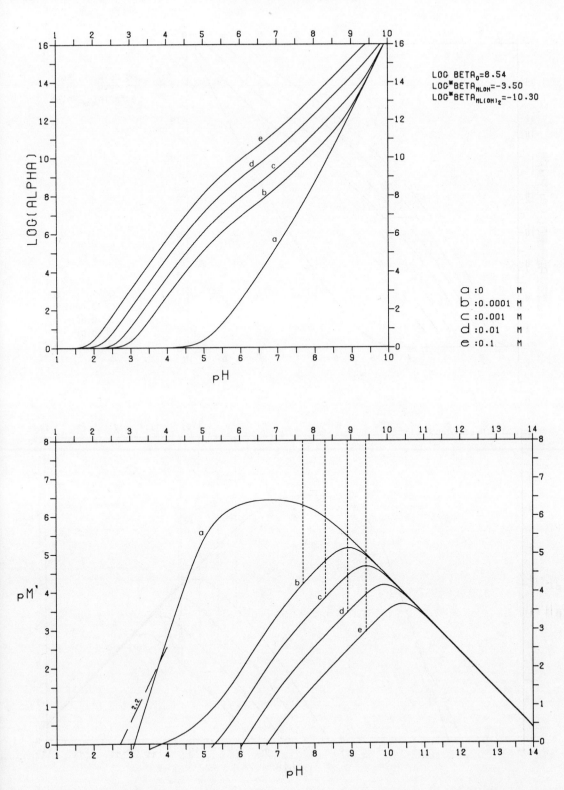

LOG BETA$_0$=8.54
LOG*BETA$_{MLOH}$=-3.50
LOG*BETA$_{ML(OH)_2}$=-10.30

a : 0   M
b : 0.0001   M
c : 0.001   M
d : 0.01   M
e : 0.1   M

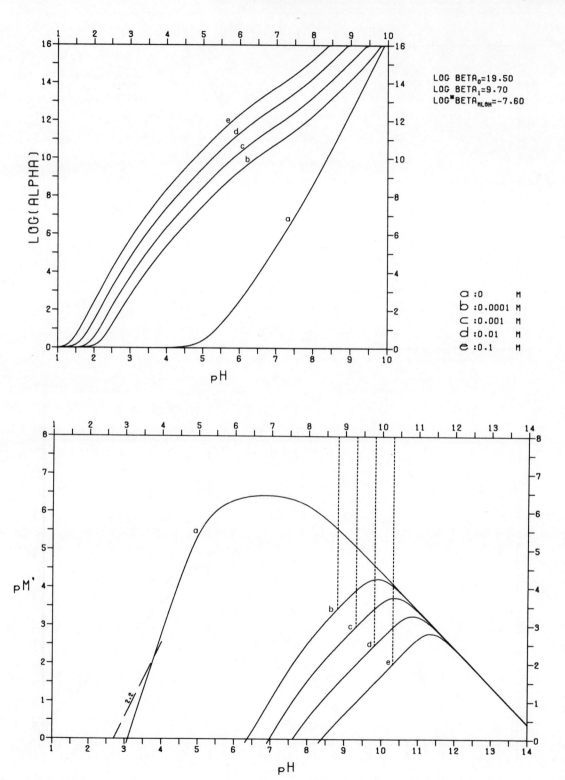

Aluminium(III)-DTPA

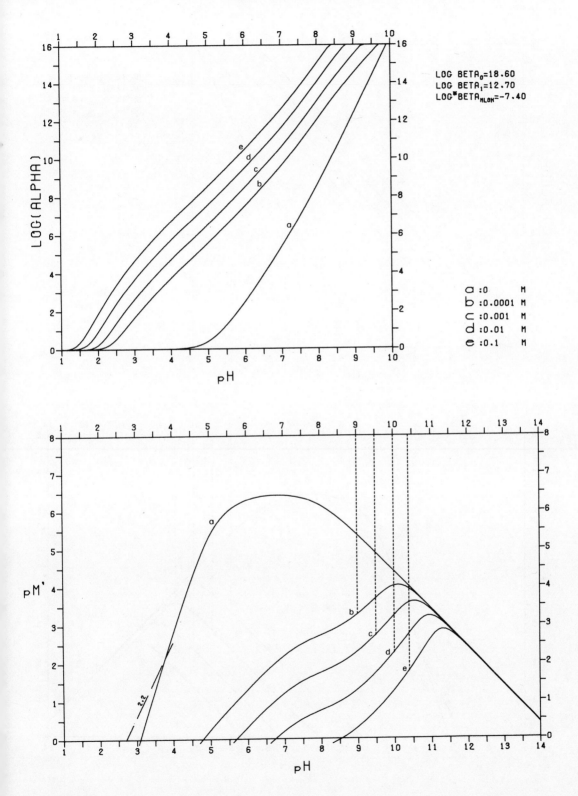

Aluminium(III)-EDTA [Ch. 3

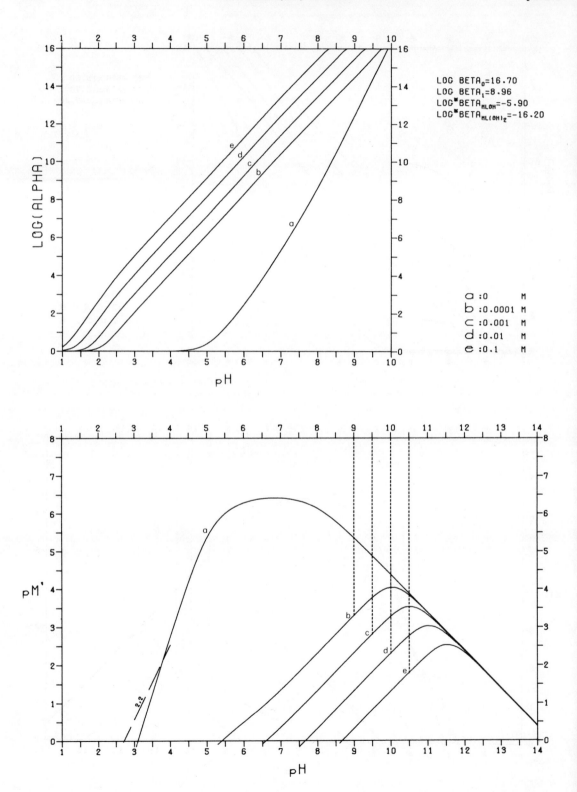

Aluminium(III)–Fluoride

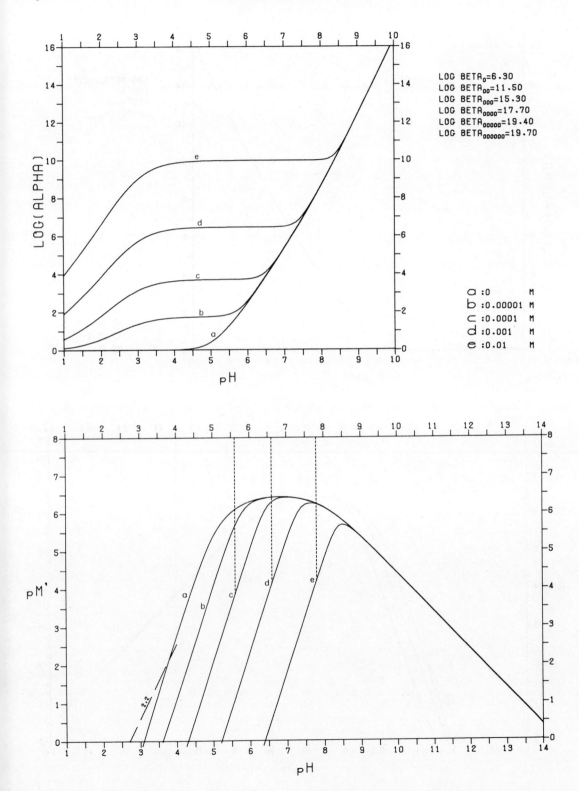

LOG BETA$_0$=6.30
LOG BETA$_{00}$=11.50
LOG BETA$_{000}$=15.30
LOG BETA$_{0000}$=17.70
LOG BETA$_{00000}$=19.40
LOG BETA$_{000000}$=19.70

a : 0 M
b : 0.00001 M
c : 0.0001 M
d : 0.001 M
e : 0.01 M

Aluminium(III)-Formate

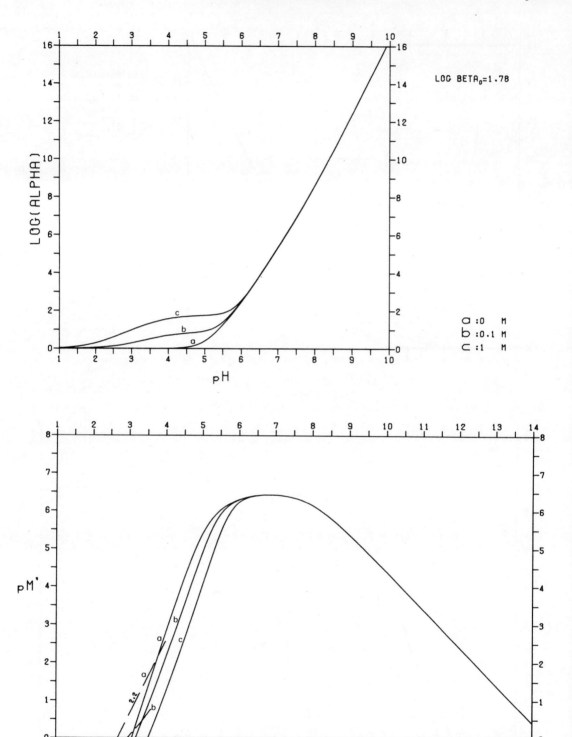

Aluminium(III)-Iminodiacetate

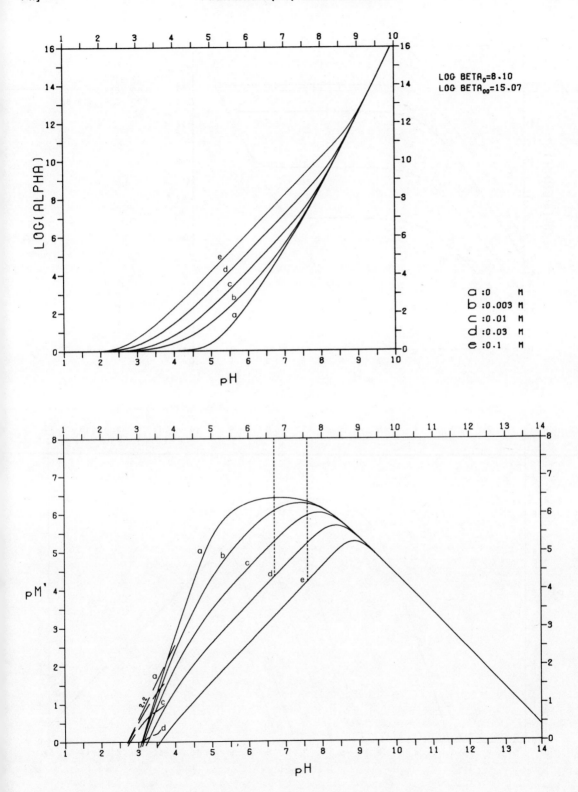

Aluminium(III)-Oxalate [Ch. 3

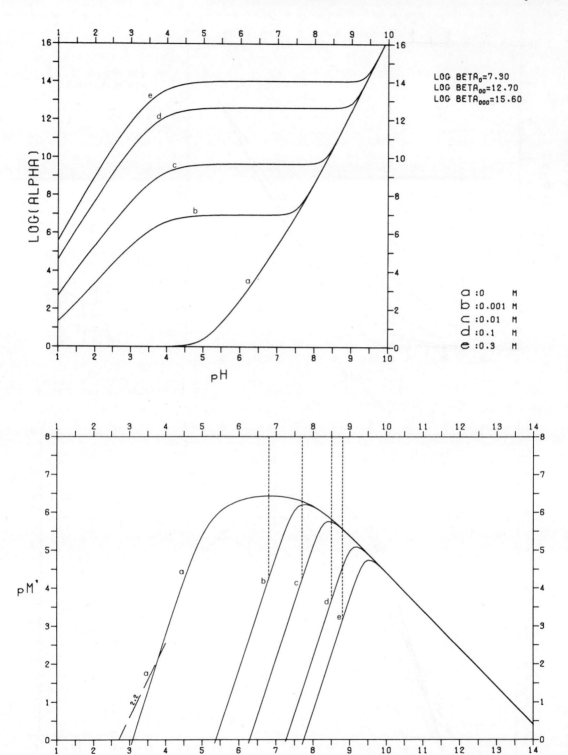

Aluminium(III)-Pyridine-2,6-dicarboxylate

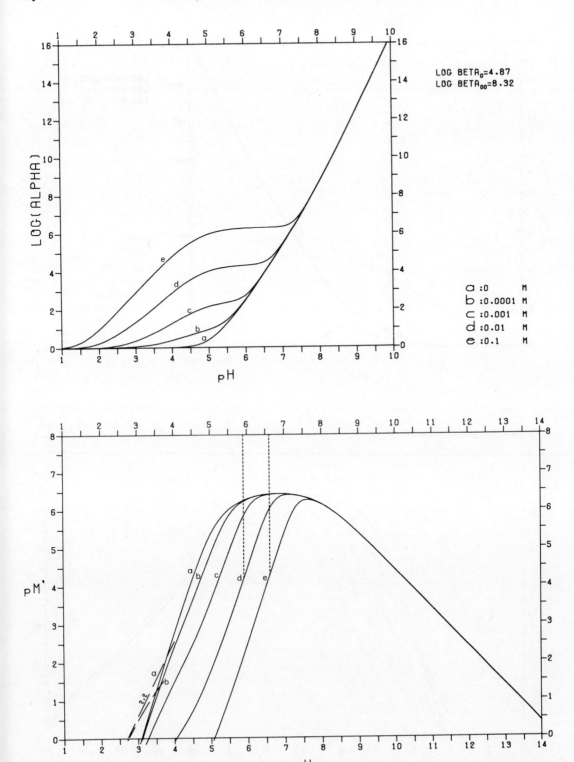

LOG BETA$_0$=4.87
LOG BETA$_{00}$=8.32

a : 0   M
b : 0.0001 M
c : 0.001 M
d : 0.01   M
e : 0.1   M

Aluminium(III)-Salicylate [Ch. 3]

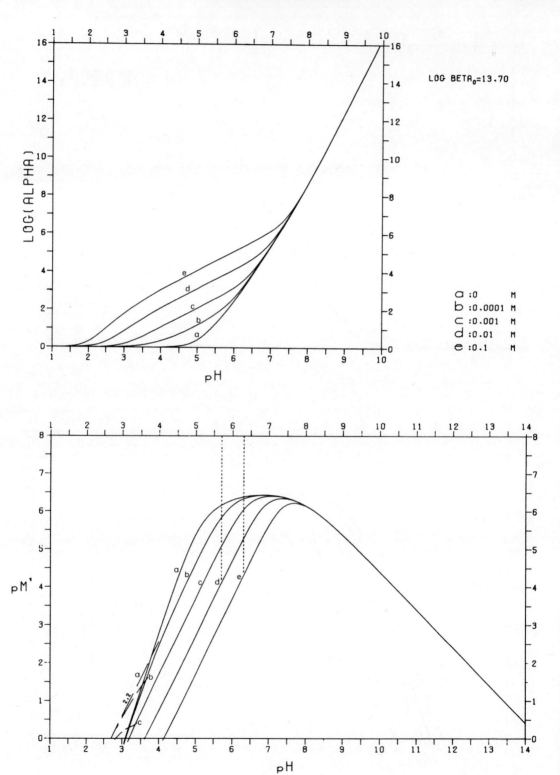

Aluminium(III)-Sulphate

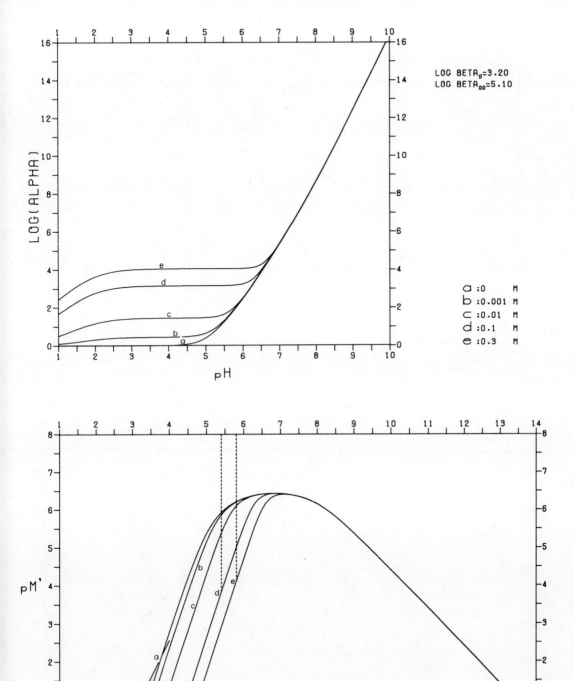

66 Aluminium(III)-5-Sulphosalicylate [Ch. 3

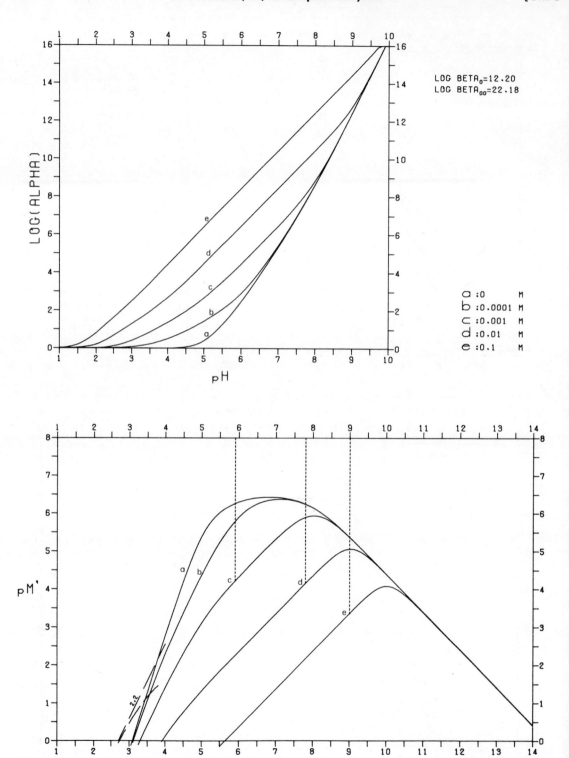

Aluminium(III)-Tartrate

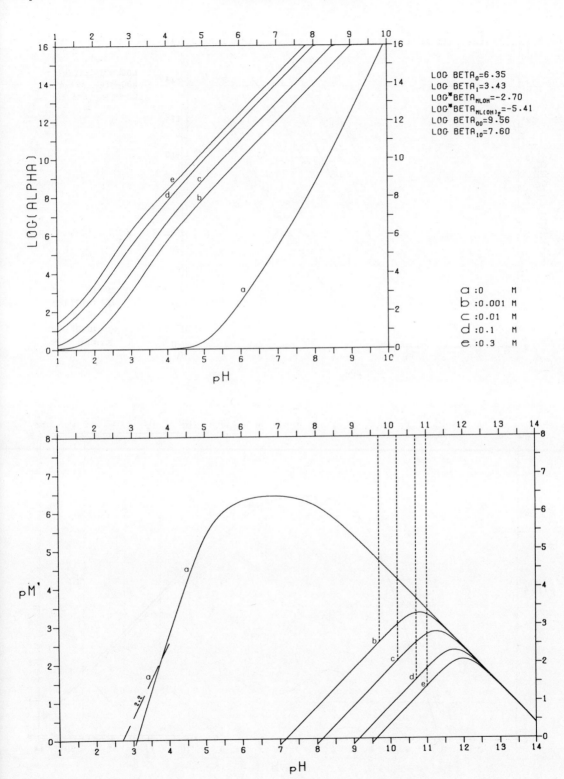

Aluminium(III)-Tiron

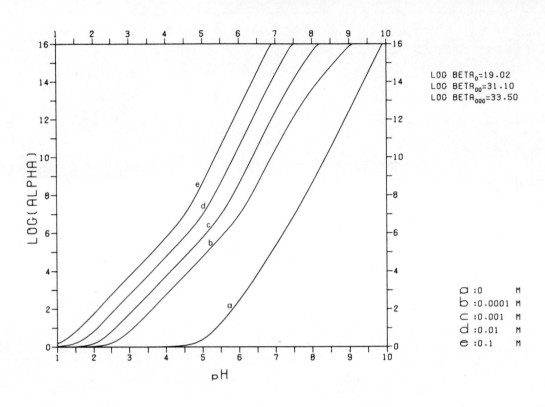

LOG BETA$_0$=19.02
LOG BETA$_{00}$=31.10
LOG BETA$_{000}$=33.50

a : 0   M
b : 0.0001 M
c : 0.001 M
d : 0.01   M
e : 0.1   M

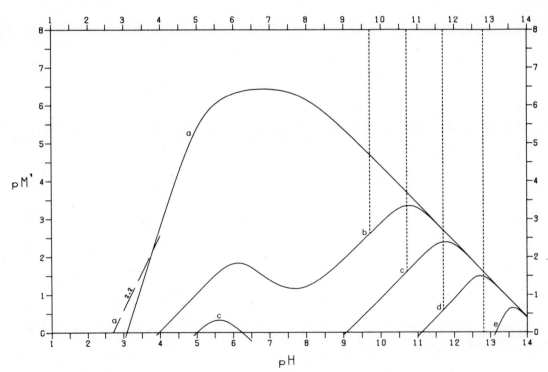

CHAPTER 4

Barium (II)   Ba

Barium shows a very slight tendency to hydrolyse. Log $*\beta_1 = -13.2$ seems to be a reliable value for the formation constant of $Ba(OH)^+$, but there are no values in the literature for log $*\beta_2$ nor for the solubility product, $*K_{s0}$, of barium hydroxide. By analogy with calcium, an approximate value of 24.0 for log $*K_{s0}$ can be assumed. The resulting plots, however, are of little practical use, because even a slight amount of CO_2 leads to the formation of the much less soluble barium carbonate.

Since the equilibria involving $BaCO_3$ are more relevant to practice, the plots in this chapter have been constructed for air-saturated solutions, for which an apparent constant log $K_{carb} = 12.3$ can be derived from the solubility product (log $K_{s0} = -9.4$) of barium carbonate.

Ba-Carbonate

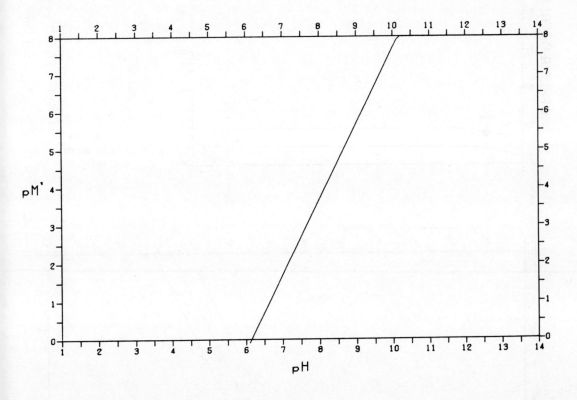

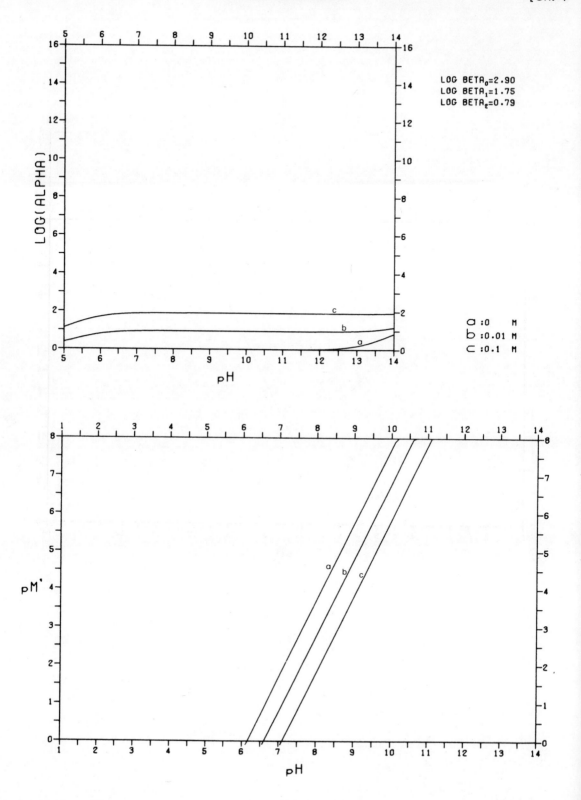

Ba-Carbonate-DCTA

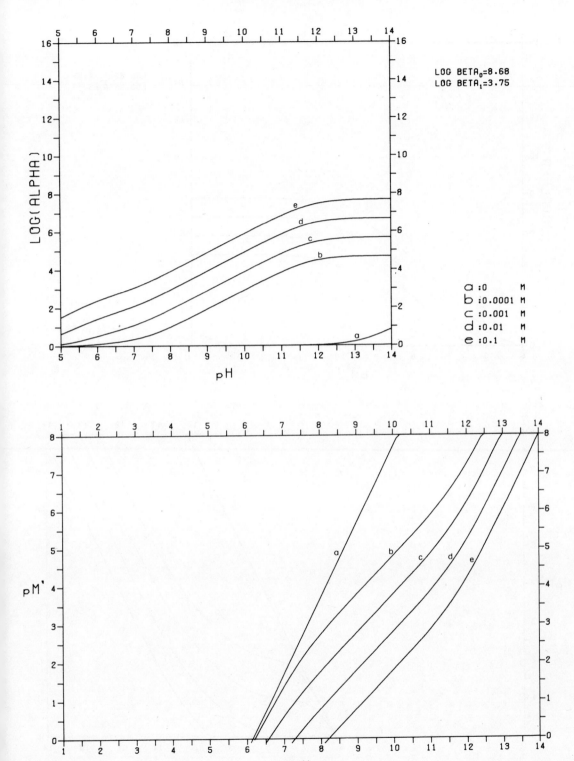

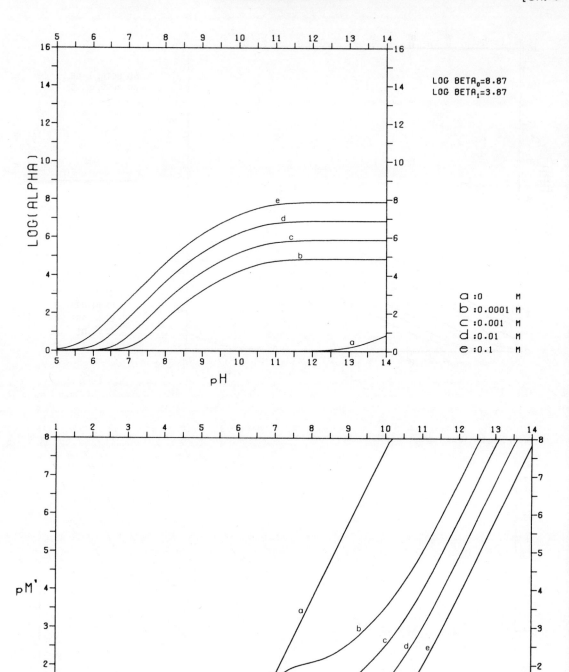

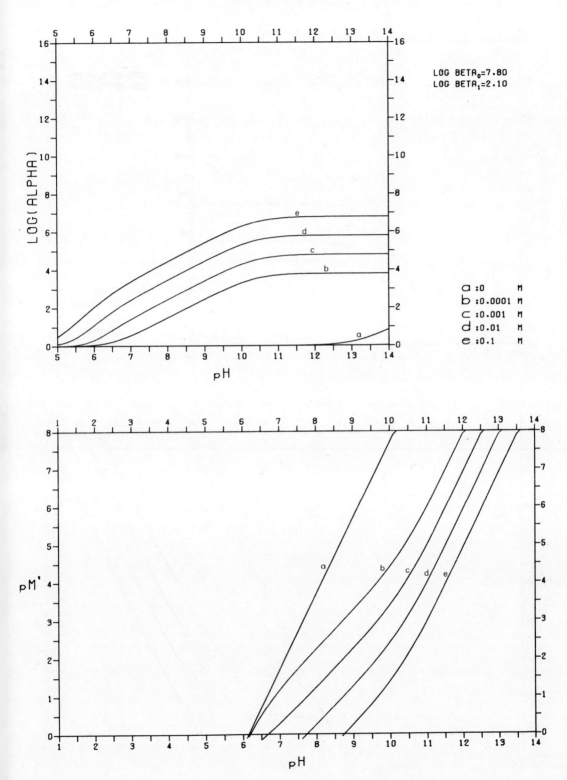

Ba-Carbonate-EGTA [Ch. 4

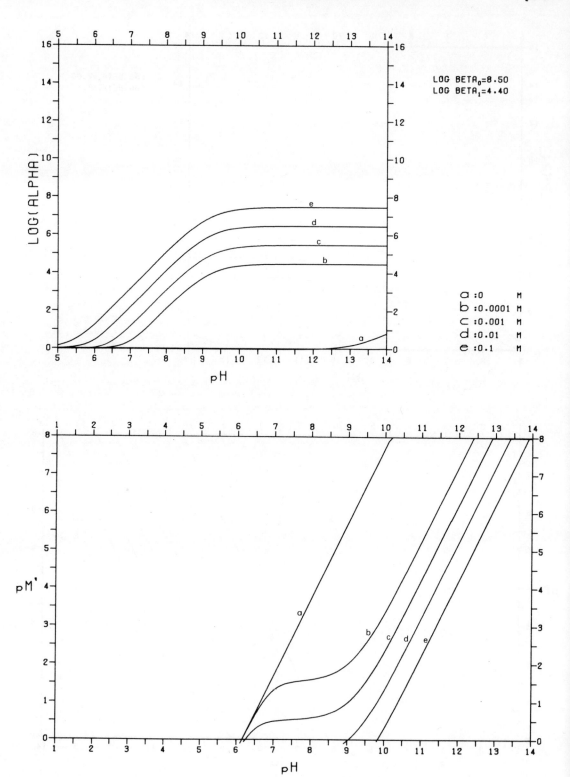

Ba-Carbonate-Pyridine-2,6-dicarboxylate

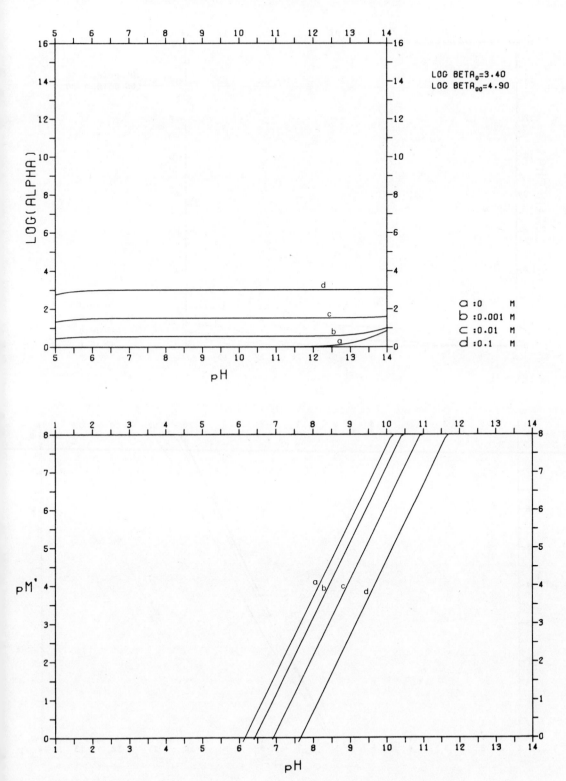

78   **Ba-Carbonate-Tiron**   [Ch. 4

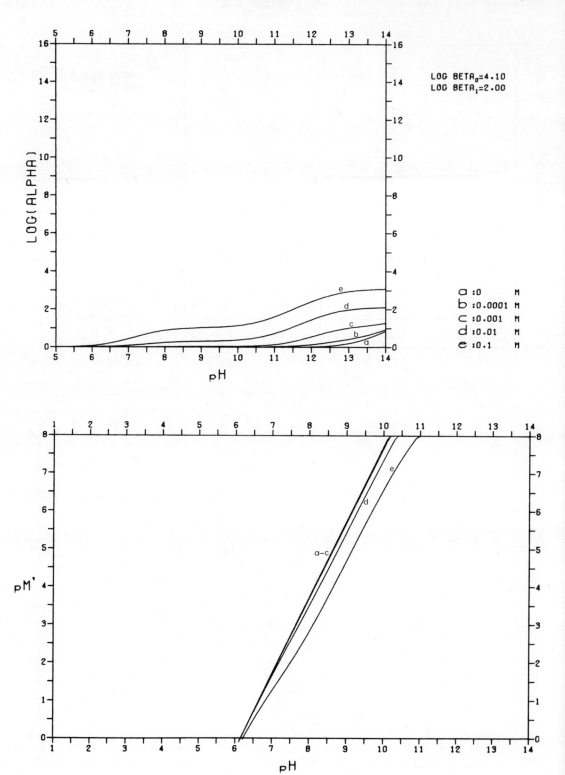

CHAPTER 5

Beryllium (II)   Be

Beryllium has a great tendency to hydrolyse because of the very small size of Be^{2+}. Of the three postulated polynuclear species, Be_2OH^{3+} and $Be_3(OH)_3^{3+}$ definitely exist, but the species $Be_6(OH)_8^{4+}$ is a minor one and its existence remains to be confirmed. Because the formation of the mononuclear species can be studied only in dilute solution, the hydrolysis constants are less accurately known than the formation constants of the polycomplexes. The effect of this inaccuracy on the positions of the lines in the pM' – pH plot is small, however.

The hydroxide is one of the least soluble beryllium compounds. It exists in three forms; the freshly precipitated amorphous material, the α-form produced on standing, and the stable β-form produced after prolonged heating. The difference in solubility between the amorphous material and the α-form is small compared with the dependence of the solubility on temperature and ionic strength. The constant, $*K_{s0}$, for the α-form was used for the plots.

The following constants were used.

$$\log *\beta_1 = -5.7 \qquad \log *\beta_{12} = -3.22$$
$$\log *\beta_2 = -13.9 \qquad \log *\beta_{33} = -8.85$$
$$\log *\beta_3 = -23.8 \qquad \log *\beta_{86} = -27.5$$
$$\log *\beta_4 = -37.8 \qquad \log *K_{s0} = 6.9$$

These values agree closely with those given by Baes and Mesmer [1].

It should be noted that beryllium hydroxide has a tendency to form colloidal solutions and that the amorphous precipitate has a slight tendency to absorb CO_2 from the air.

[1] C. F. Baes and R. E. Mesmer, *The Hydrolysis of Cations*, Wiley, New York, 1976.

Beryllium(III)

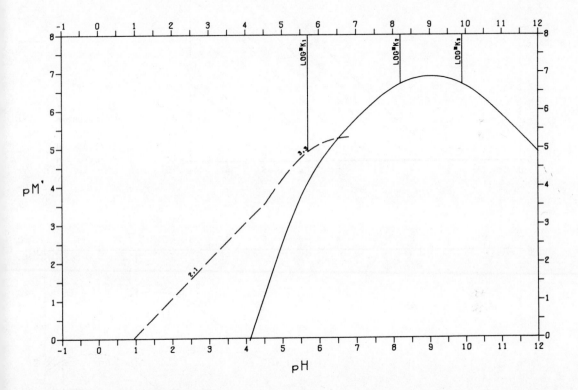

Beryllium(II)-Acetylacetone [Ch. 5

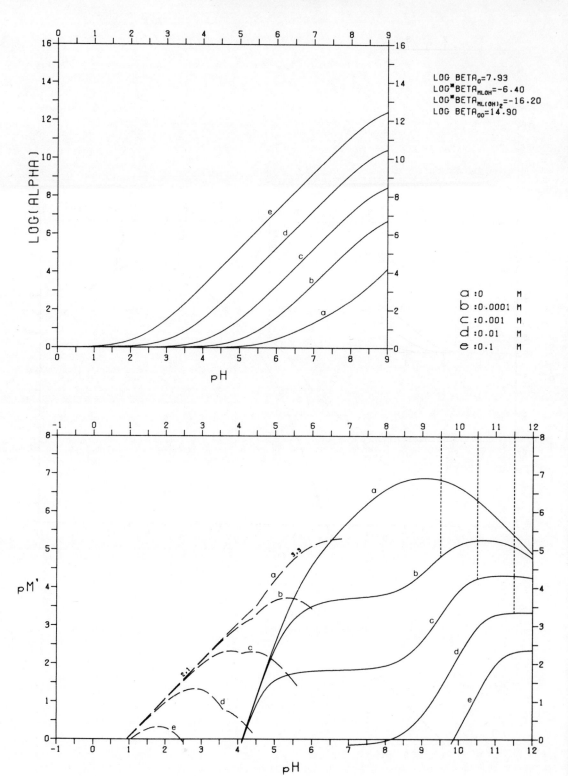

Beryllium(II)-Citrate

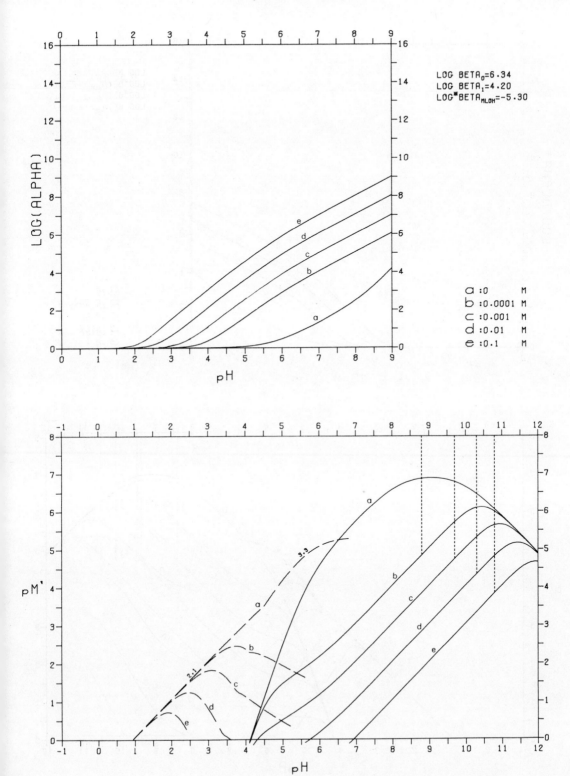

Beryllium(II)–DCTA [Ch. 5]

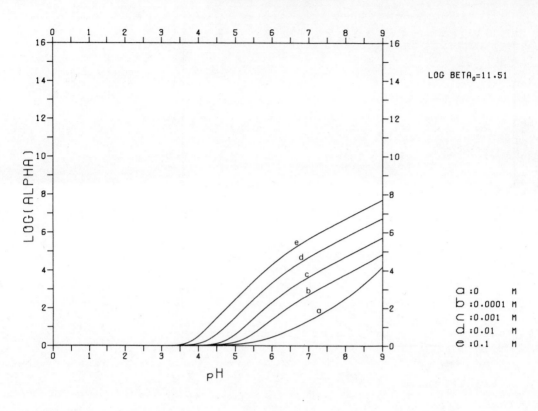

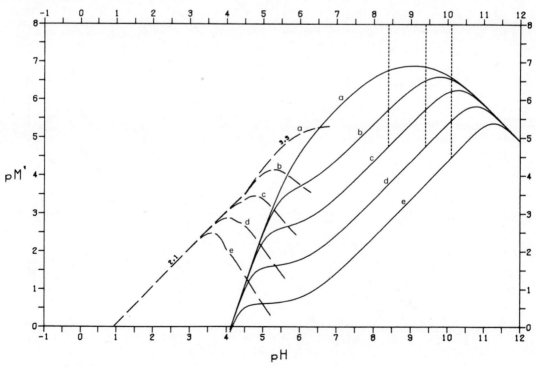

Beryllium(II)-EDTA

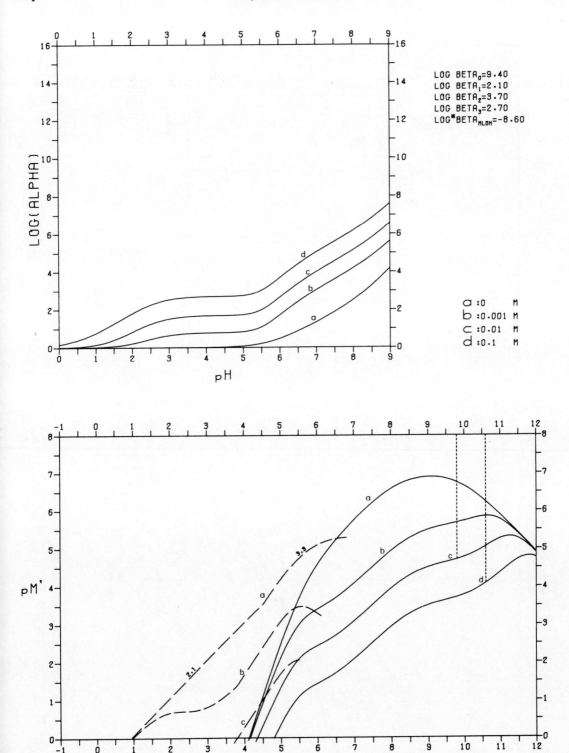

Beryllium(II)-Fluoride [Ch. 5

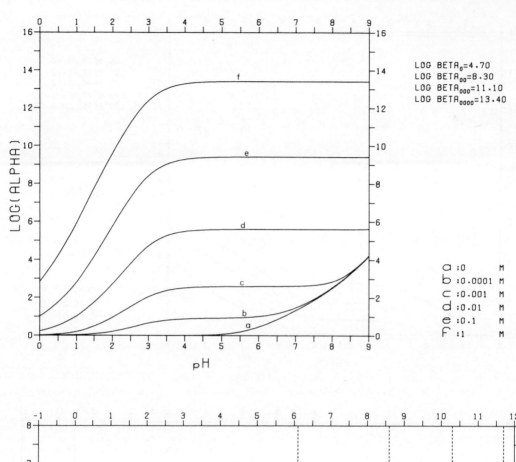

LOG BETA$_0$=4.70
LOG BETA$_{00}$=8.30
LOG BETA$_{000}$=11.10
LOG BETA$_{0000}$=13.40

a : 0   M
b : 0.0001 M
c : 0.001   M
d : 0.01   M
e : 0.1   M
f : 1   M

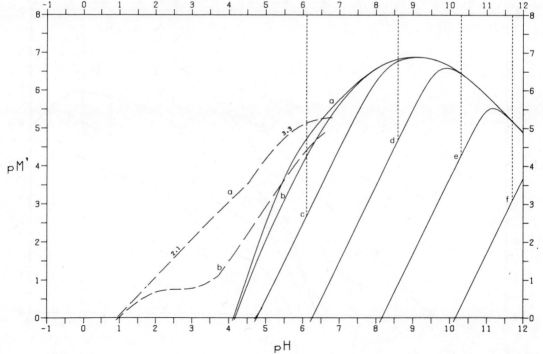

Beryllium(II)-Oxalate

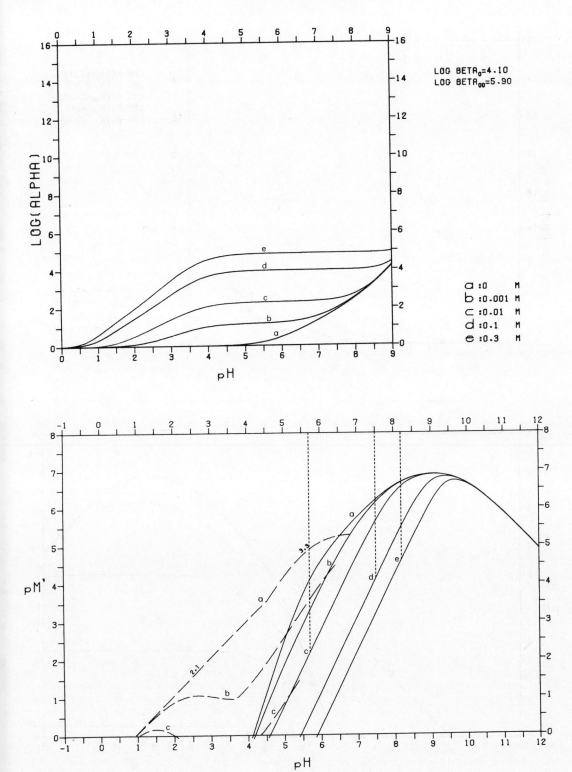

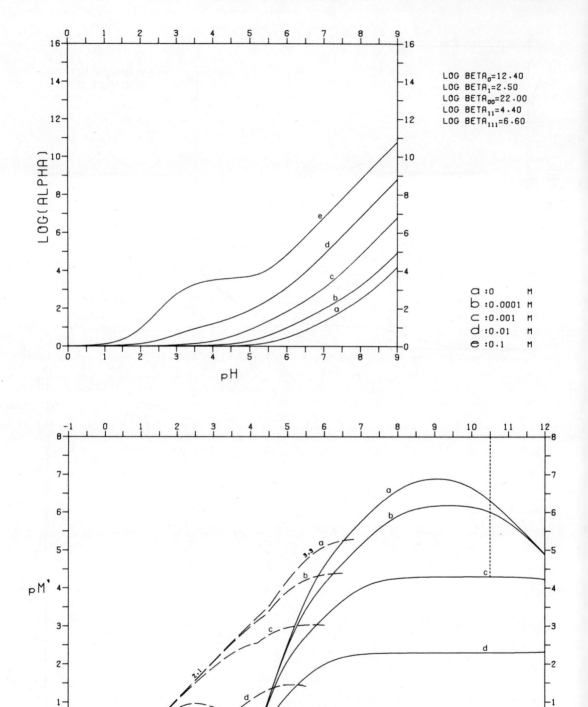

Beryllium(II)-Sulphate

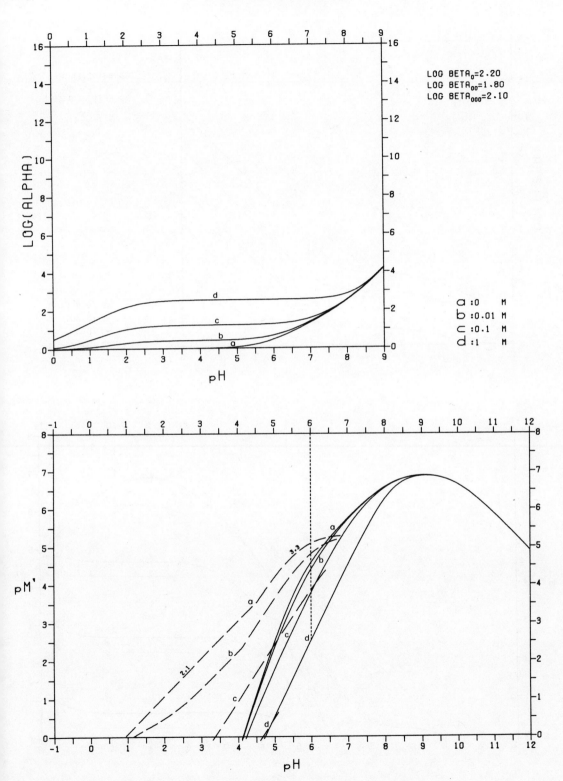

90 **Beryllium(II)-5-Sulphosalicylate** [Ch. 5

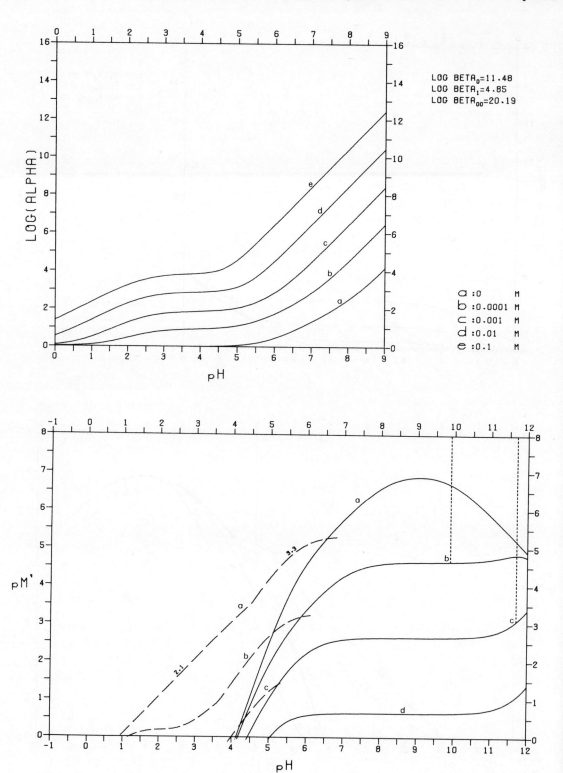

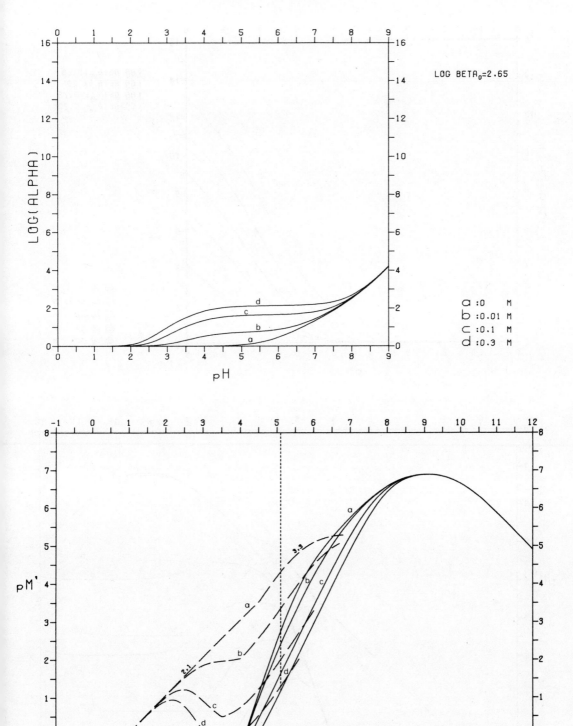

Beryllium(II)–Tiron

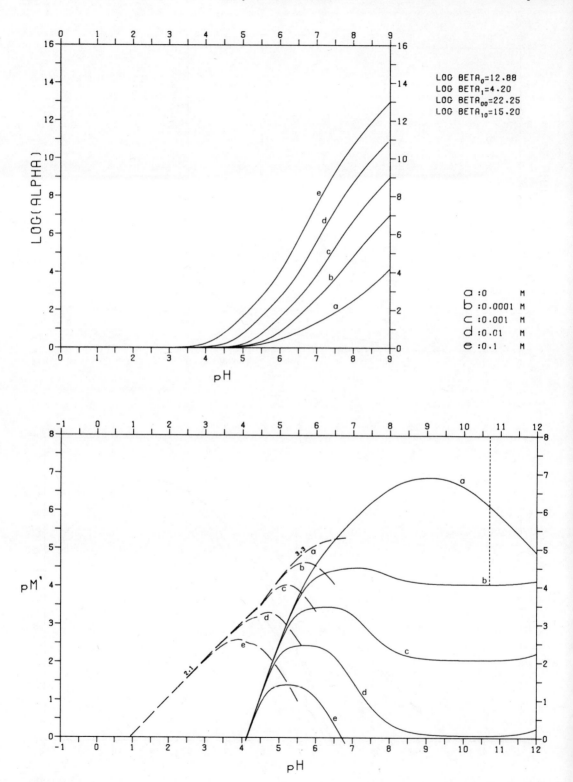

CHAPTER 6
Bismuth (III)   Bi

The hydrolysis of Bi^{3+} has been studied extensively, particularly with regard to the identification of polynuclear species. It is generally agreed that $Bi_6(OH)_{12}^{6+}$ is the principal polycomplex. A series of enneamers (Bi_9-species) has also been reported, but insufficient data are available for constants to be estimated. This leads to some uncertainty in the position of the steep side of the precipitation boderline in the pM' – pH plot.

Of the three crystalline modifications of Bi_2O_3 only the α-form is of importance. The difference between the α-form and the freshly precipitated white insoluble form can be neglected.

The following values were used for constructing the plots.

$$\log {}^*\beta_1 = -1.43 \qquad \log {}^*\beta_{12,6} = 0.33$$
$$\log {}^*\beta_2 = -4.6$$
$$\log {}^*\beta_3 = -9.44$$
$$\log {}^*\beta_4 = -22.2 \qquad \log {}^*K_{s0} = 4.04$$

It should be noted that bismuth has a tendency to form basic salts (such as BiOCl and $BiONO_3$) in concentrated solutions, and that these are not included in the plots.

Bismuth(III)

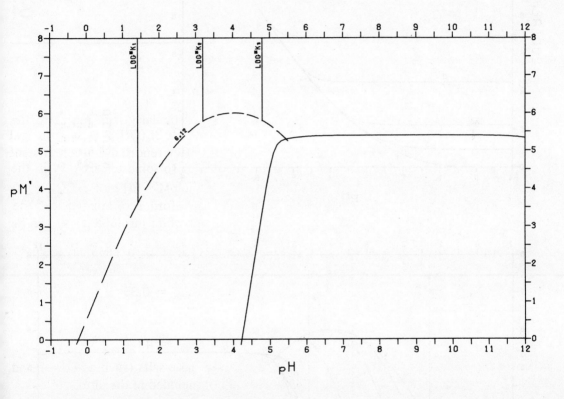

Bismuth(III)–Bromide

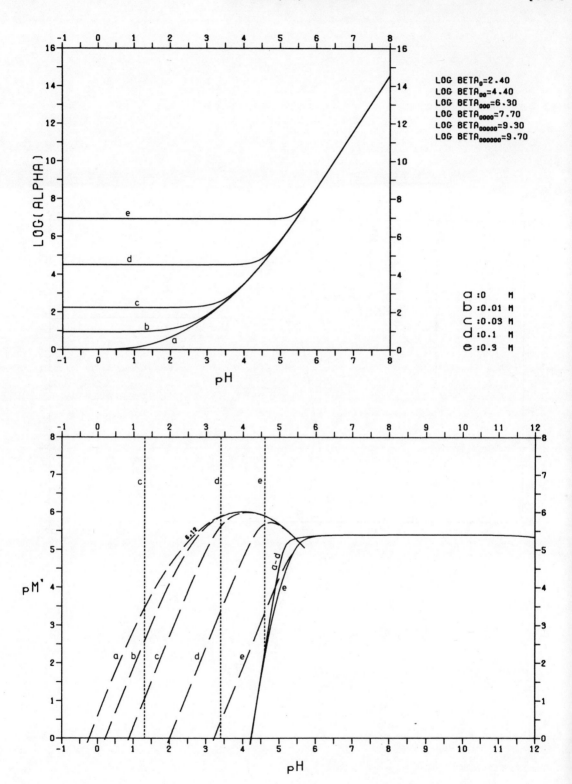

Bismuth(III)-Chloride

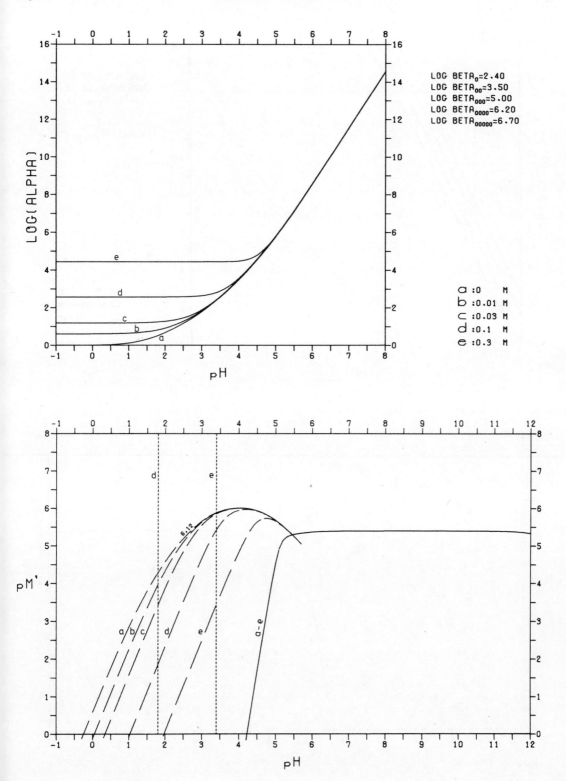

Bismuth(III)-DCTA

[Ch. 6

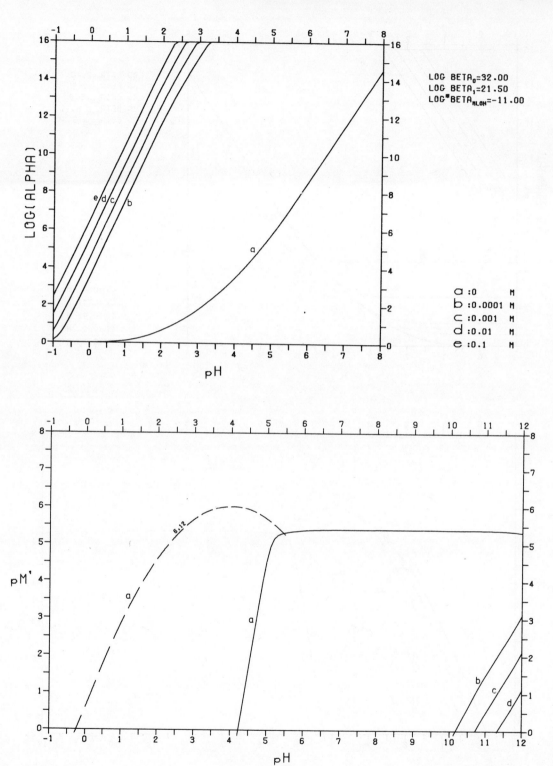

Bismuth(III)-DTPA

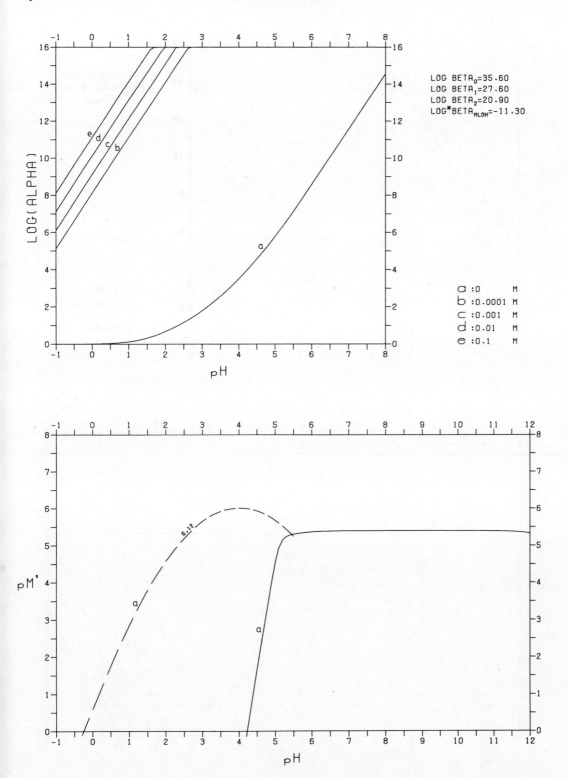

LOG BETA₀=35.60
LOG BETA₁=27.60
LOG BETA₂=20.90
LOG*BETA_MLOH=-11.30

a: 0 M
b: 0.0001 M
c: 0.001 M
d: 0.01 M
e: 0.1 M

Bismuth(III)-EDTA [Ch. 6]

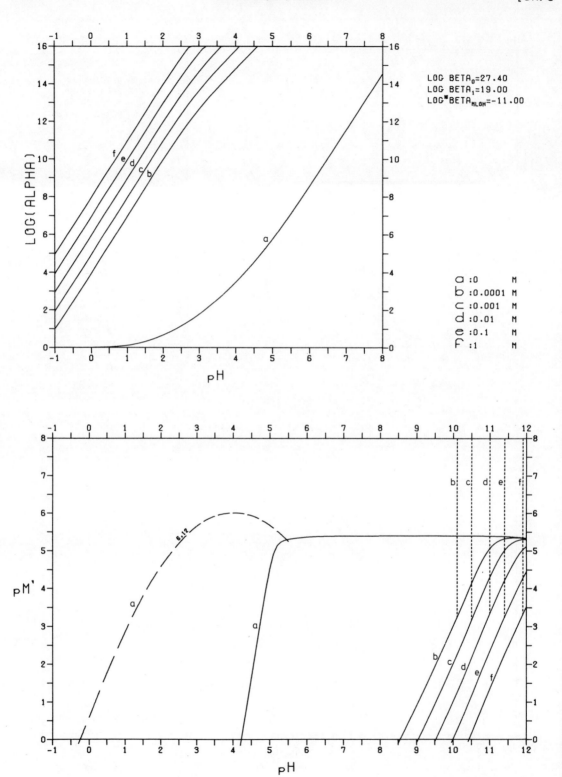

Bismuth(III)-Fluoride

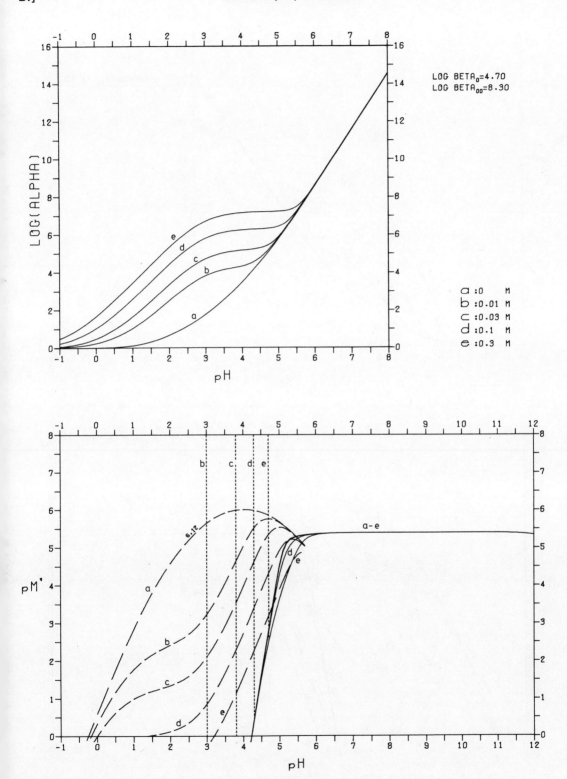

102 Bismuth(III)-Tartrate [Ch. 6

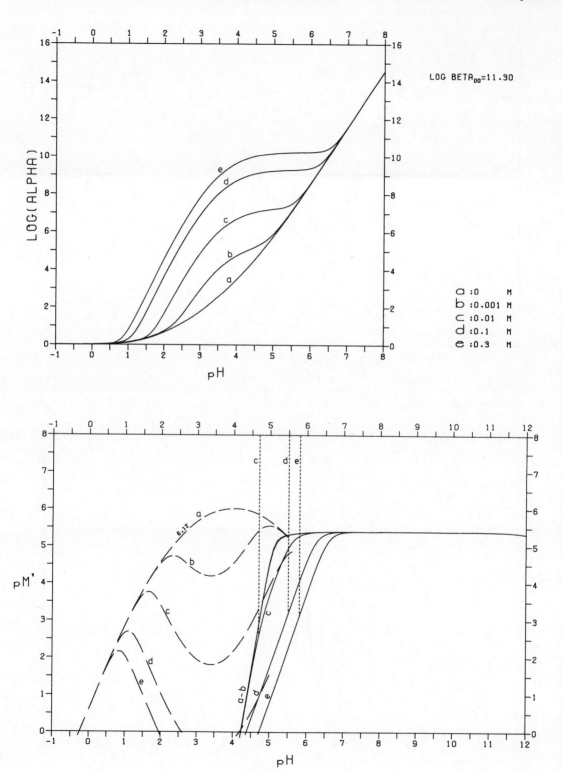

CHAPTER 7

Calcium (II)   Ca

Calcium has a slight tendency to hydrolyse, and the values log $*\beta_1 = -12.6$ and log $*K_{s0} = 23.0$ can be found in the literature. No value has been found for log $*\beta_2$. The plots resulting from these values are not very useful in practice because even a slight amount of CO_2 absorption leads to the formation of the much more insoluble calcium carbonate. For practical purposes, it is much more appropriate to consider this compound, so the plots have been constructed with the apparent constant log $K_{carb} = 13.4$, which has been derived from the solubility product (log $K_{s0} = -8.3$) and holds for air-saturated solutions. The constants used for construction of the plots are:

$$\log *\beta_1 = -12.6$$
$$\log K_{carb} = 13.4$$

Ca-Carbonate

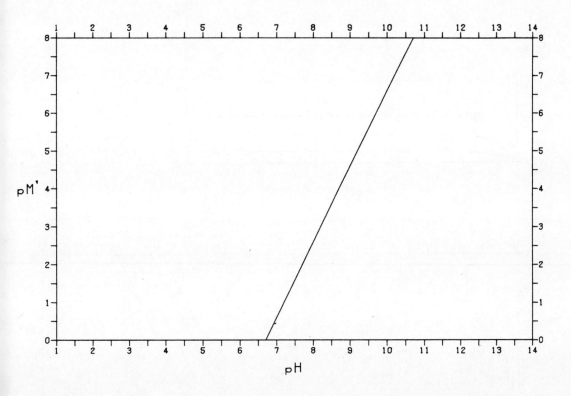

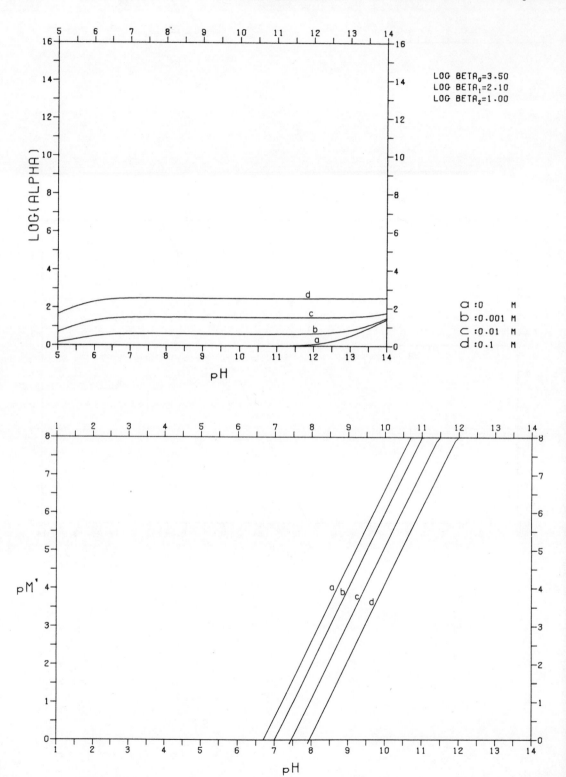

Ca-Carbonate-DCTA

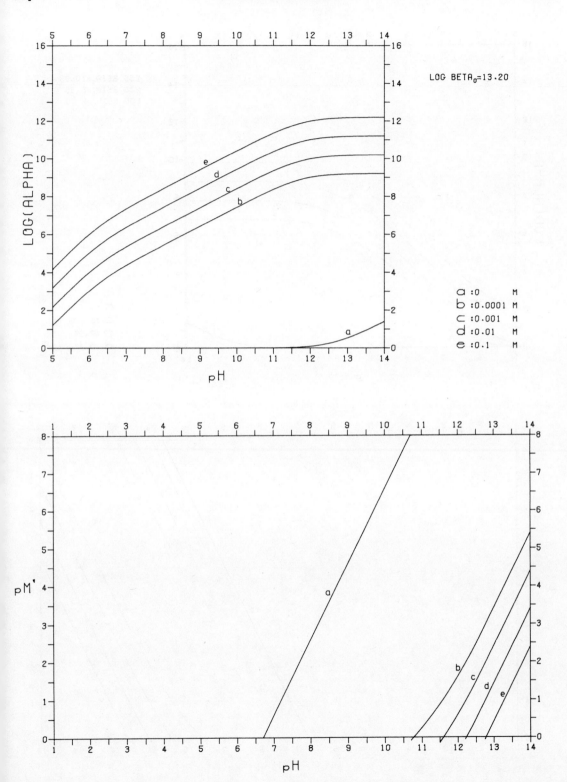

108 Ca-Carbonate-DTPA [Ch. 7

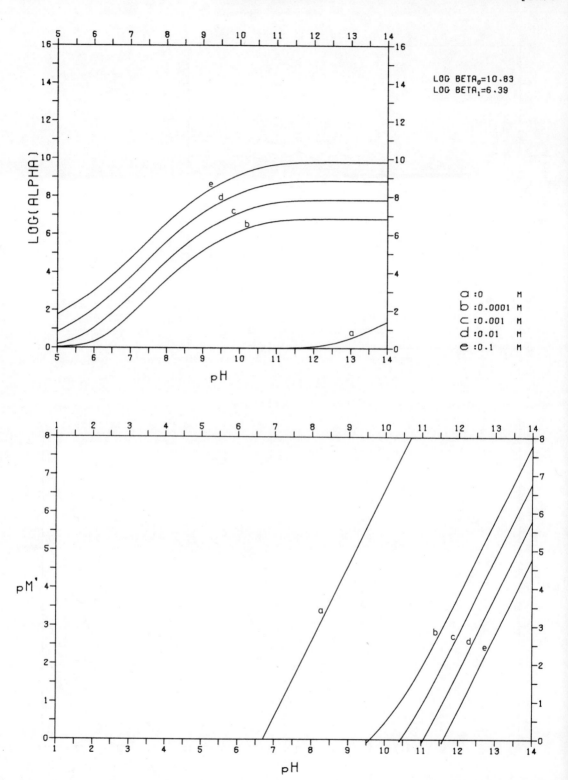

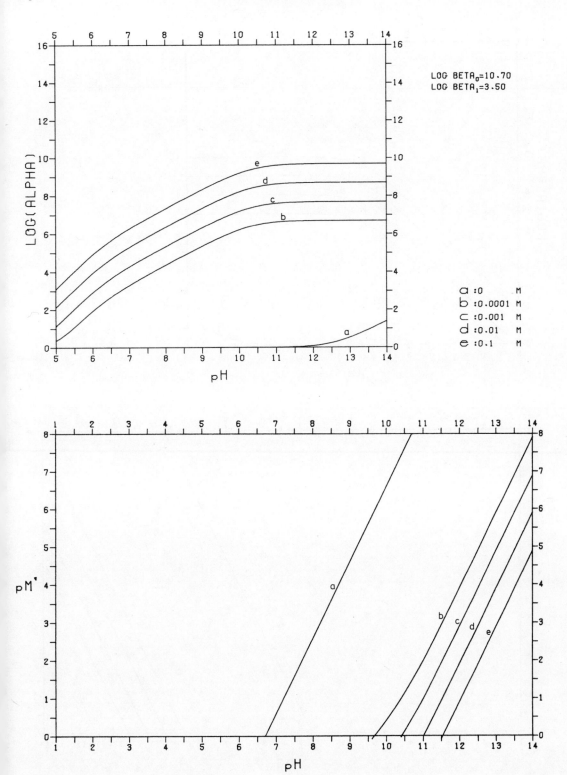

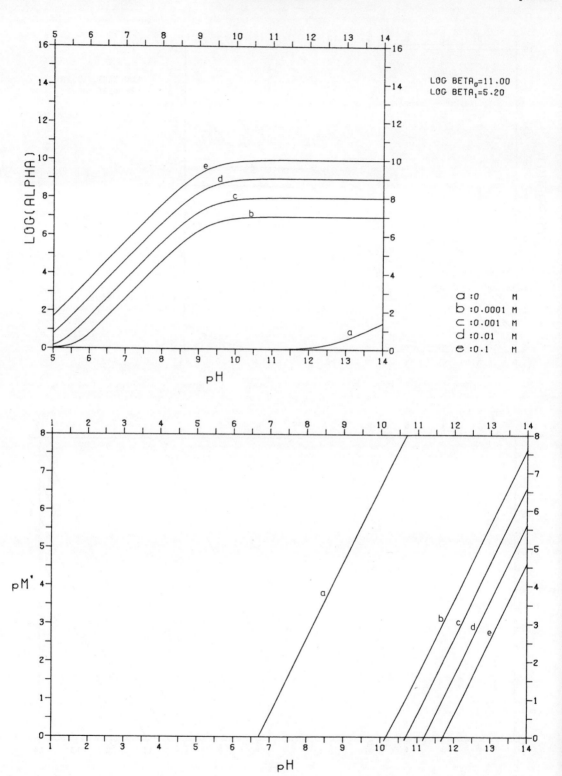

Ca-Carbonate-Iminodiacetate

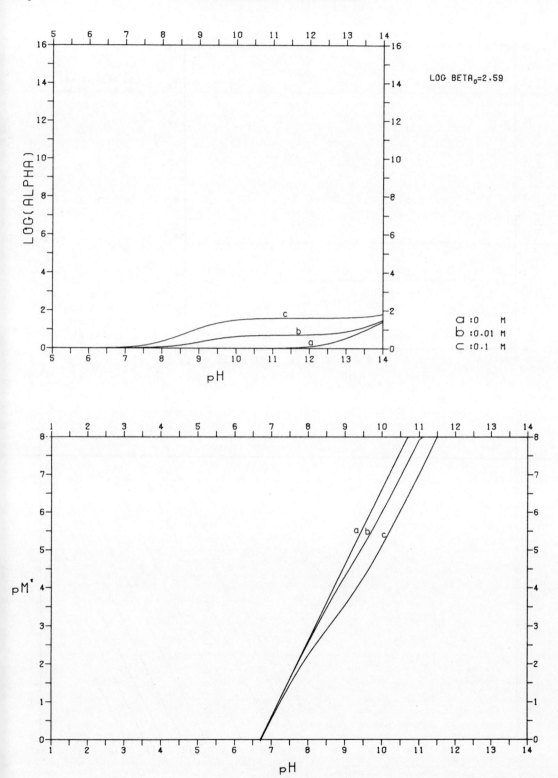

Ca-Carbonate-Pyridine-2,6-dicarboxylate [Ch. 7

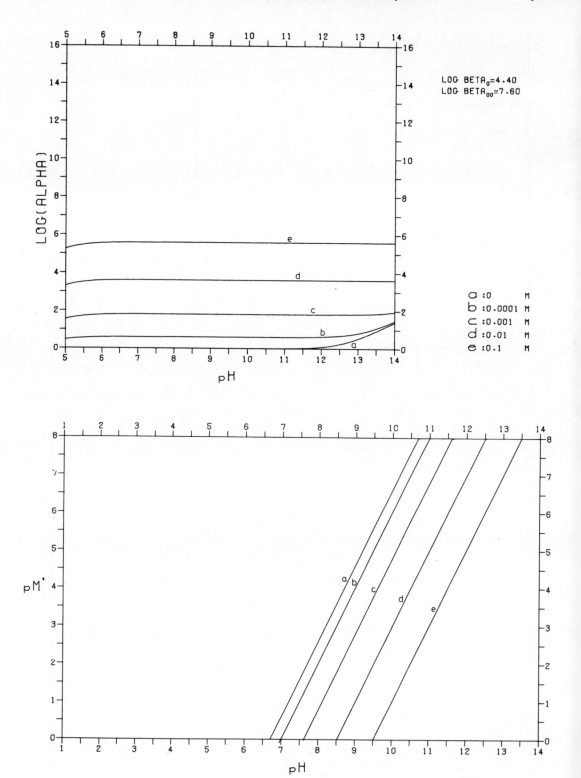

Ca-Carbonate-Tiron

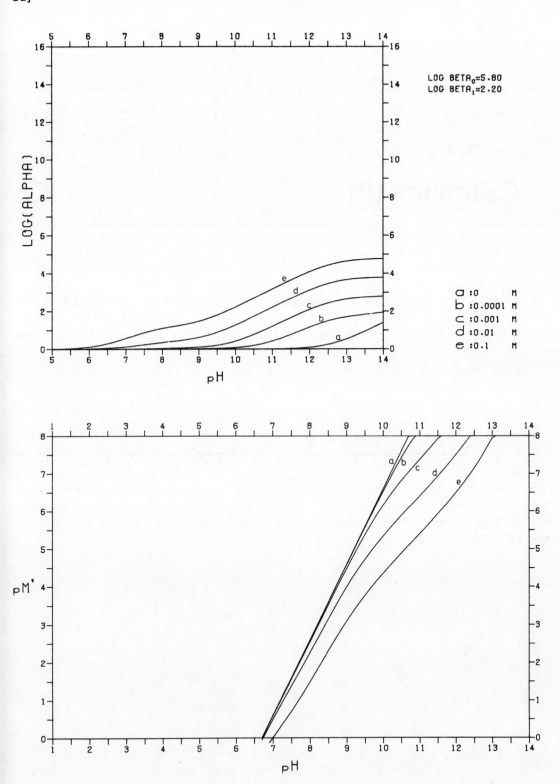

CHAPTER 8

Cadmium (II)   Cd

Cadmium forms polycomplexes in a small range of concentrated solutions (>0.1M) just before $Cd(OH)_2$ is precipitated. The species $Cd_2(OH)^{3+}$ definitely exists; in addition $Cd_4(OH)_4^{4+}$ and/or $Cd_4(OH)_5^{5+}$ have been proposed, but, if they exist, they are of minor importance.

The following set of constants was selected for the first set of plots:

$$\log {}^*\beta_1 = -10.3 \qquad \log {}^*\beta_{12} = -9.2$$
$$\log {}^*\beta_2 = -20.6 \qquad \log {}^*\beta_{44} = -32.4$$
$$\log {}^*\beta_3 = -33.8$$
$$\log {}^*\beta_4 = -46.9 \qquad \log {}^*K_{s0} = 13.65$$

Study of the formation of cadmium hydroxide requires careful exclusion of carbonate. Cadmium carbonate is extremely insoluble and it precipitates from air-saturated aqueous solution long before $Cd(OH)_2$. Plots for the cadmium carbonate system are given on pages 137–158.

Cadmium(II)

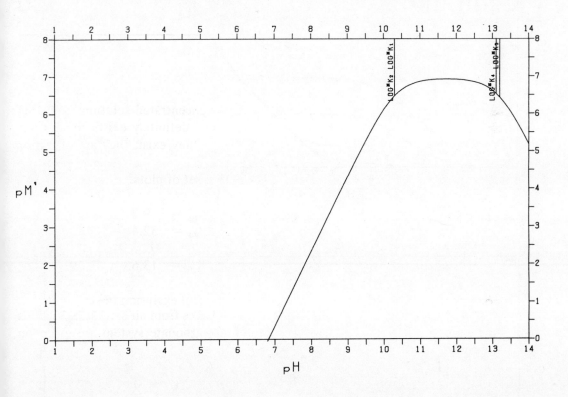

Cadmium(II)-Acetylacetone [Ch. 8

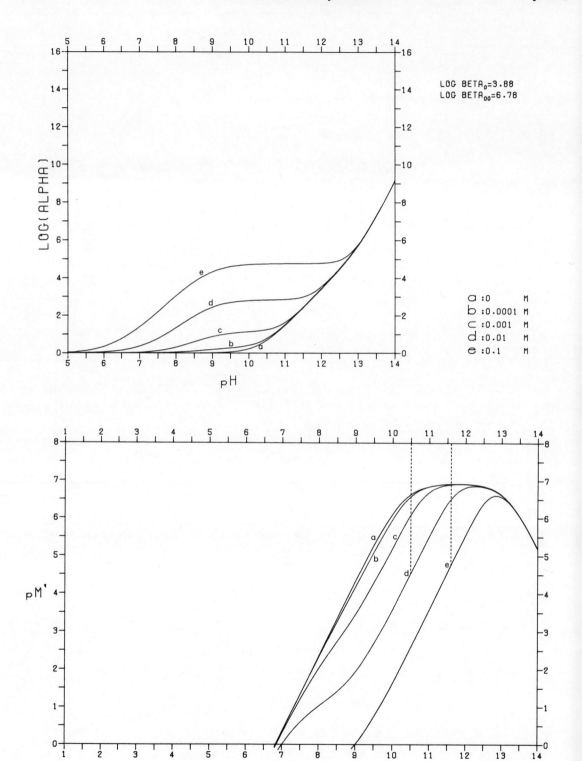

Cadmium(II)-Ammonia

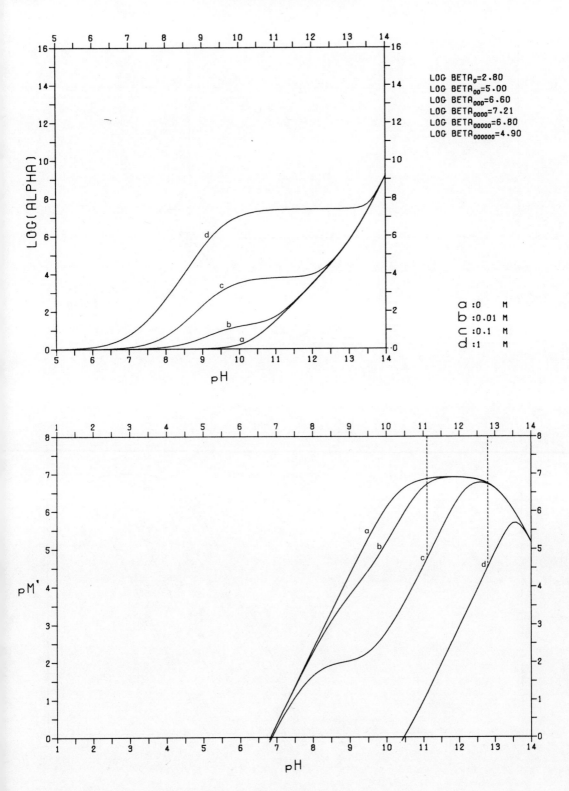

LOG BETA$_0$=2.80
LOG BETA$_{00}$=5.00
LOG BETA$_{000}$=6.60
LOG BETA$_{0000}$=7.21
LOG BETA$_{00000}$=6.80
LOG BETA$_{000000}$=4.90

a : 0 M
b : 0.01 M
c : 0.1 M
d : 1 M

Cadmium(II)–Bromide

[Ch. 8

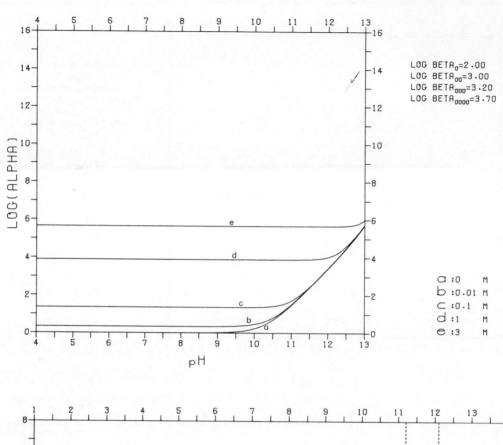

LOG BETA$_0$=2.00
LOG BETA$_{00}$=3.00
LOG BETA$_{000}$=3.20
LOG BETA$_{0000}$=3.70

a : 0   M
b : 0.01 M
c : 0.1 M
d : 1   M
e : 3   M

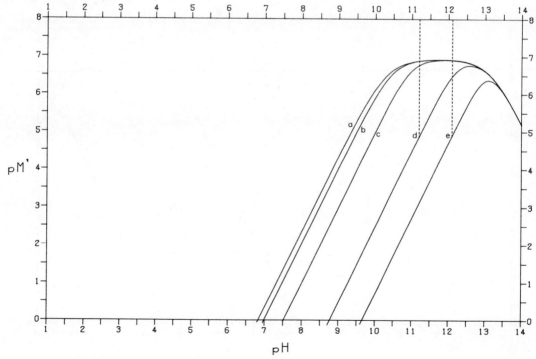

Cadmium(II)-Chloride

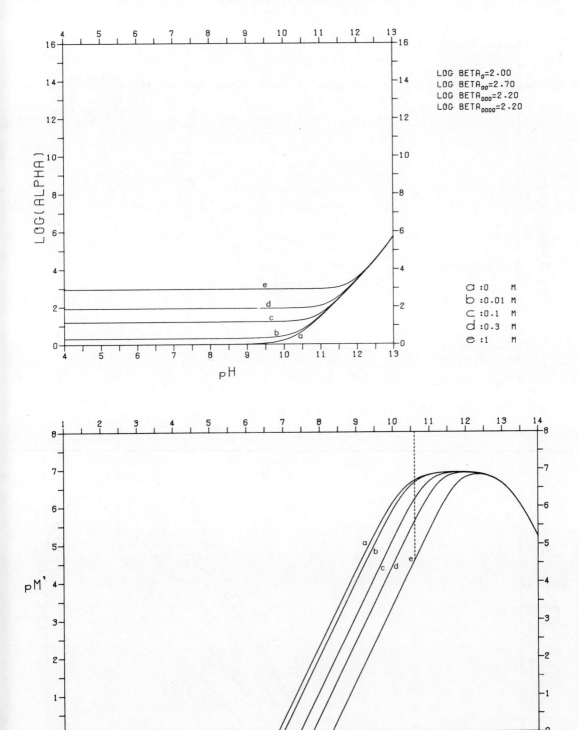

LOG BETA$_0$=2.00
LOG BETA$_{00}$=2.70
LOG BETA$_{000}$=2.20
LOG BETA$_{0000}$=2.20

a : 0   M
b : 0.01 M
c : 0.1 M
d : 0.3 M
e : 1   M

Cadmium(II)-Citrate [Ch. 8]

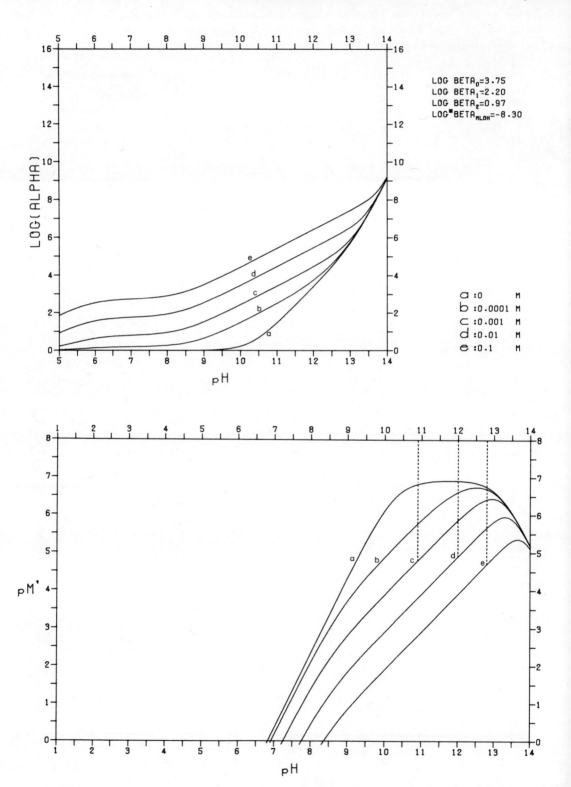

Cadmium(II)-Cyanide

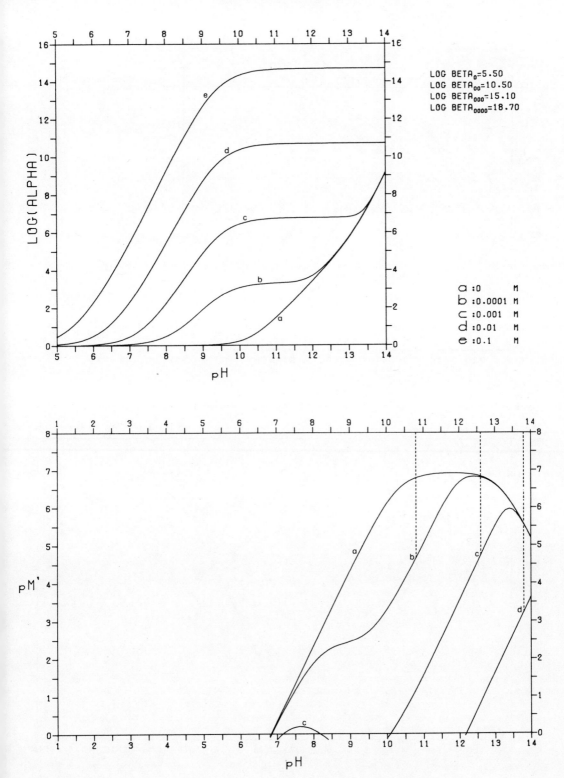

LOG BETA$_0$=5.50
LOG BETA$_{00}$=10.50
LOG BETA$_{000}$=15.10
LOG BETA$_{0000}$=18.70

a : 0 M
b : 0.0001 M
c : 0.001 M
d : 0.01 M
e : 0.1 M

122 Cadmium(II)-DCTA [Ch. 8

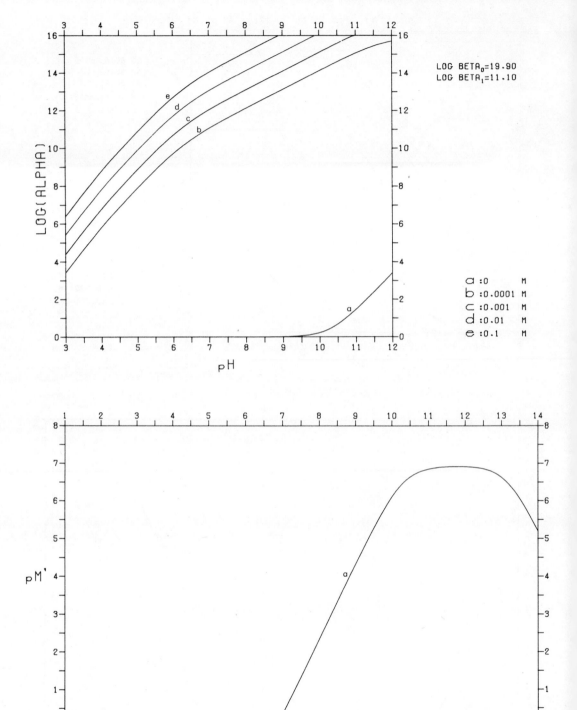

Cadmium(II)-DTPA

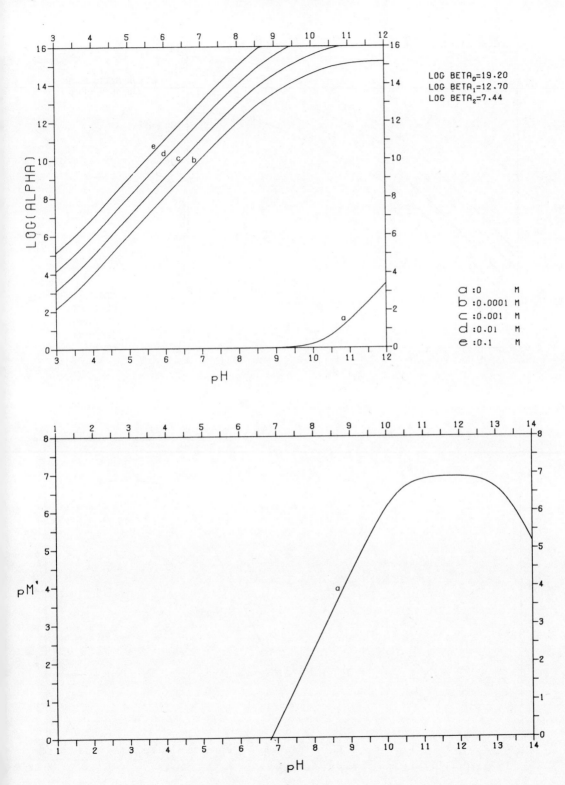

Cadmium(II)-EDTA

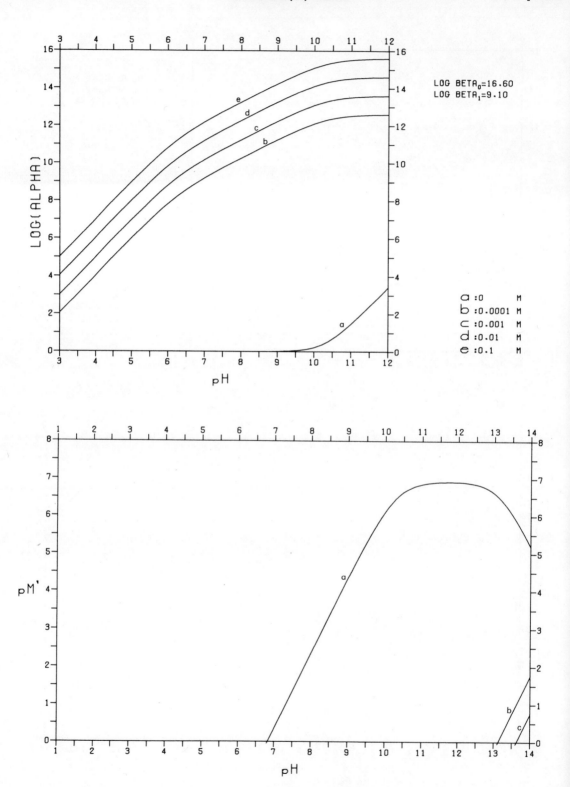

Cadmium(II)-EGTA

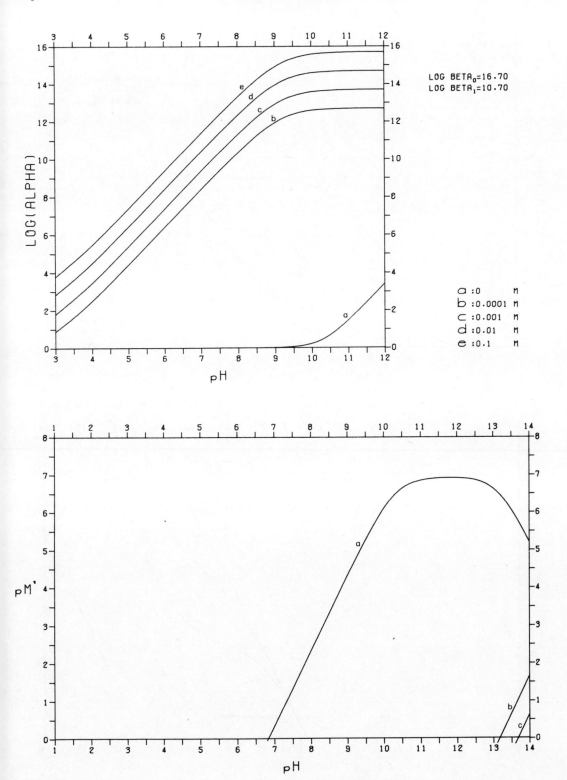

LOG BETA$_0$=16.70
LOG BETA$_1$=10.70

a : 0 M
b : 0.0001 M
c : 0.001 M
d : 0.01 M
e : 0.1 M

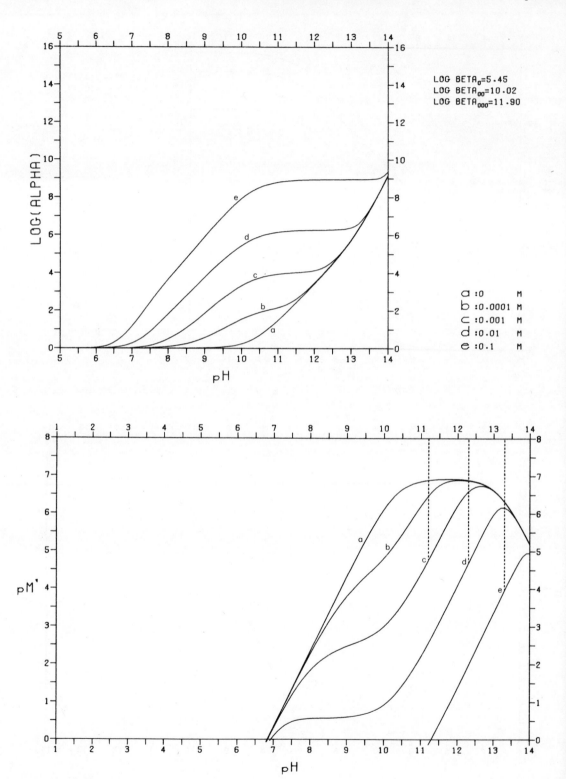

Cadmium(II)-Glycine

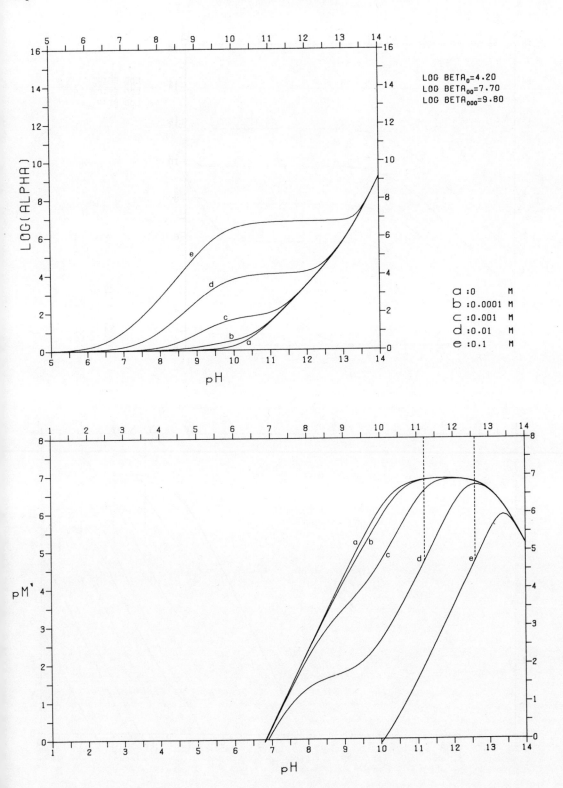

LOG BETA₀=4.20
LOG BETA₀₀=7.70
LOG BETA₀₀₀=9.80

a : 0   M
b : 0.0001 M
c : 0.001   M
d : 0.01   M
e : 0.1   M

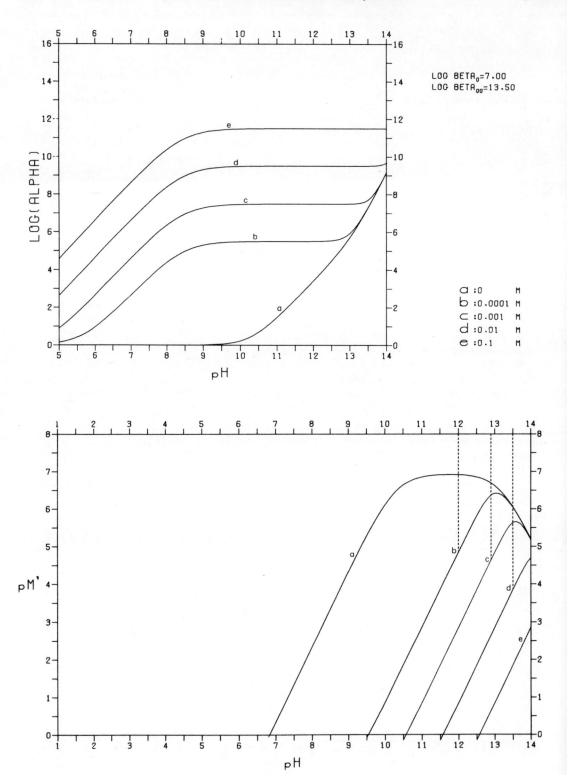

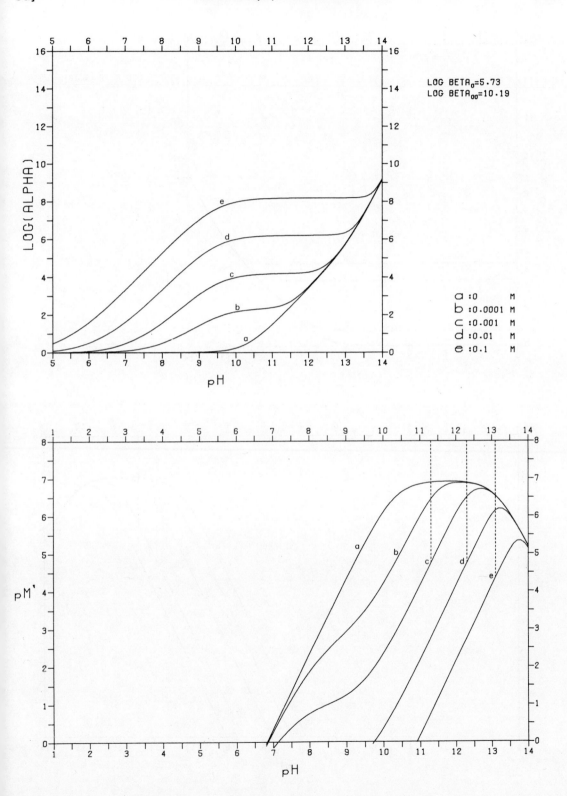

Cadmium(II)-Oxalate

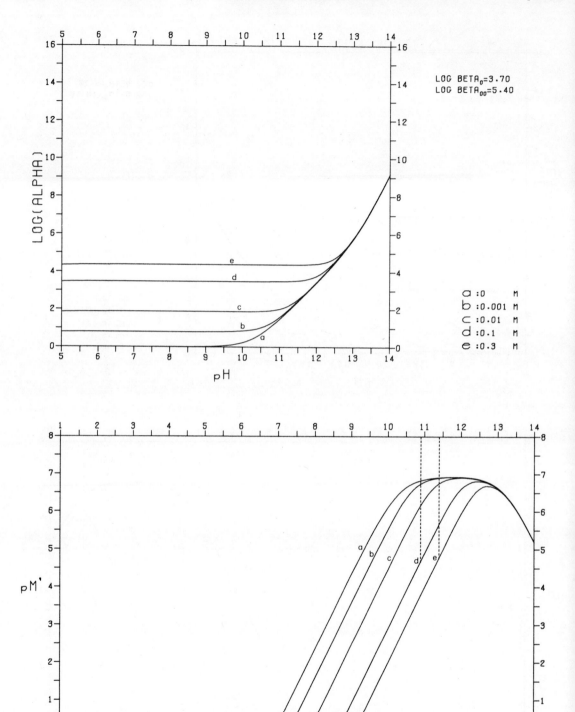

Cadmium(II)-Pyridine-2,6-dicarboxylate

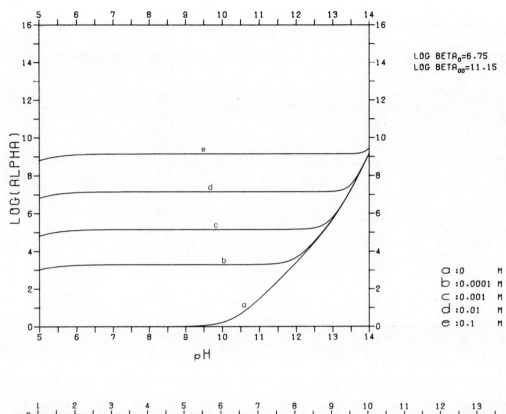

LOG BETA$_0$=6.75
LOG BETA$_{00}$=11.15

a : 0 M
b : 0.0001 M
c : 0.001 M
d : 0.01 M
e : 0.1 M

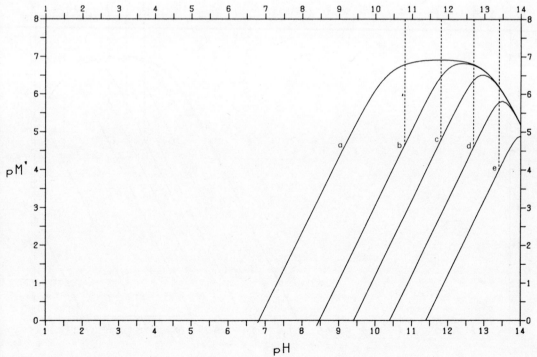

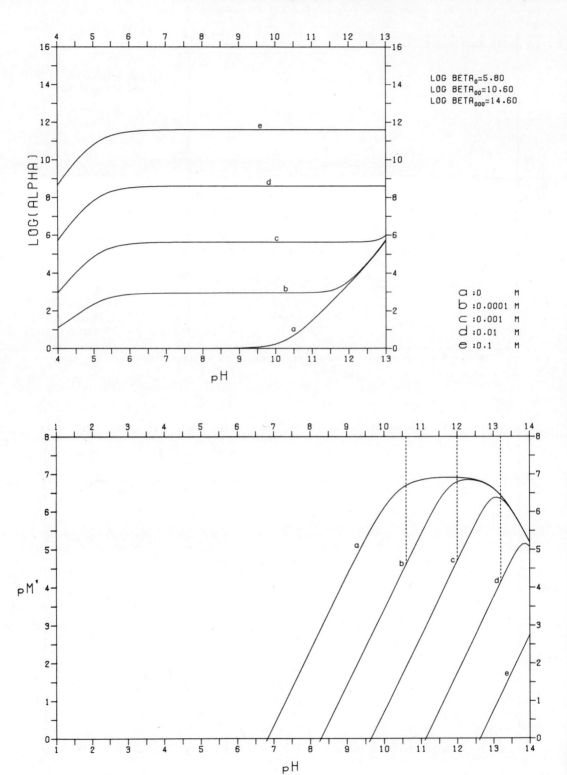

Cadmium(II)-Sulphate

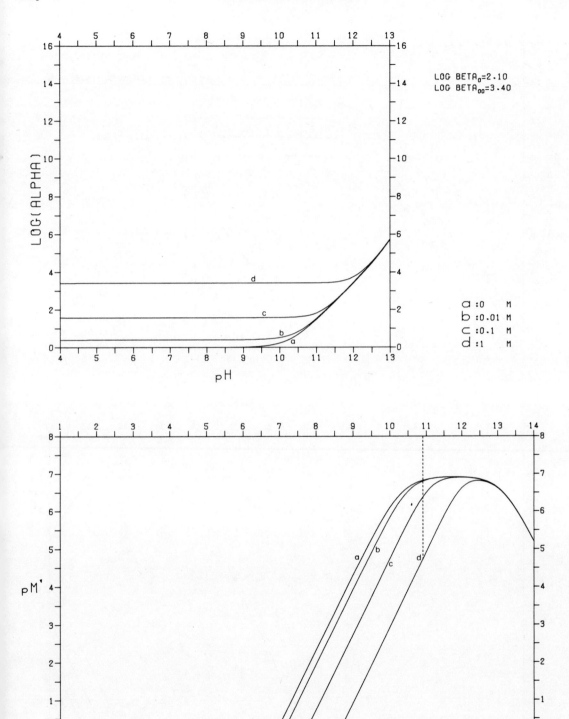

Cadmium(II)-Tetren

[Ch. 8

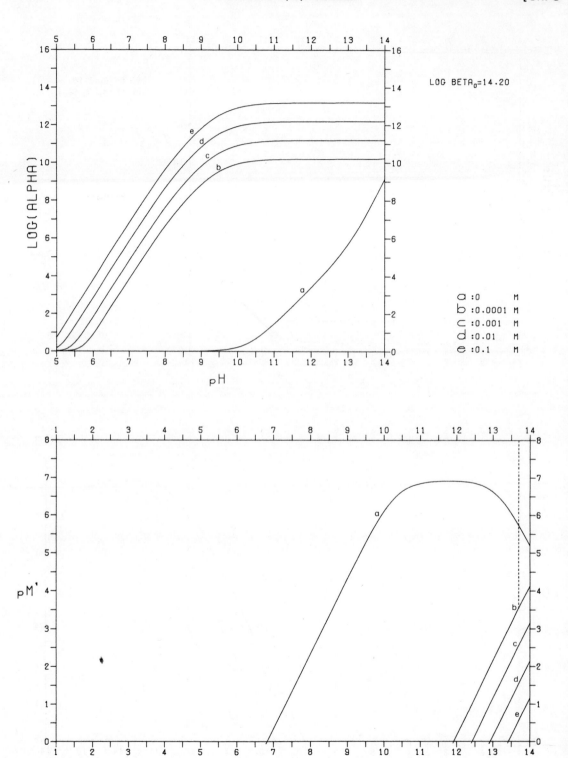

Cadmium(II)-Tiron

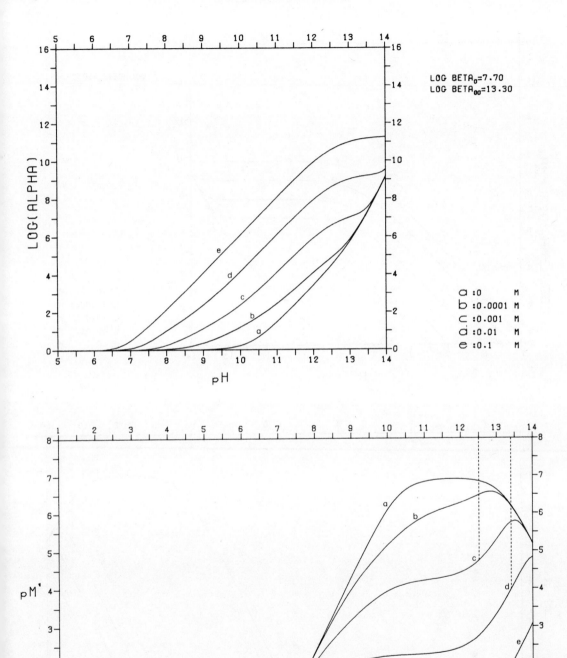

LOG BETA₀=7.70
LOG BETA₀₀=13.30

a : 0 M
b : 0.0001 M
c : 0.001 M
d : 0.01 M
e : 0.1 M

Cadmium(II)-Trien

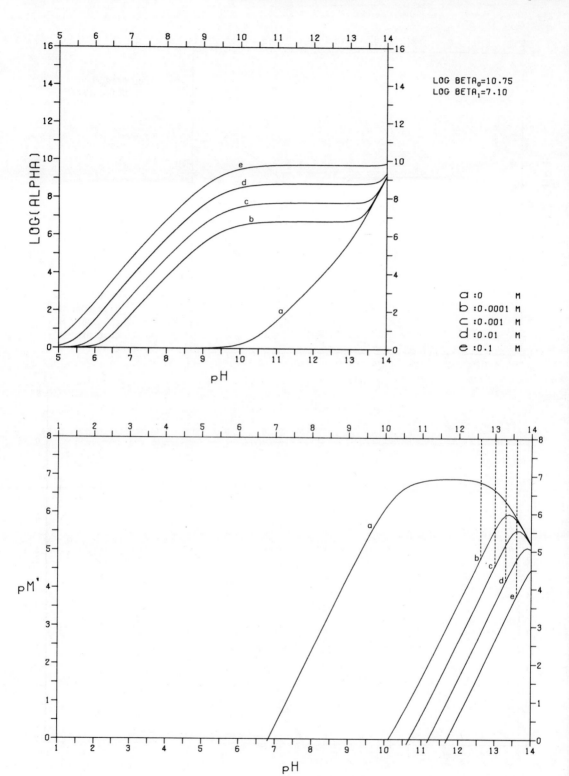

Cd — carbonate system

The partial pressure of CO_2 in air corresponds to log P_{CO_2} = −3.52 (P in atmospheres) and this is sufficient to cause precipitation of the extremely insoluble cadmium carbonate. From this value and the solubility product of $CdCO_3$ (log K_{so} = −11.8) an apparent constant log K_{carb} = 9.9 can be derived. The following set of constants was used for construction of the plots which follow.

$$\log {}^*\beta_1 = -10.3 \qquad \log {}^*\beta_{12} = -9.2$$
$$\log {}^*\beta_2 = -20.6 \qquad \log {}^*\beta_{44} = -32.4$$
$$\log {}^*\beta_3 = -33.8$$
$$\log {}^*\beta_4 = -46.9 \qquad \log K_{carb} = 9.9$$

The presence of CO_2 does not affect the log α plots, but, for convenience, they have been repeated.

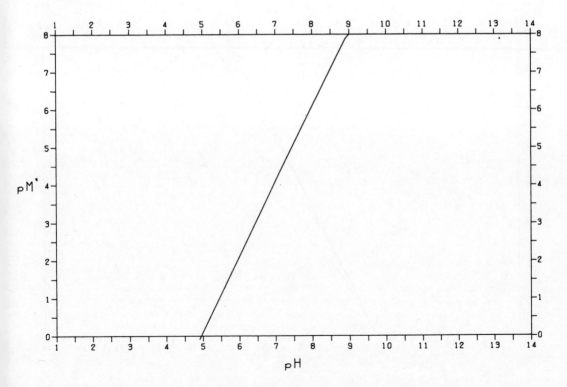

Cd–Carbonate–Acetylacetone [Ch. 8

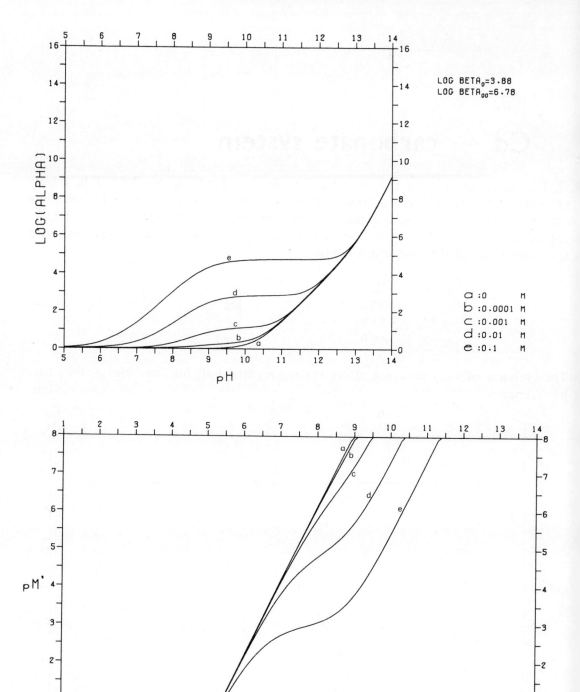

Cd-Carbonate-Ammonia

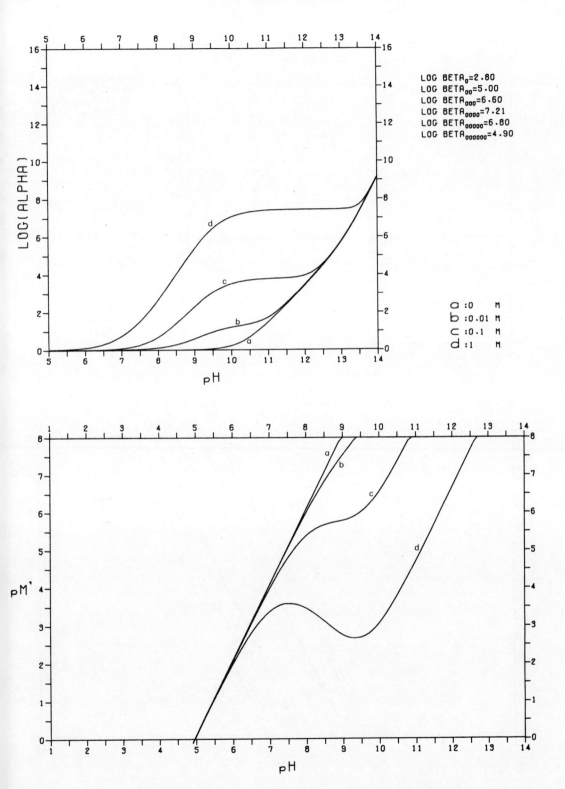

Cd–Carbonate–Bromide [Ch. 8

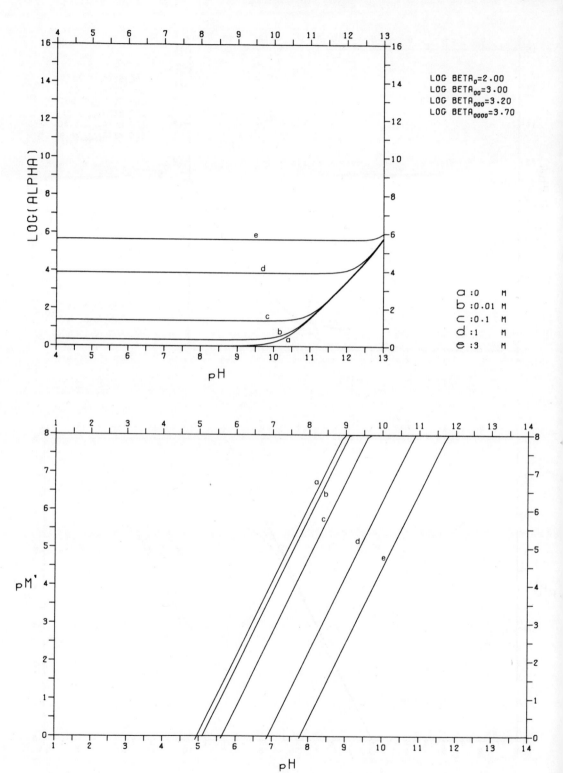

Cd-Carbonate-Chloride

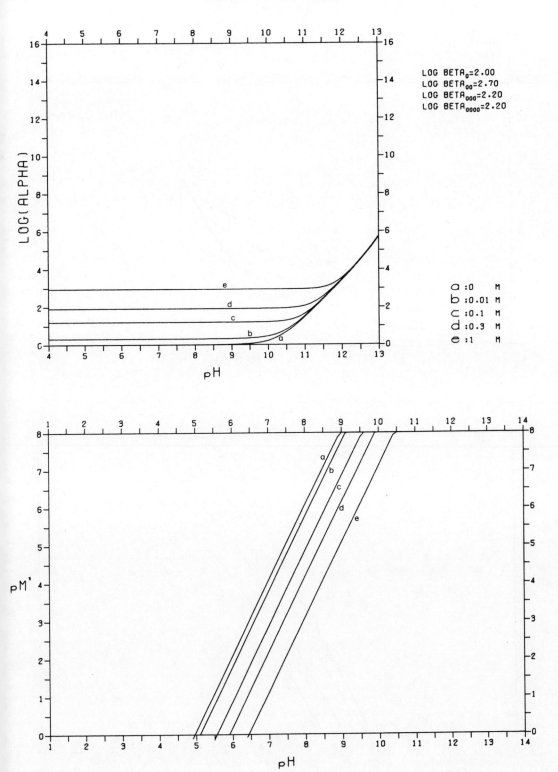

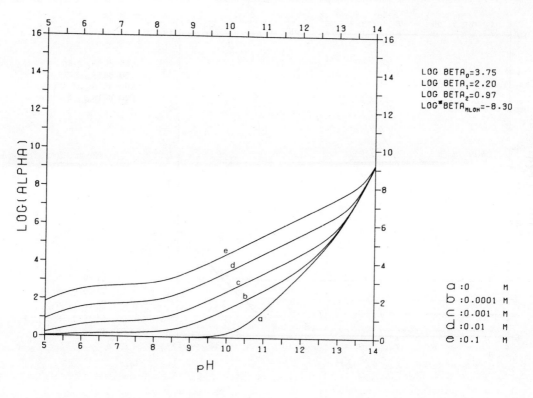
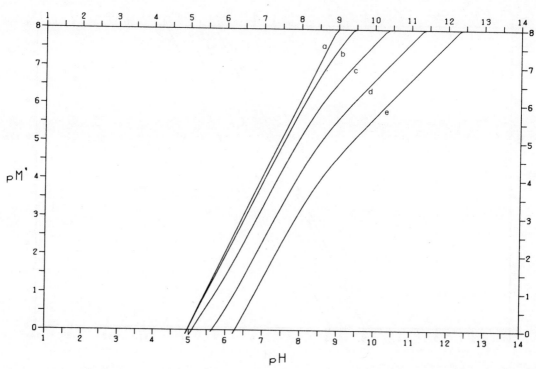

Cd-Carbonate-Cyanide

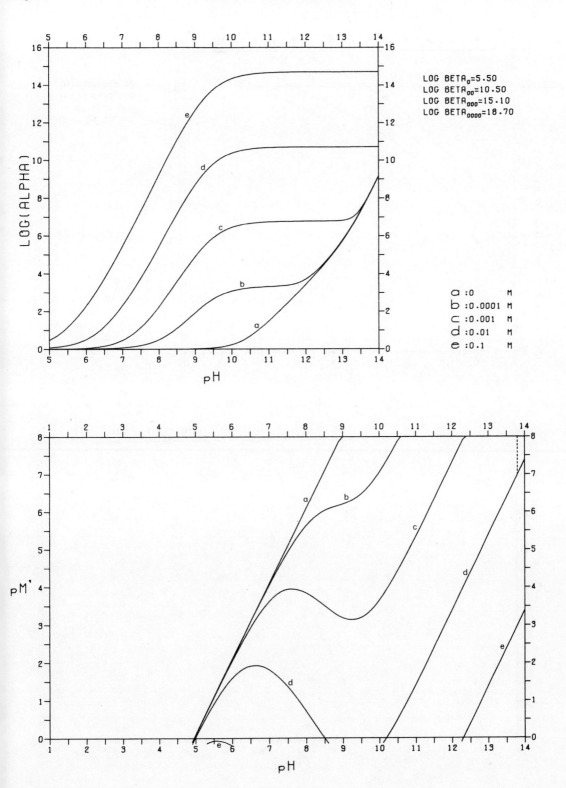

Cd--Carbonate--DCTA [Ch. 8]

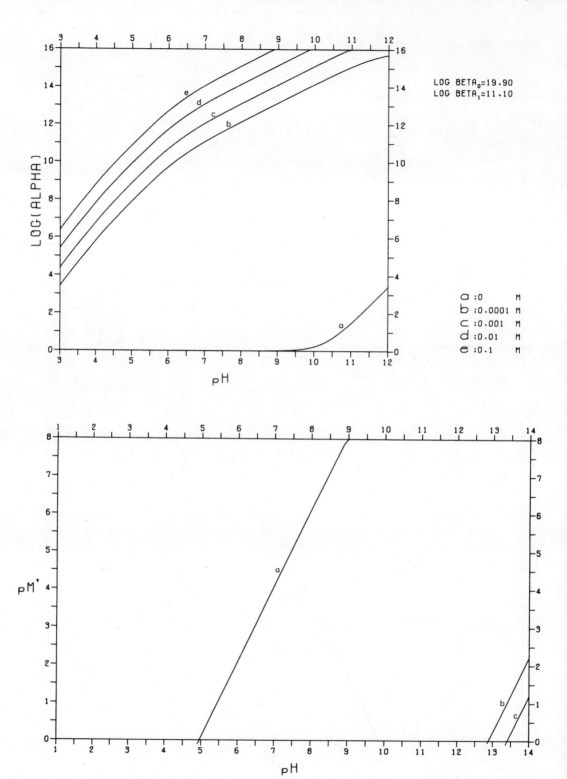

Cd-Carbonate-DTPA

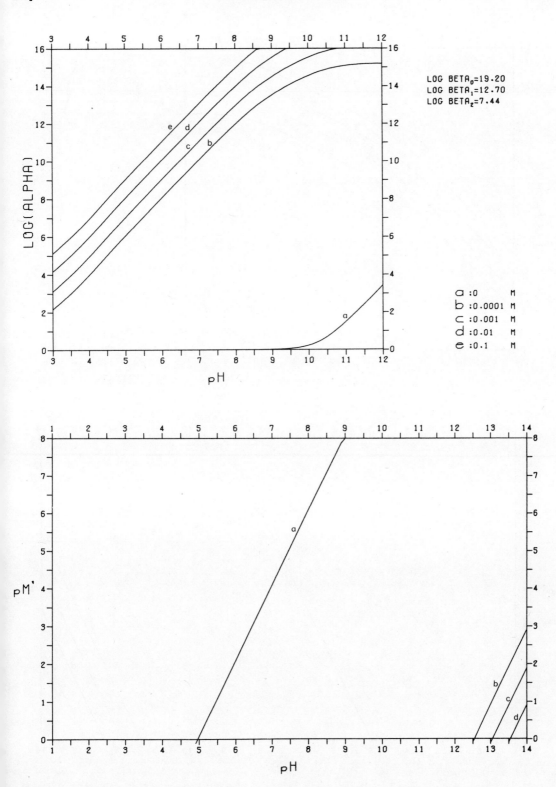

LOG BETA$_0$=19.20
LOG BETA$_1$=12.70
LOG BETA$_2$=7.44

a : 0   M
b : 0.0001 M
c : 0.001   M
d : 0.01   M
e : 0.1   M

Cd–Carbonate–EDTA [Ch. 8

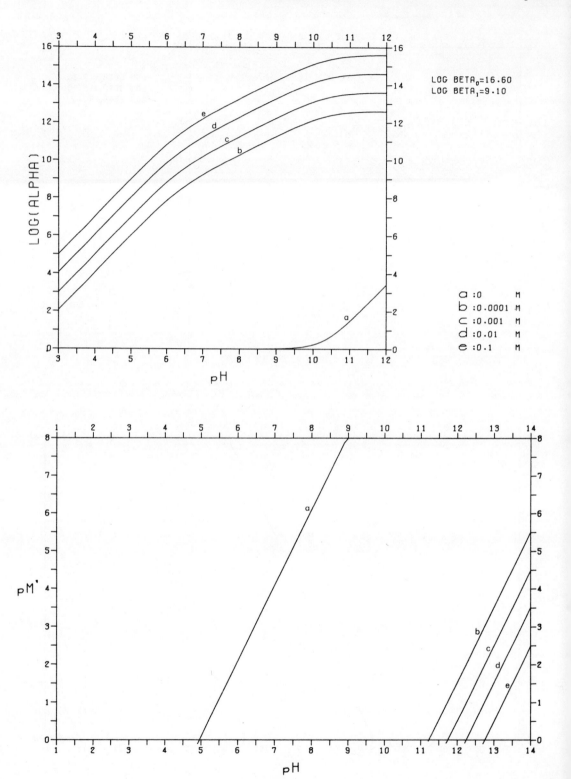

Cd-Carbonate-EGTA

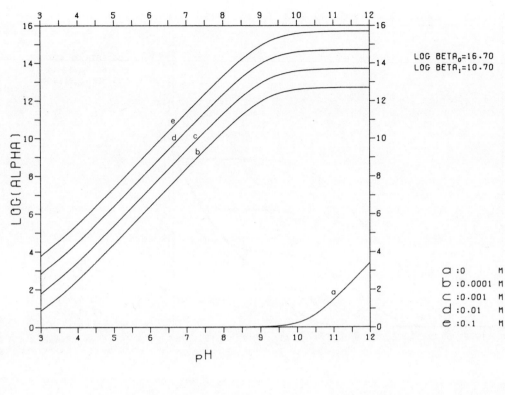

LOG BETA$_0$=16.70
LOG BETA$_1$=10.70

a : 0   M
b : 0.0001 M
c : 0.001 M
d : 0.01   M
e : 0.1   M

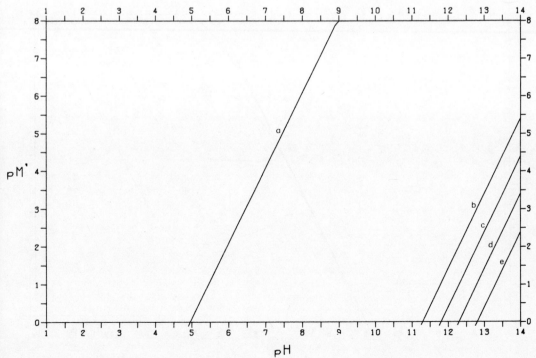

148 Cd-Carbonate-Ethylenediamine [Ch. 8

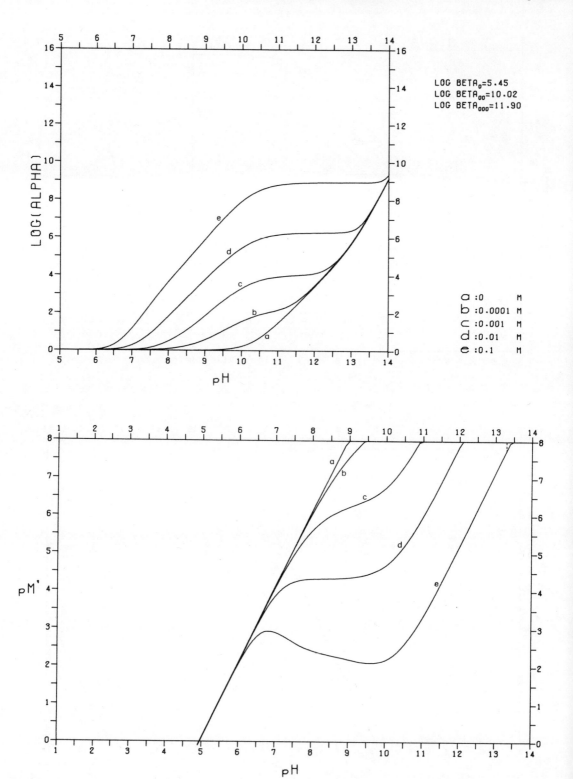

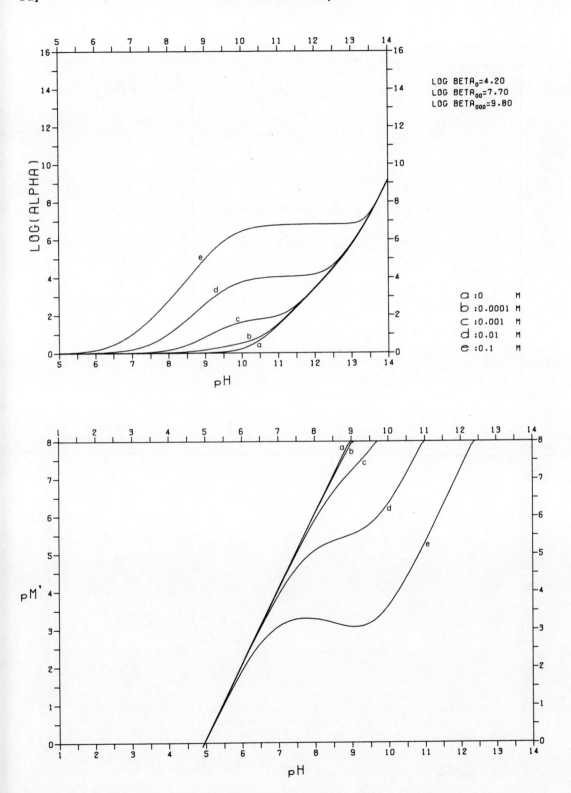

Cd–Carbonate–(OH)–Quinolinesulphonate [Ch. 8

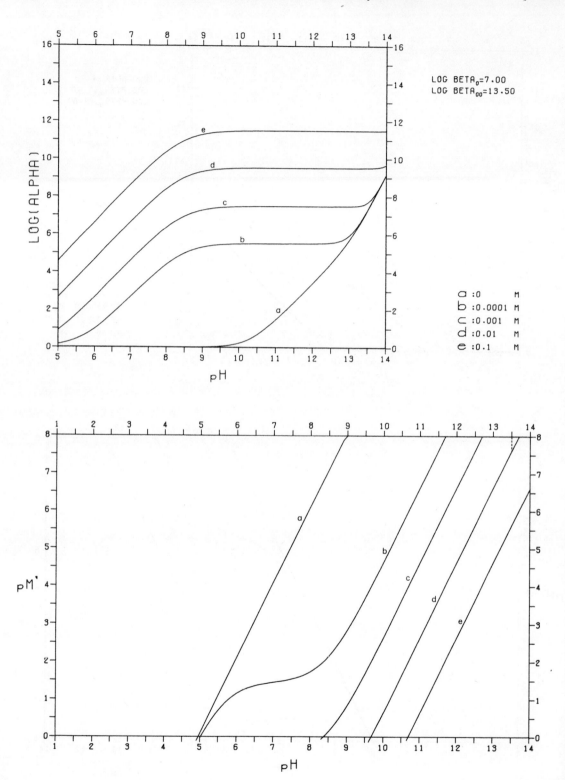

Cd-Carbonate-Iminodiacetate

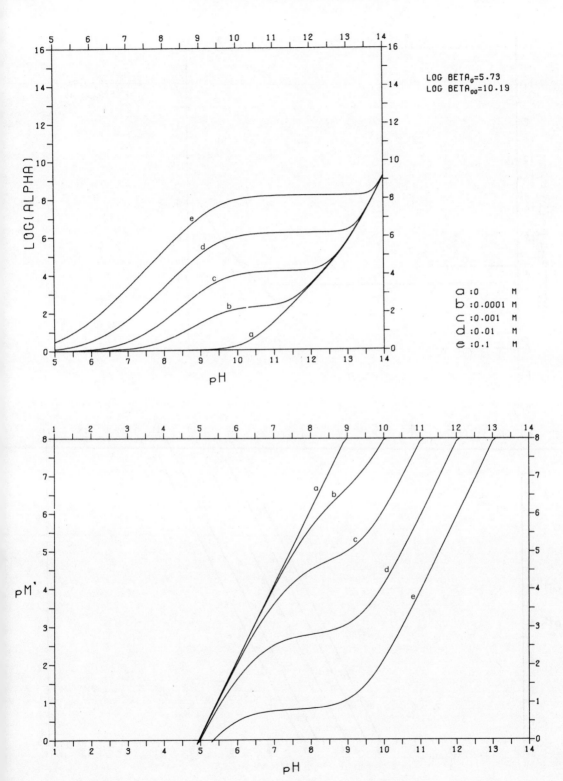

LOG BETA$_0$=5.73
LOG BETA$_{00}$=10.19

a : 0   M
b : 0.0001 M
c : 0.001   M
d : 0.01   M
e : 0.1   M

Cd-Carbonate-Oxalate

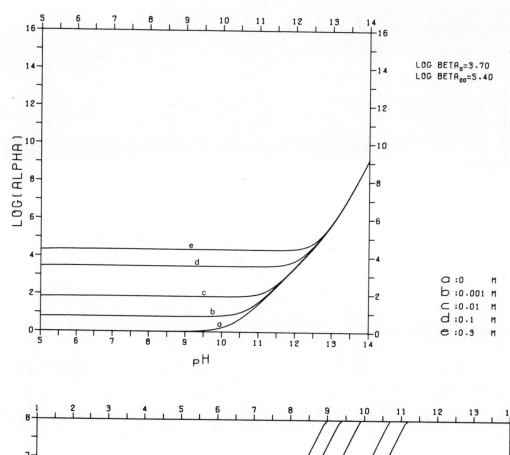

LOG BETA$_0$=3.70
LOG BETA$_{00}$=5.40

a : 0   M
b : 0.001 M
c : 0.01   M
d : 0.1   M
e : 0.3   M

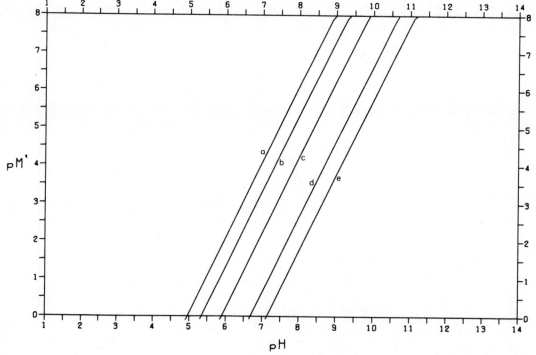

Cd–Carbonate–Pyridine-2,6-dicarboxylate

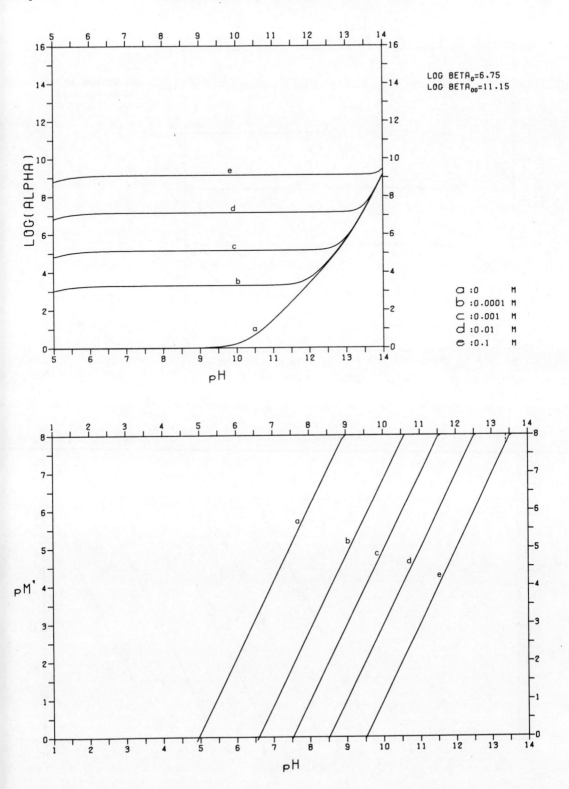

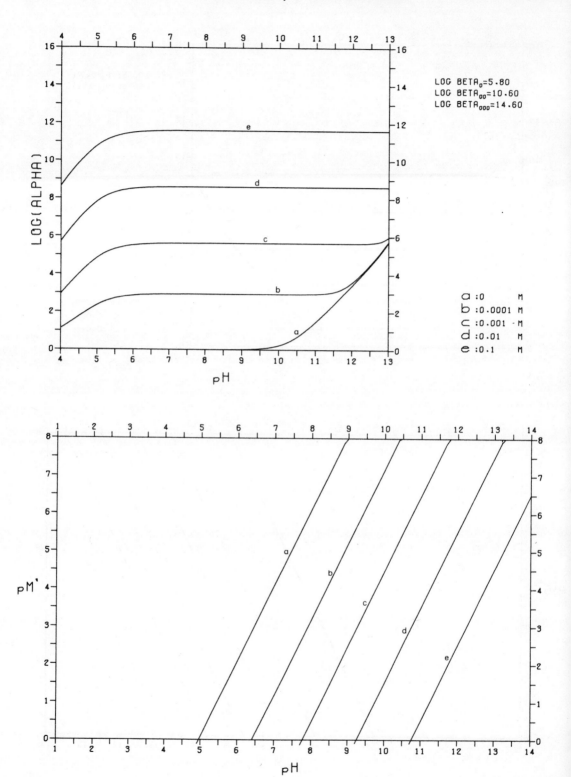

Cd-Carbonate-Sulphate

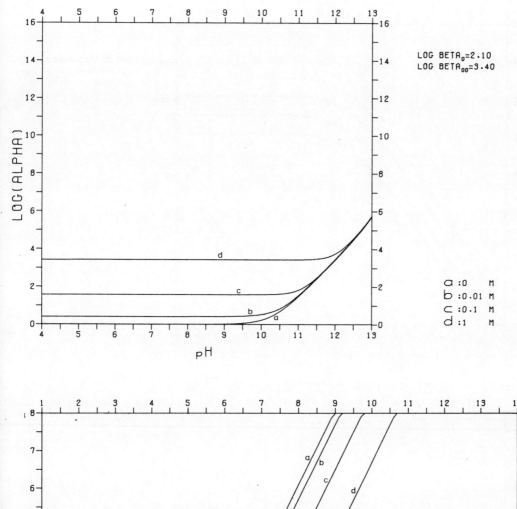

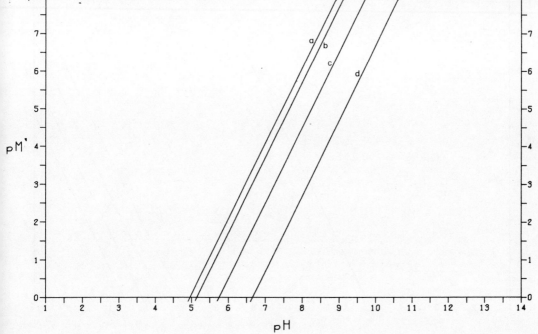

156   **Cd–Carbonate–Tetren**   [Ch. 8

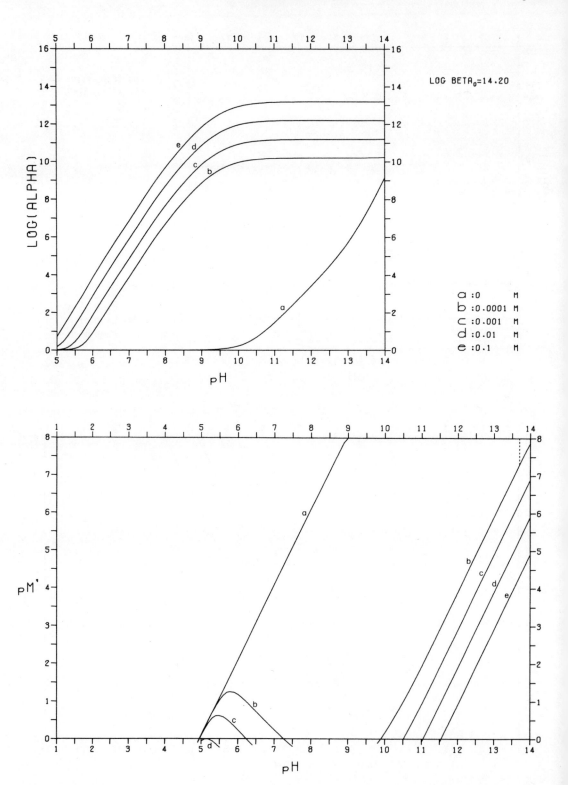

Cd-Carbonate-Tiron

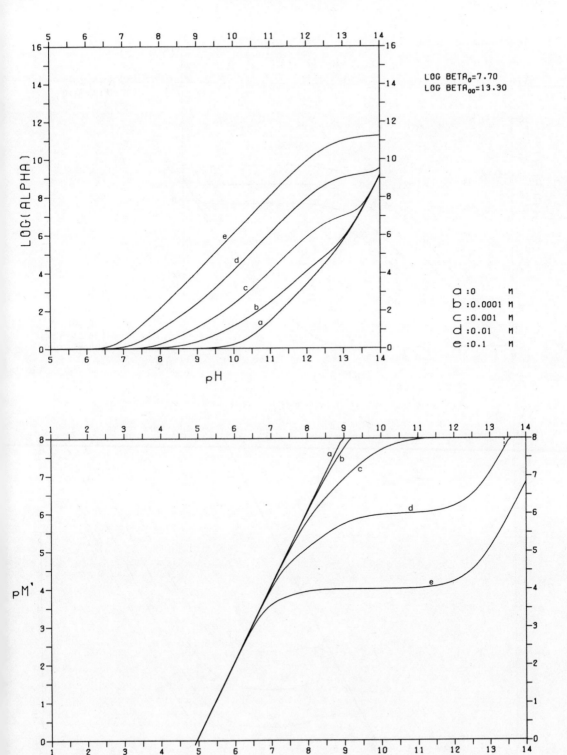

LOG BETA₀=7.70
LOG BETA₀₀=13.30

a : 0   M
b : 0.0001   M
c : 0.001   M
d : 0.01   M
e : 0.1   M

158 Cd-Carbonate-Trien [Ch. 8

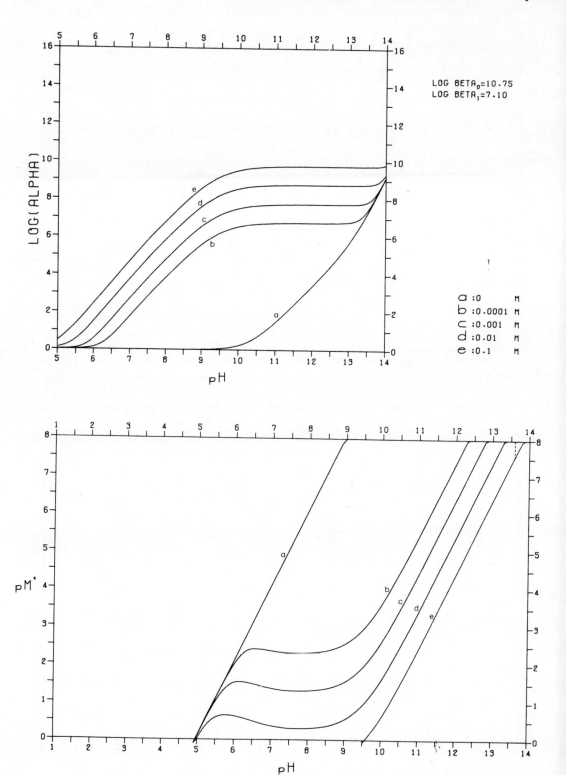

CHAPTER 9
Cerium (III)   Ce

Hydrolysis of the rather large Ce^{3+} ion does not become appreciable until fairly high pH values (>6). Hydrolysis studies are rather difficult because, at ordinary concentration levels, hydroxide precipitation invariably occurs when the average coordination number is still small, but there is conclusive evidence that hydroxide complexes are formed at pH values just before the onset of precipitation. The limited amount of fairly reliable published information suggests that the usual product is $Ce_3(OH)_5^{4+}$ and that $Ce(OH)^{2+}$ is only present as a minor species.

Studies of hydroxide-precipitate formation are difficult because of supersaturation, the great dependence of the solubility product on particle size, the changes in pH caused by rapid partial recrystallization, aerial oxidation to yellow $CeO_2 \cdot nH_2O$, and carbonate formation. These phenomena also make it difficult to define an equilibrium situation which is of practical use for as many scientists as possible.

The data used were determined from solubility measurements [1]; they agree with those compiled by Baes and Mesmer [2]. They refer to solutions, from which O_2 and CO_2 have been excluded, after removal by filtration (under nitrogen) of a one-hour-old precipitate. The value of log $*\beta_4$ is estimated from the increase in the solubility of the precipitate above pH 11.

$$\log *\beta_1 = -\ 8.1 \qquad \log *\beta_{53} = -32.8$$
$$\log *\beta_2 = -16.3$$
$$\log *\beta_3 = -26.0$$
$$\log *\beta_4 = -38.0 \qquad \log *K_{s0} = \ 20.1$$

The influence of air ($O_2 + CO_2$) on the system would be small compared with the uncertainties in the data. Its introduction should not lead to significantly different plots and thus has been omitted.

[1] J. Kragten and L. G. Decnop-Weever, *Talanta*, in press.
[2] C. F. Baes and R. E. Mesmer, *The Hydrolysis of Cations*, Wiley, New York, 1976.

Cerium(III)

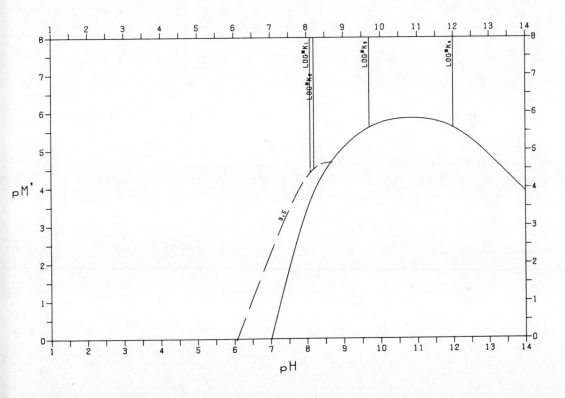

Cerium(III)-Acetylacetone [Ch. 9]

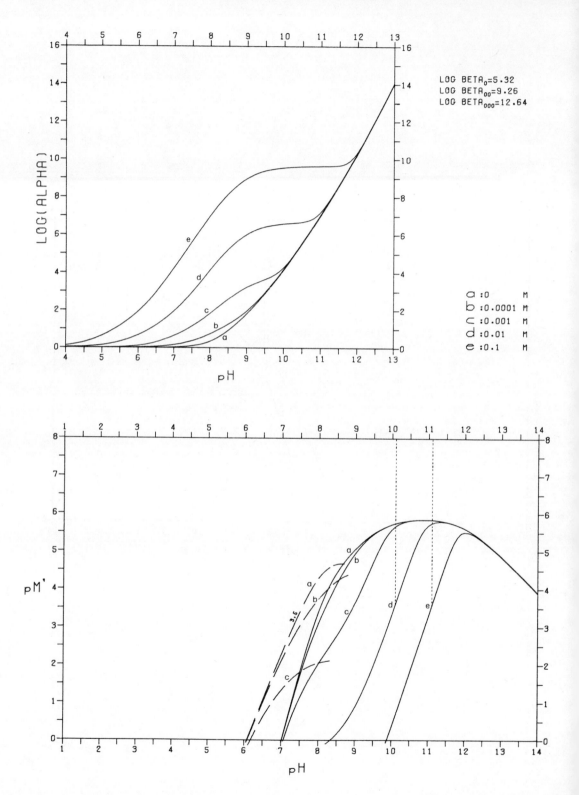

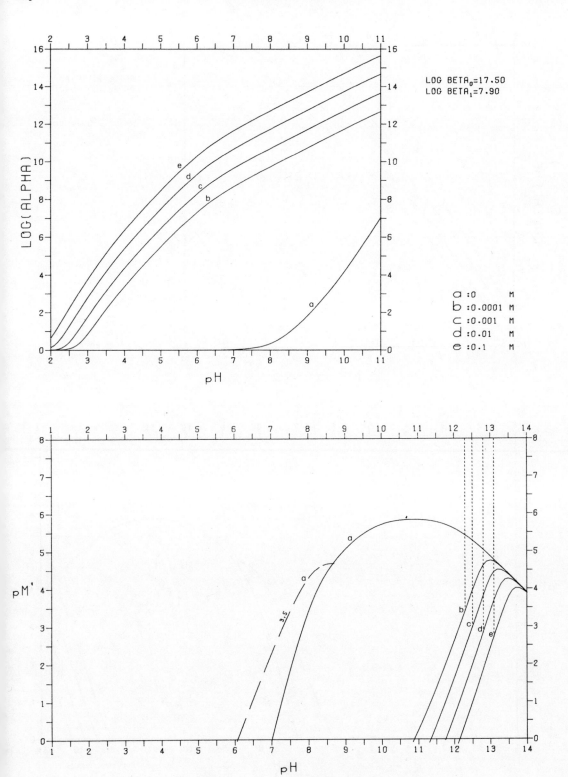

Cerium(III)–DTPA

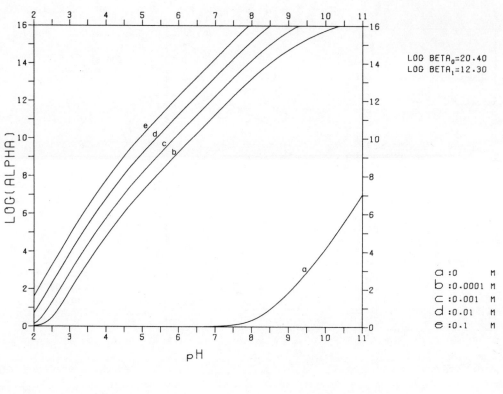

LOG BETA$_0$=20.40
LOG BETA$_1$=12.30

a : 0   M
b : 0.0001 M
c : 0.001 M
d : 0.01   M
e : 0.1   M

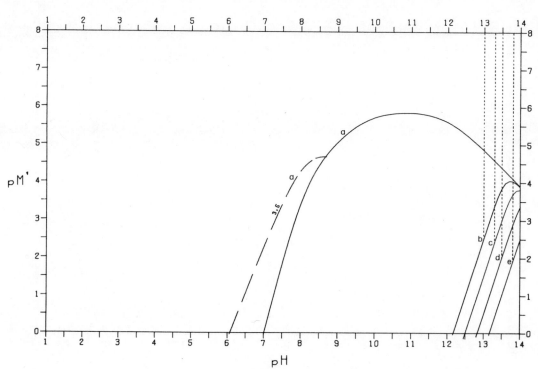

Cerium(III)-EDTA

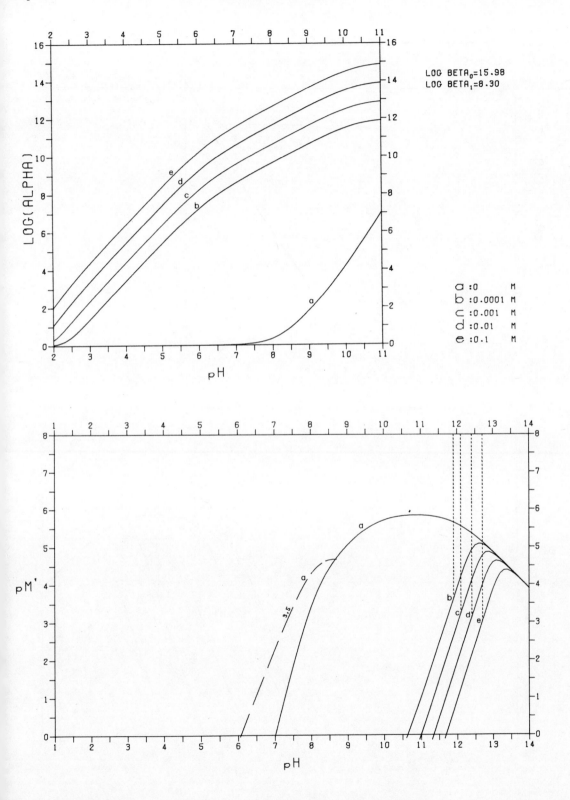

166 Cerium(III)-(OH)-Quinolinesulphonate [Ch. 9

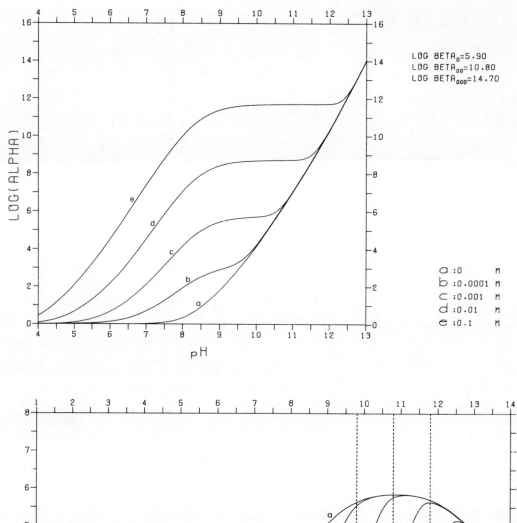

Cerium(III)-Iminodiacetate

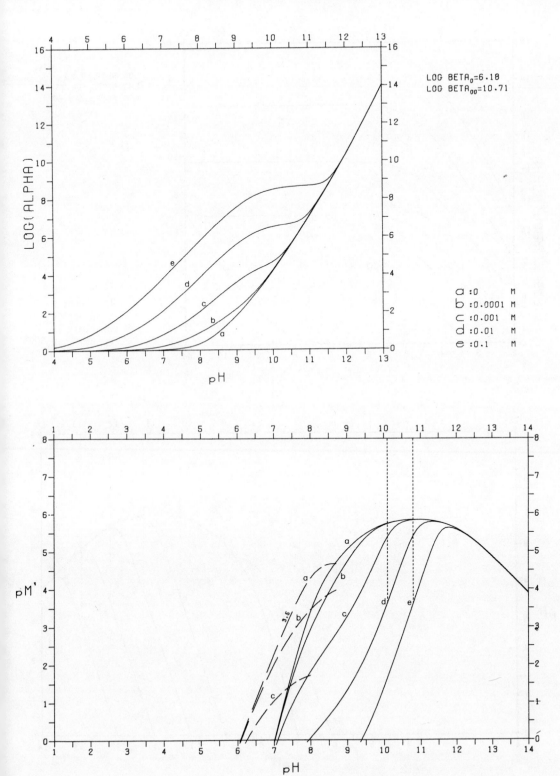

LOG BETA$_0$=6.18
LOG BETA$_{00}$=10.71

a : 0 M
b : 0.0001 M
c : 0.001 M
d : 0.01 M
e : 0.1 M

Cerium(III)-Pyridine-2,6-dicarboxylate [Ch.9]

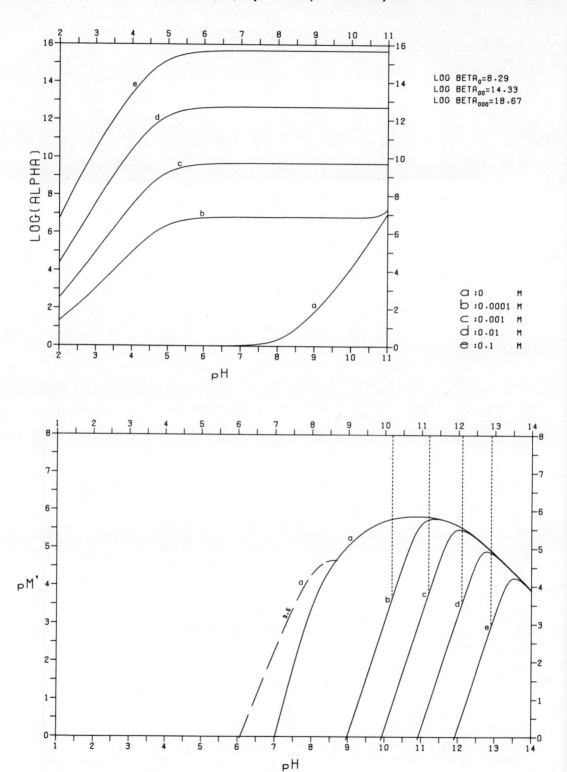

Cerium(III)-5-Sulphosalicylate

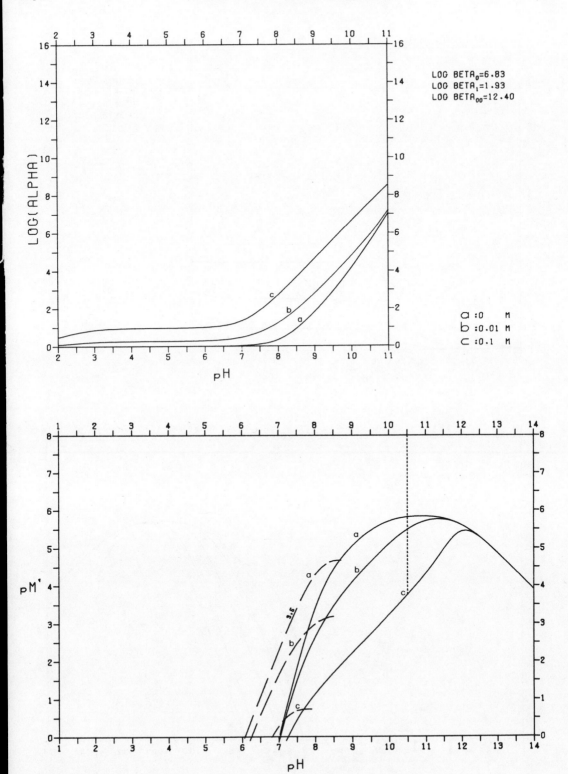

LOG BETA$_0$=6.83
LOG BETA$_1$=1.93
LOG BETA$_{00}$=12.40

a : 0 M
b : 0.01 M
c : 0.1 M

Cerium(III)-Tartrate [Ch. 9

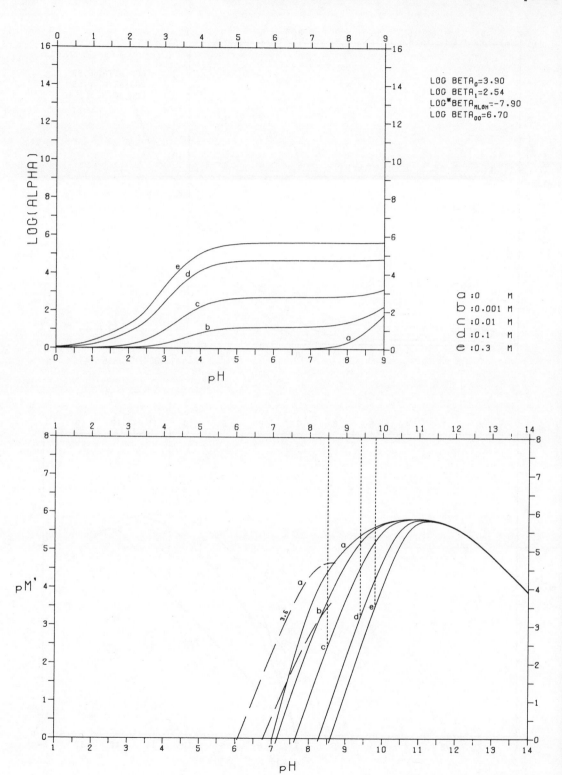

CHAPTER 10
Cobalt (II)   Co

The Co^{2+} ion has been shown to form mononuclear hydrolysis products ranging from $Co(OH)^+$ to $Co(OH)_4^{2-}$. There is some doubtful evidence for the formation of polynuclear complexes, but the relevant data do not lead to a meaningful polycomplex line on the pM'–pH plot. Precipitated $Co(HO)_2$ exists as blue and pink forms, of which the latter is the more stable. The difference between the solubility products of the two forms is small in comparison with experimental variables; the mean value is listed.

$$\log {}^*\beta_1 = -\ 9.90 \qquad \log {}^*\beta_{12} = -11.0$$
$$\log {}^*\beta_2 = -18.8 \qquad \log {}^*\beta_{44} = -30.1$$
$$\log {}^*\beta_3 = -31.5 \qquad \log {}^*K_{s0} = \ \ 12.8$$

The study of hydroxide formation requires exclusion of carbonate and oxygen. The plots for the $CoCO_3$ system are given on pages 193–212. Oxygen may interfere, because the reaction $Co^{2+} + 2H_2O + \frac{1}{2}O_2 \rightarrow Co(OH)_3(s) + H^+$ proceeds rapidly. In the absence of complexing ligands in air-saturated aqueous solutions, the borderline of $Co(OH)_3$ precipitation is given by $p[Co^{2+}] = -8.8 + 2\ pH$. Reliable information about the complexation of Co(III) ions is not available. Ligand-exchange reactions of Co^{3+} proceed very slowly. It may be assumed that when the redox potential is below $E = (993 - 59.1\ pH)$ mV, Co^{2+} will always be in equilibrium with $Co(OH)_2$ rather than $Co(OH)_3$.

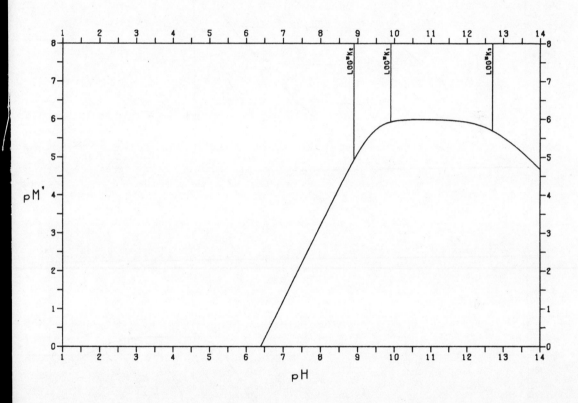

Cobalt(II)-Acetylacetone [Ch. 10

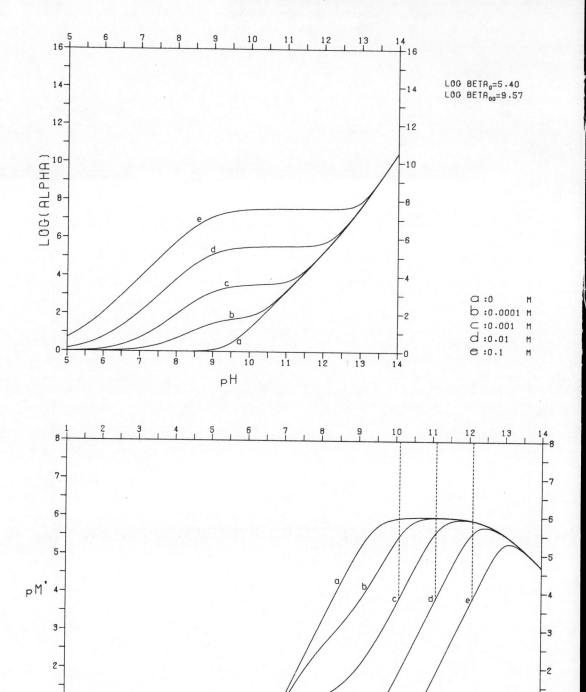

Cobalt(II)–Ammonia

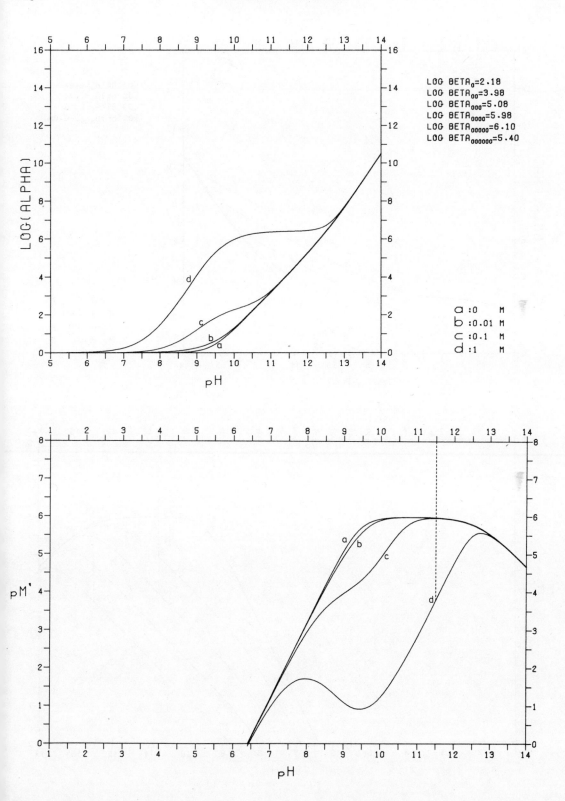

LOG BETA$_0$=2.18
LOG BETA$_{00}$=3.98
LOG BETA$_{000}$=5.08
LOG BETA$_{0000}$=5.98
LOG BETA$_{00000}$=6.10
LOG BETA$_{000000}$=5.40

a : 0 M
b : 0.01 M
c : 0.1 M
d : 1 M

Cobalt(II)–Citrate [Ch. 10]

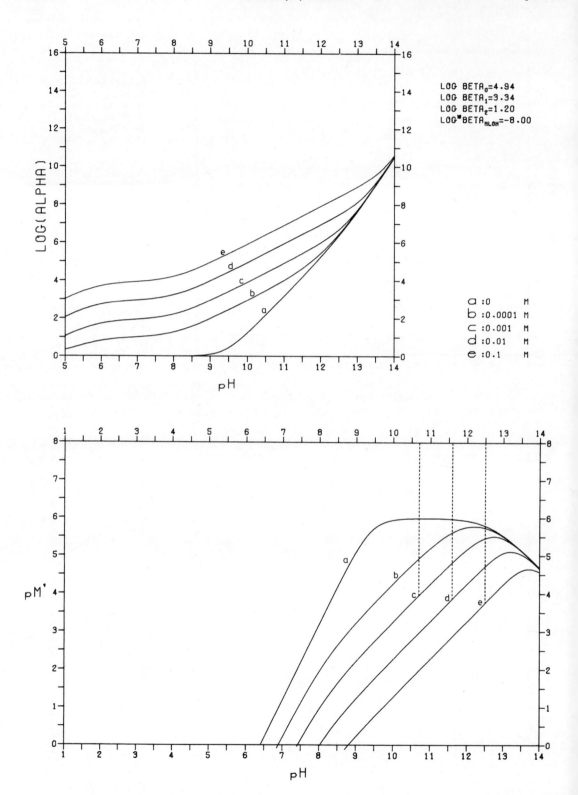

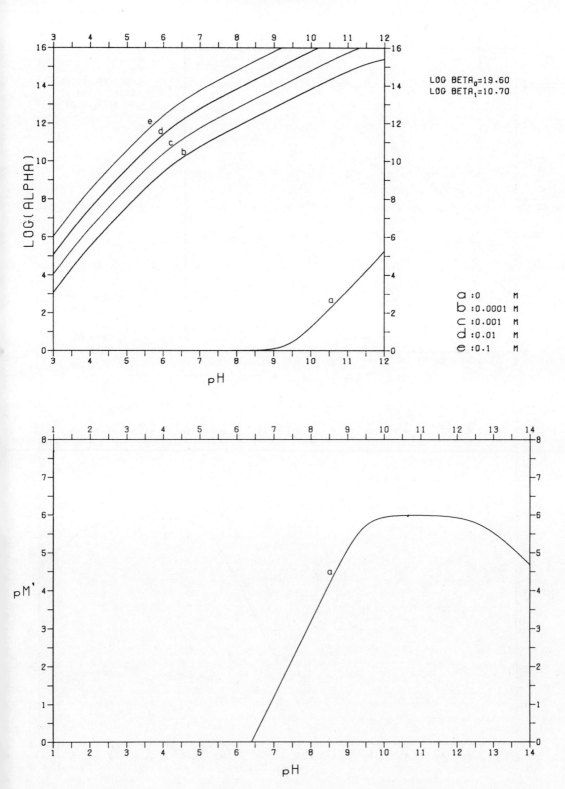

Cobalt(II)-DTPA

[Ch. 10

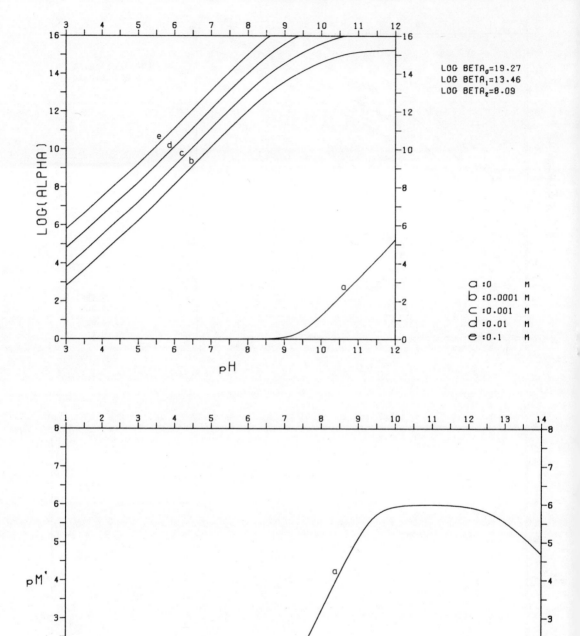

Cobalt(II)-EDTA

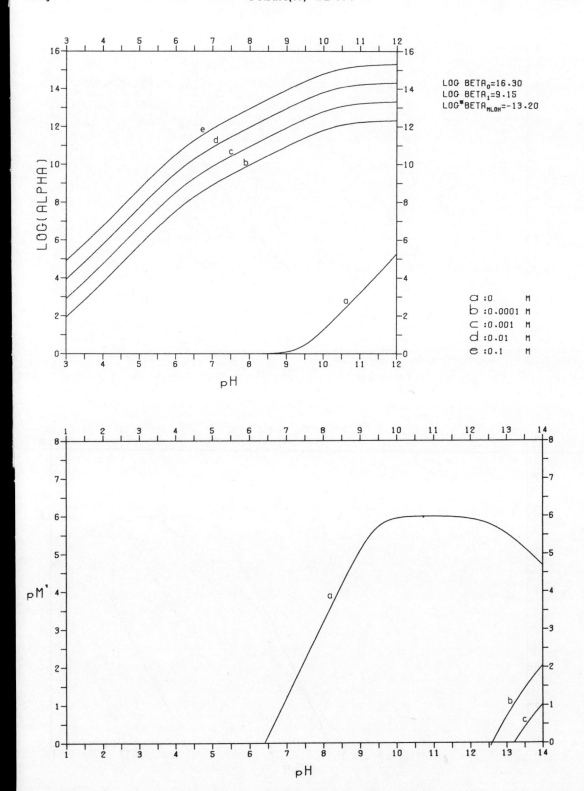

Cobalt(II)-Ethylenediamine [Ch. 10]

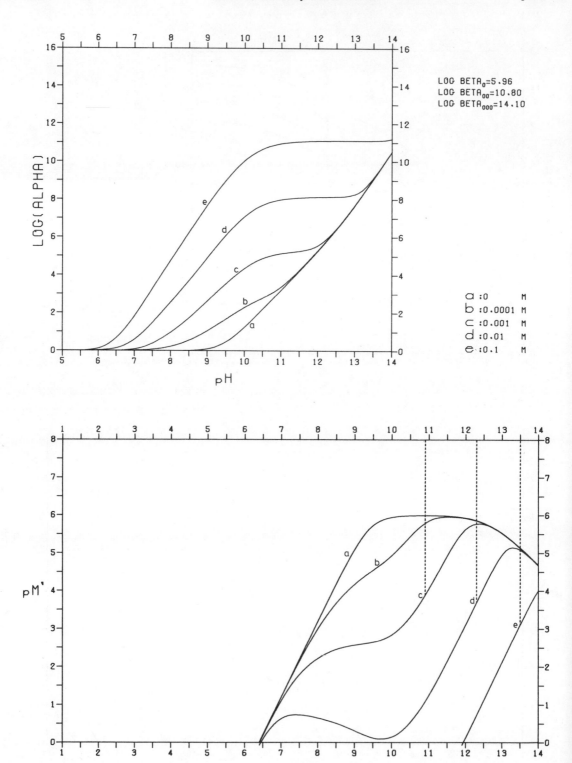

LOG BETA$_0$=5.96
LOG BETA$_{00}$=10.80
LOG BETA$_{000}$=14.10

a : 0   M
b : 0.0001 M
c : 0.001   M
d : 0.01   M
e : 0.1   M

Cobalt(II)–Glycine

LOG BETA$_0$=4.64
LOG BETA$_{00}$=8.46
LOG BETA$_{000}$=10.81

a : 0 M
b : 0.0001 M
c : 0.001 M
d : 0.01 M
e : 0.1 M

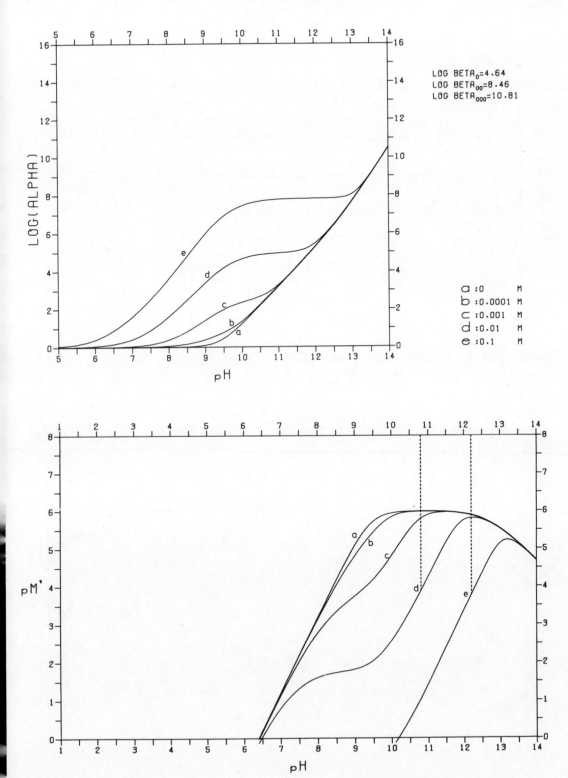

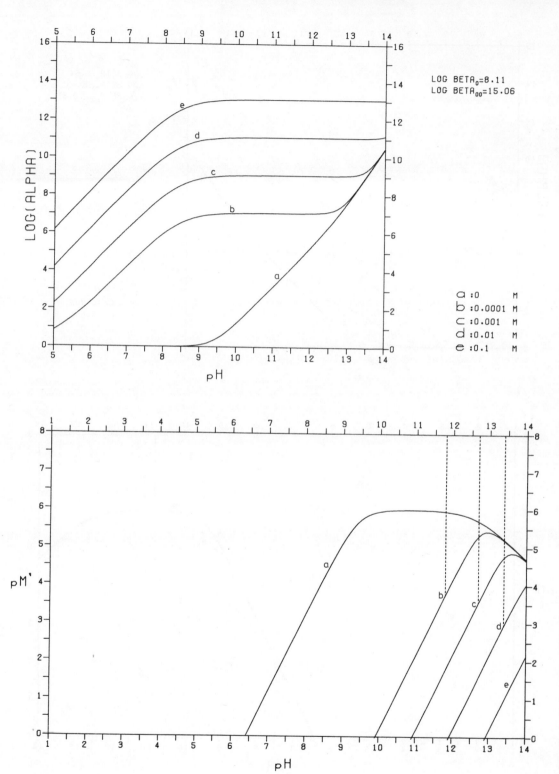

Cobalt(II)-Iminodiacetate

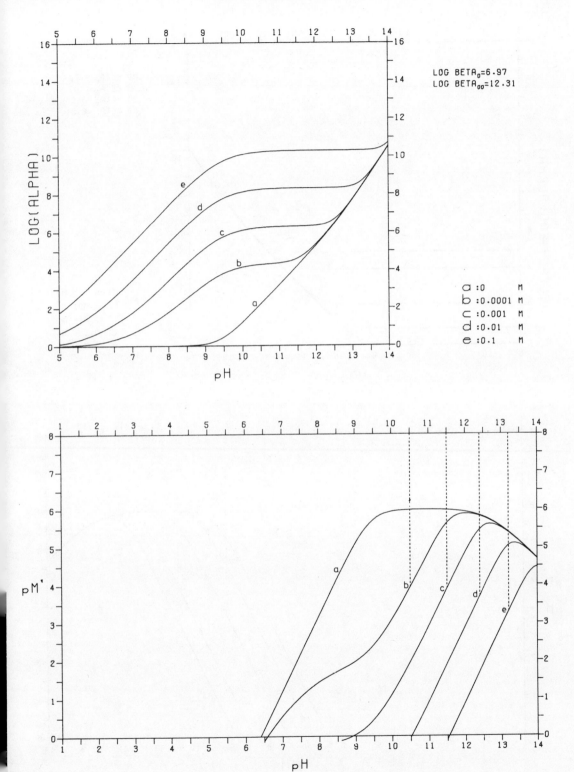

Cobalt(II)–Oxalate [Ch. 10]

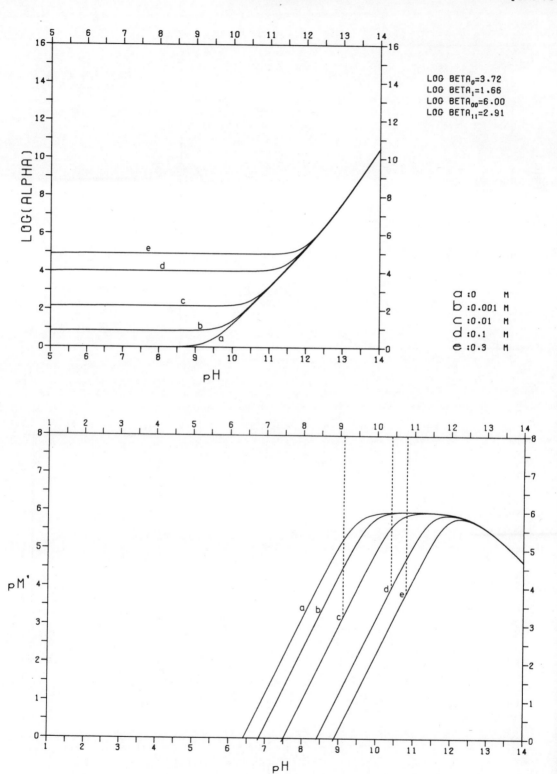

Cobalt(II)-Pyridine-2,6-dicarboxylate

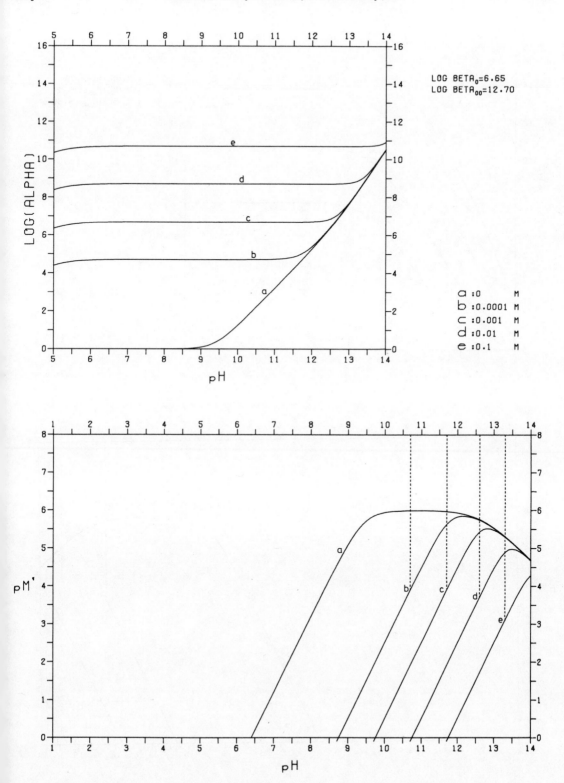

Cobalt(II)-1,10-Phenanthroline

[Ch. 10

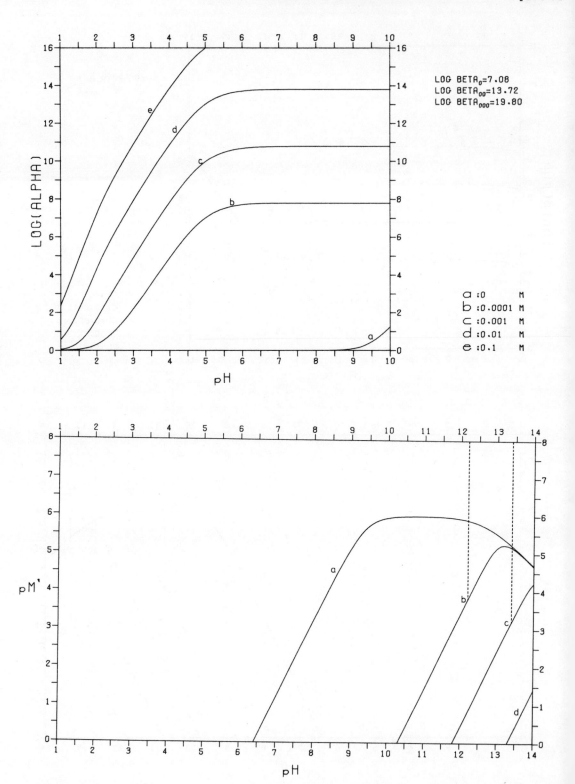

LOG BETA$_0$=7.08
LOG BETA$_{00}$=13.72
LOG BETA$_{000}$=19.80

a : 0   M
b : 0.0001 M
c : 0.001   M
d : 0.01   M
e : 0.1   M

Cobalt(II)-Sulphate

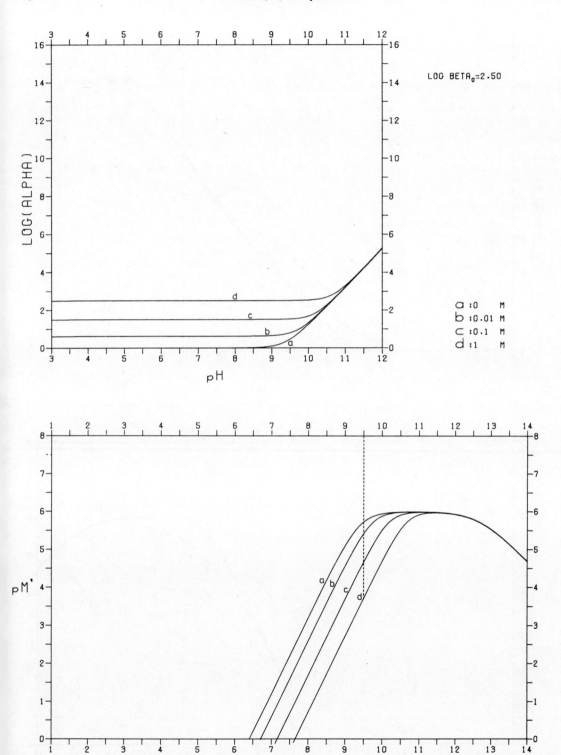

LOG BETA$_0$=2.50

a : 0   M
b : 0.01 M
c : 0.1 M
d : 1   M

188 Cobalt(II)-5-Sulphosalicylate [Ch. 10

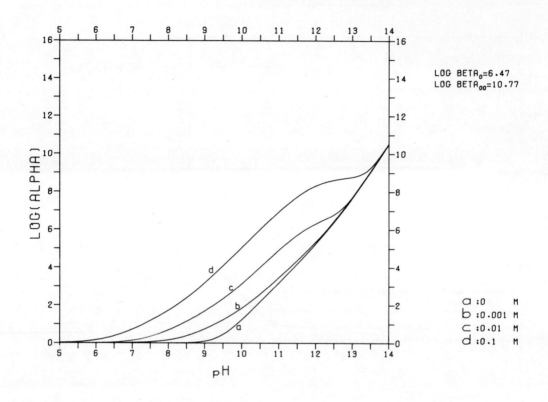

LOG BETA$_0$=6.47
LOG BETA$_{00}$=10.77

a :0   M
b :0.001 M
c :0.01 M
d :0.1   M

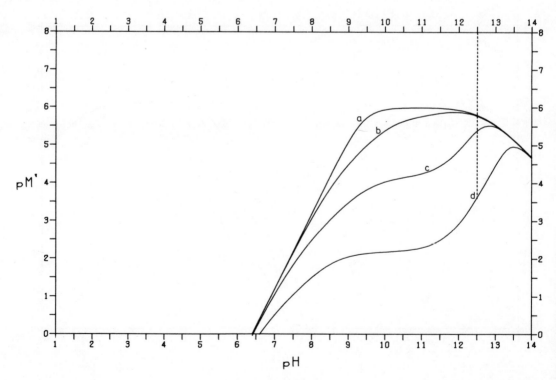

Cobalt(II)-Tartrate

189

LOG BETA$_0$=3.00
LOG BETA$_{00}$=3.70

a : 0   M
b : 0.001 M
c : 0.01   M
d : 0.1   M
e : 0.3   M

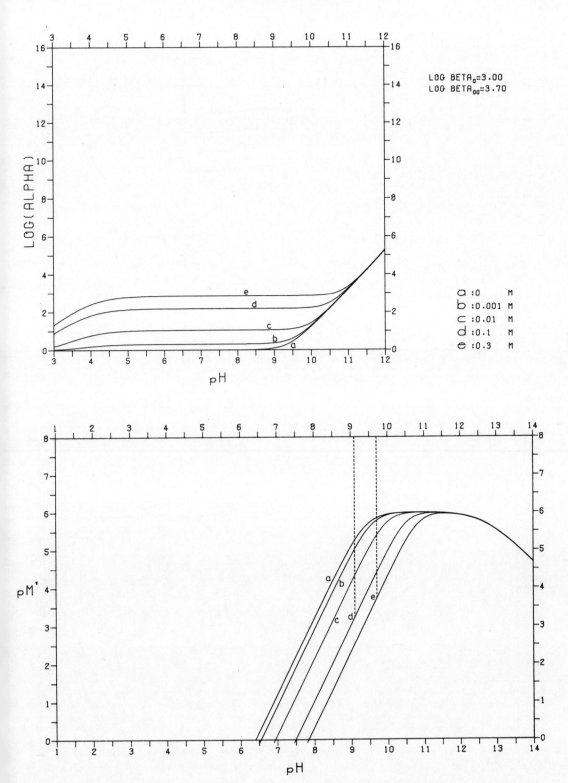

190 Cobalt(II)-Tetren [Ch. 10]

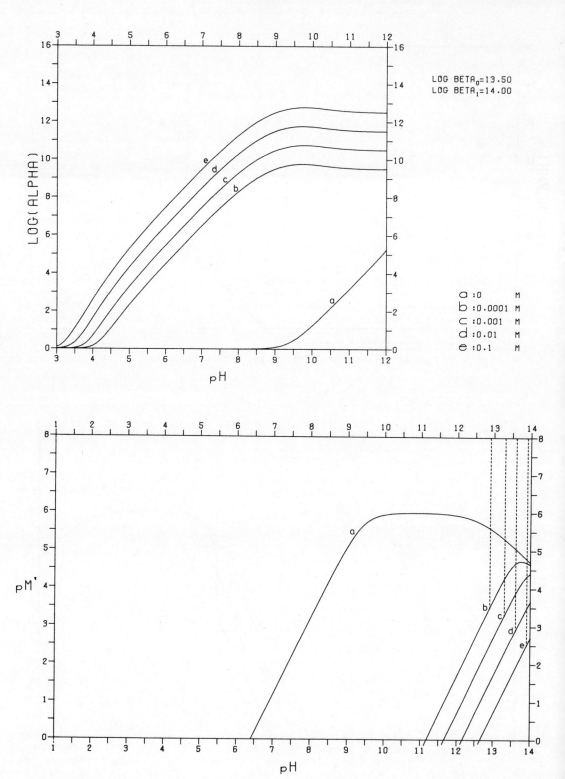

Cobalt(II)-Tiron

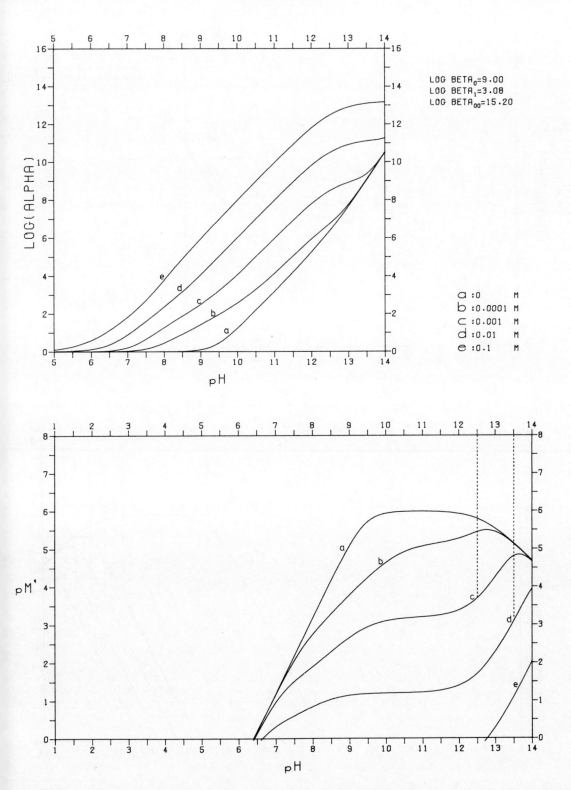

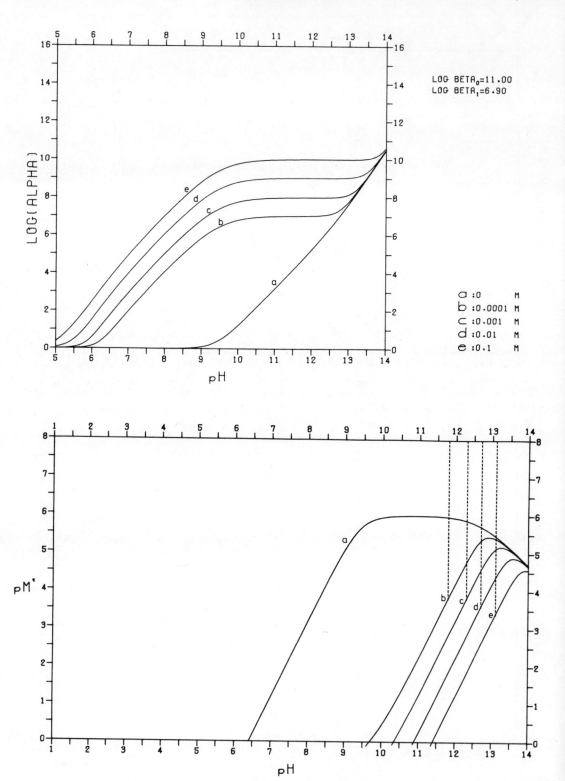

Co — carbonate system

From the solubility of $CoCO_3$ ($\log K_{s0} = -10.0$) and the partial pressure of CO_2 in air ($10^{-3.5}$ atmospheres) an apparent constant $\log K_{carb} = 11.7$ may be derived. The following set of constants was used for construction of the plots.

$$\log {}^*\!\beta_1 = -9.90 \qquad \log K_{carb} = 11.7$$
$$\log {}^*\!\beta_2 = -18.8$$
$$\log {}^*\!\beta_3 = -31.5$$

The presence of CO_2 has no influence on the log α plots but they have been repeated for convenience.

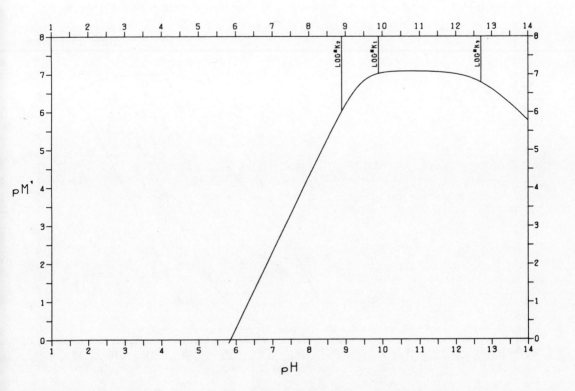

194 Co-Carbonate-Acetylacetone [Ch. 10

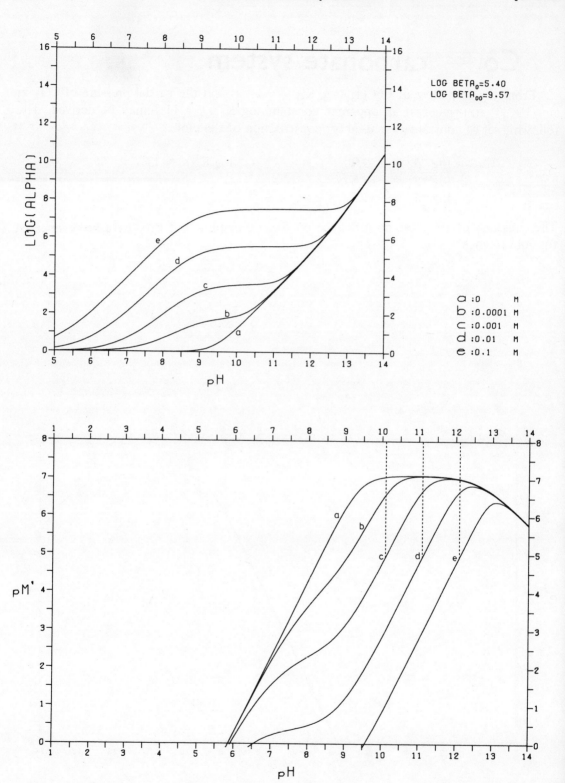

Co-Carbonate-Ammonia

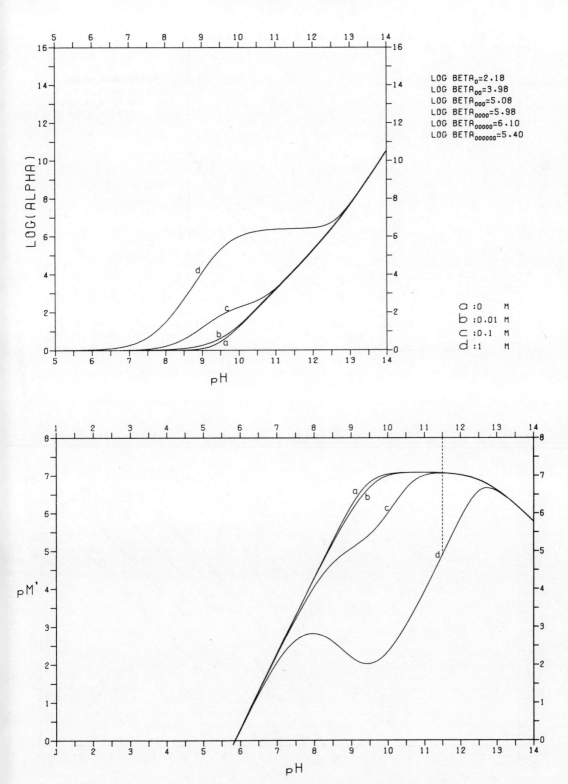

LOG BETA$_0$=2.18
LOG BETA$_{00}$=3.98
LOG BETA$_{000}$=5.08
LOG BETA$_{0000}$=5.98
LOG BETA$_{00000}$=6.10
LOG BETA$_{000000}$=5.40

a : 0 M
b : 0.01 M
c : 0.1 M
d : 1 M

196 Co-Carbonate-Citrate [Ch. 10

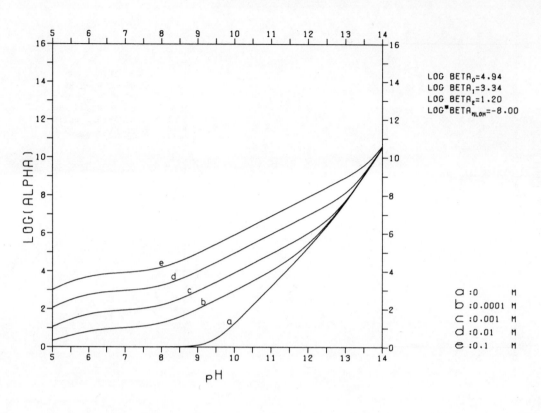

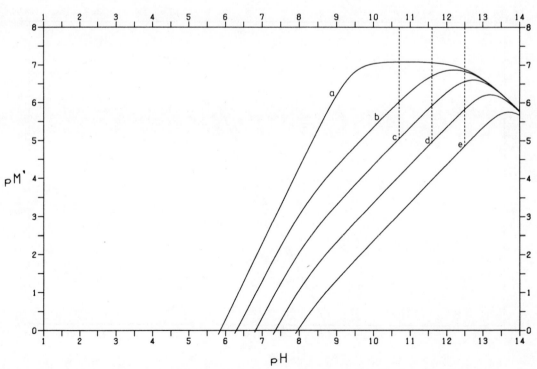

Co-Carbonate-DCTA

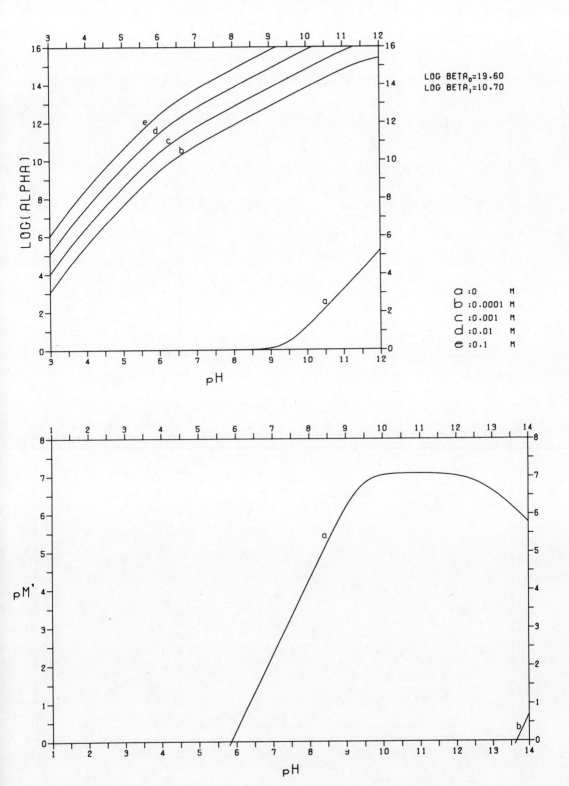

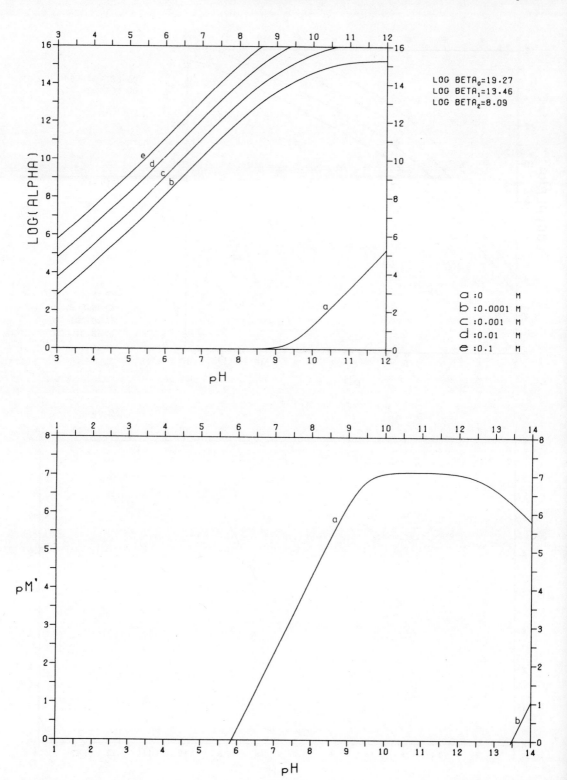

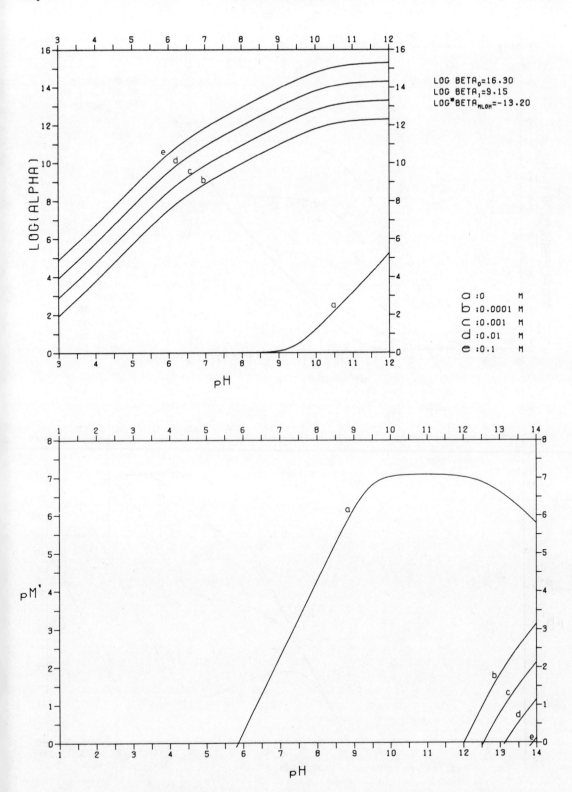

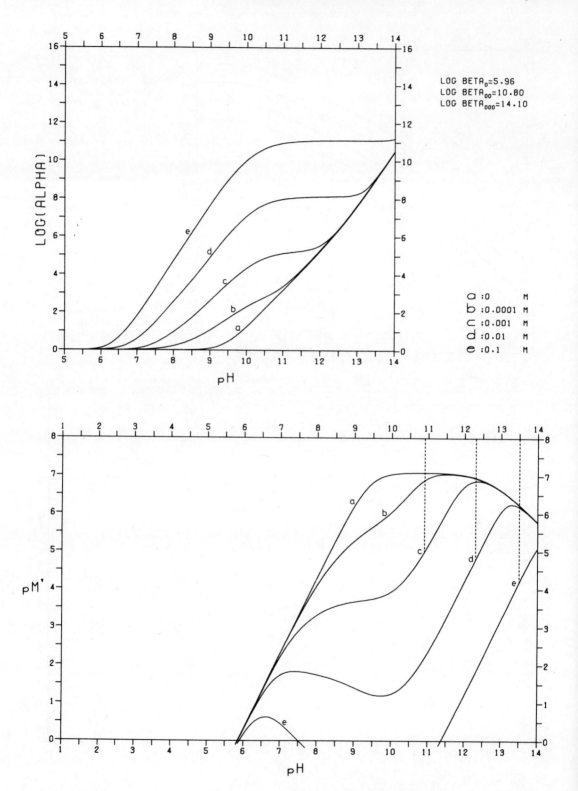

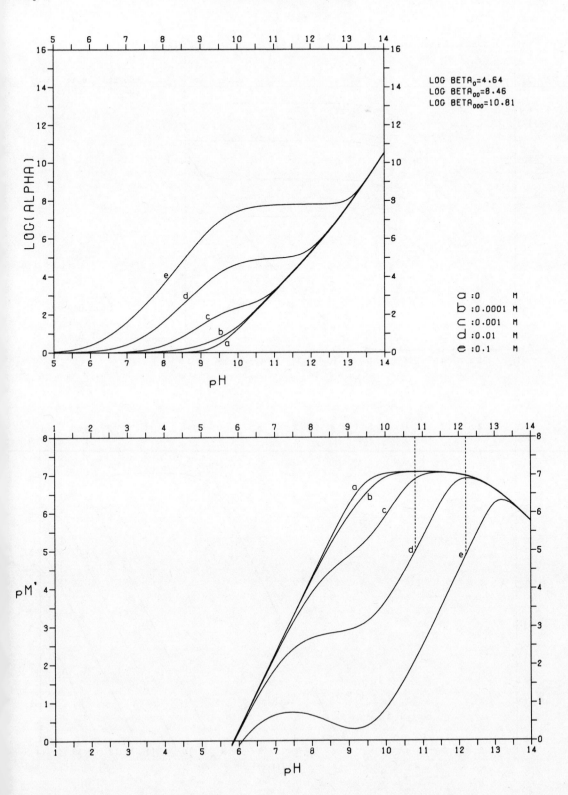

Co-Carbonate-(OH)-Quinolinesulphonate [Ch. 10]

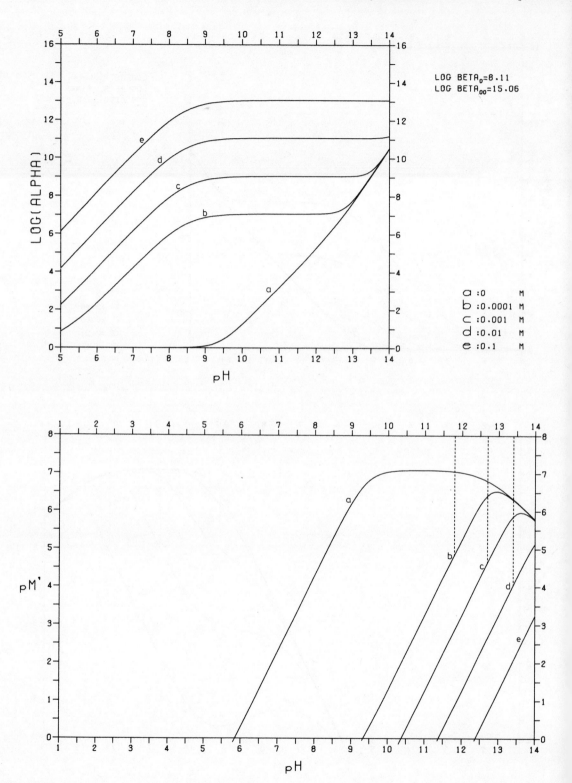

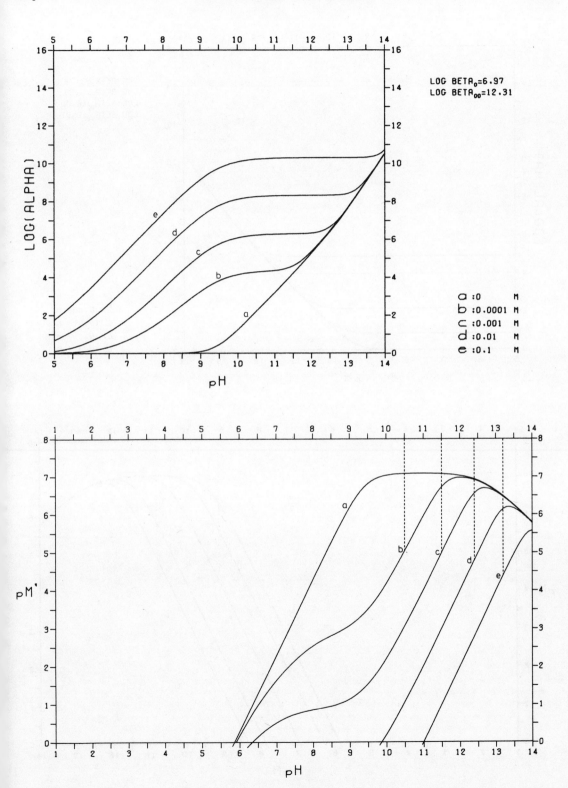

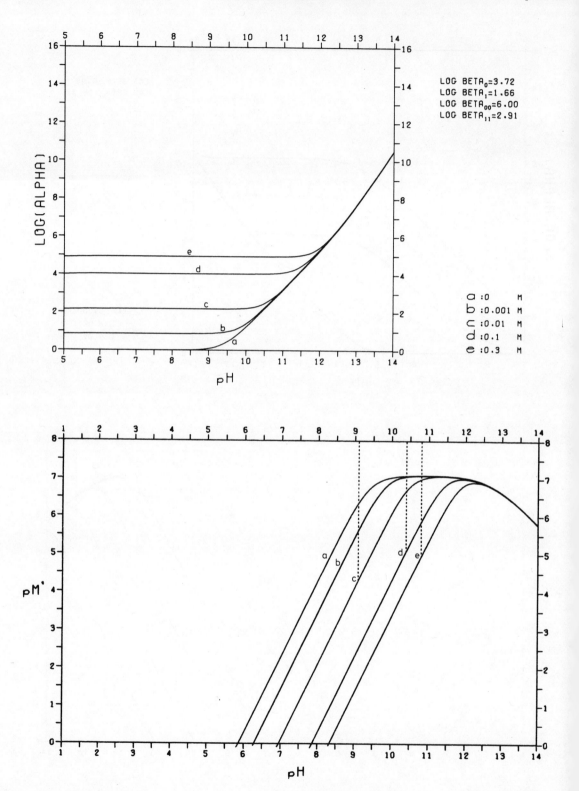

Co-Carbonate-Pyridine-2,6-dicarboxylate

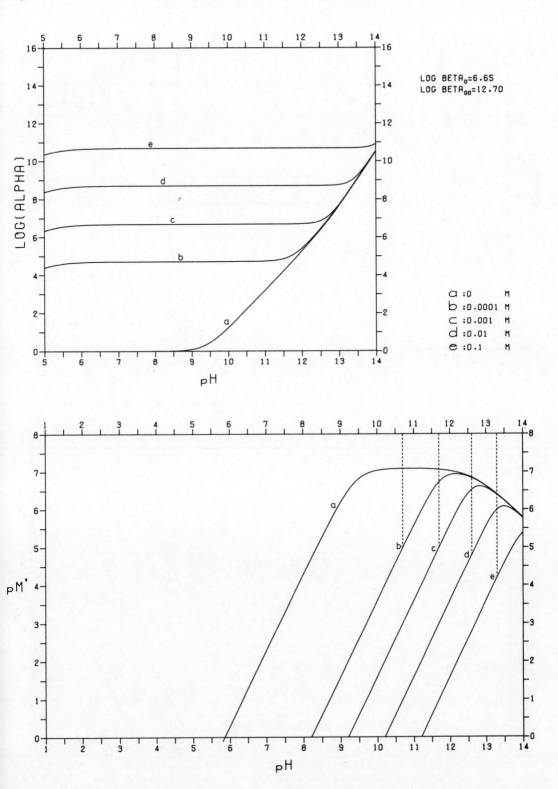

Co-Carbonate-1,10-Phenanthroline [Ch. 10]

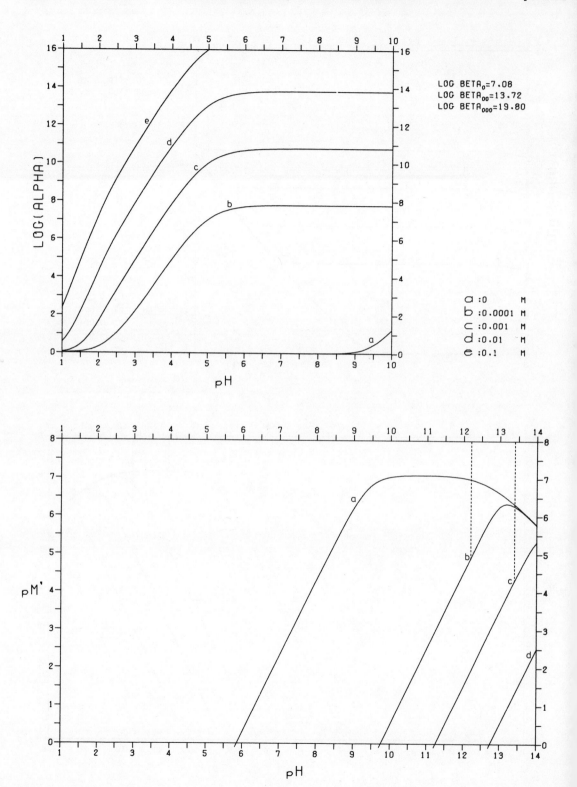

LOG BETA$_0$=7.08
LOG BETA$_{00}$=13.72
LOG BETA$_{000}$=19.80

a : 0   M
b : 0.0001 M
c : 0.001   M
d : 0.01   M
e : 0.1   M

Co-Carbonate-Sulphate

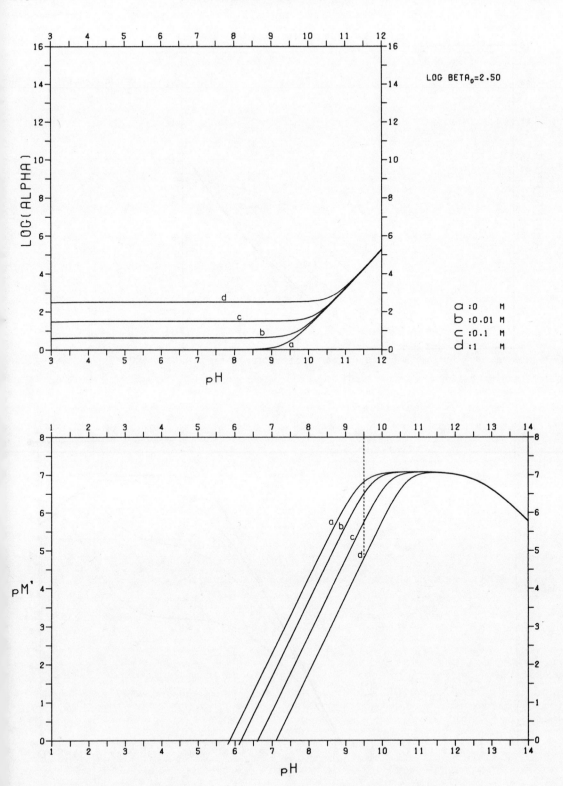

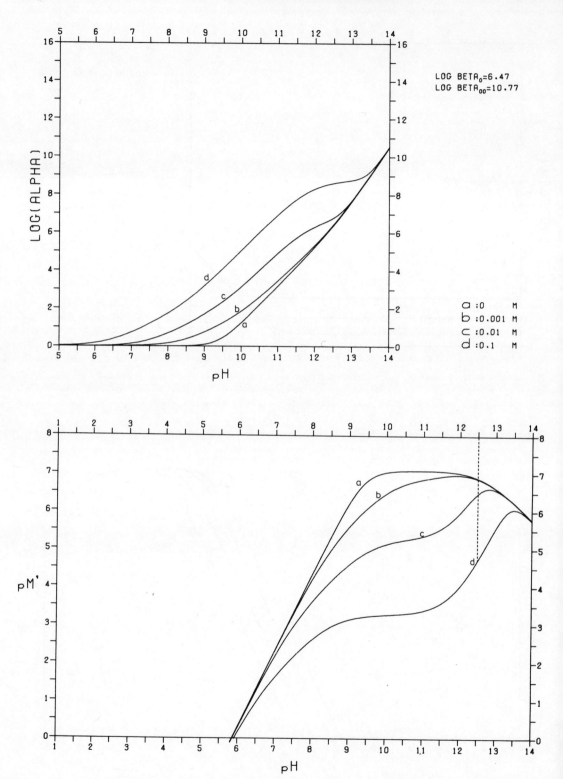

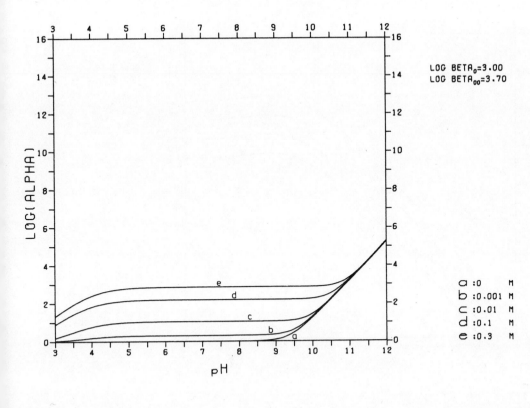

LOG BETA₀=3.00
LOG BETA₀₀=3.70

a : 0 M
b : 0.001 M
c : 0.01 M
d : 0.1 M
e : 0.3 M

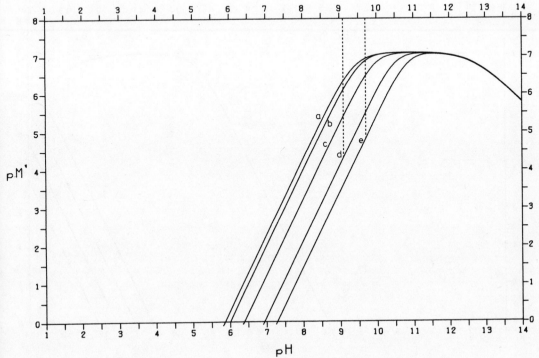

210 **Co–Carbonate–Tetren** [Ch. 10

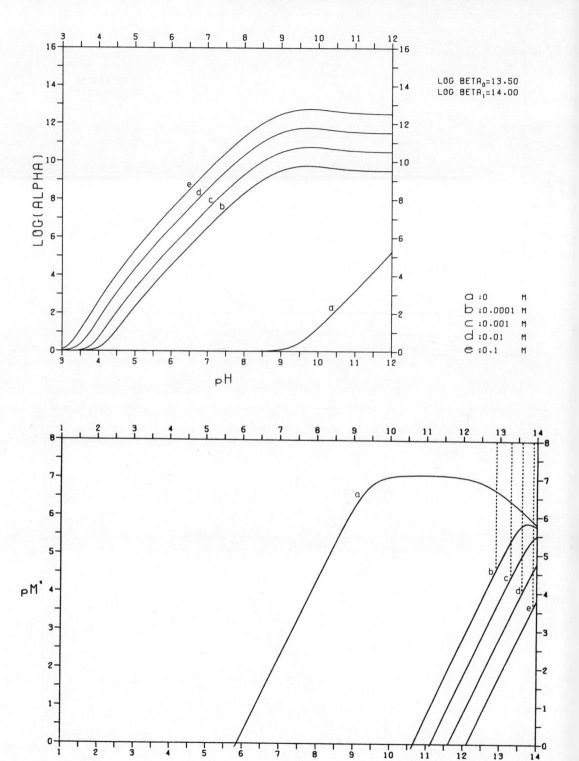

Co-Carbonate-Tiron

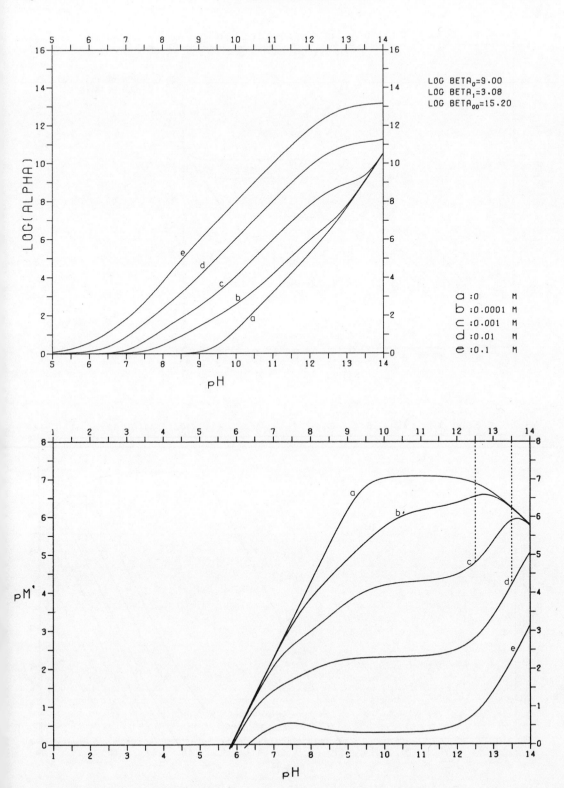

212   Co-Carbonate-Trien   [Ch. 10

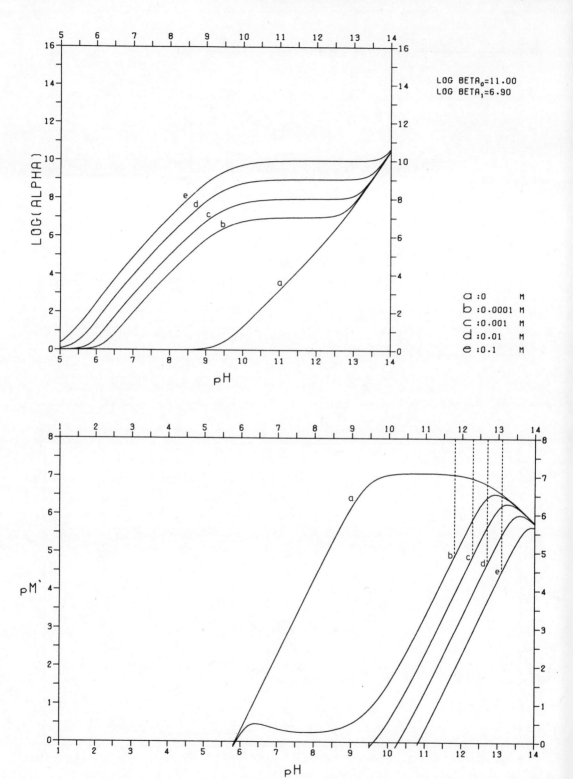

CHAPTER 11

Chromium (III)   Cr

The release of protons from the water of hydration of $Cr(H_2O)_6^{3+}$ to form mononuclear hydrolysis products occurs rapidly. The equilibrium constants have been determined by kinetic methods. Conversely, the reactions involving replacement of water molecules generally proceed very slowly. In particular, polymerization of Cr^{3+} is complicated by the exceedingly slow kinetics of the formation of even small polymers. The inertness of the chromium polyhydroxides even allows separation of the various species by ion-exchange techniques, which implies that it may take days to reach a state of equilibrium. This, together with the slowness of Cr^{3+}–ligand reactions, should be borne in mind when the plots are being used. The following data were used for construction of the plots.

$$\log {}^*\beta_1 = -4.2$$
$$\log {}^*\beta_2 = -10.4$$
$$\log {}^*\beta_3 = -18.7$$
$$\log {}^*\beta_4 = -27.8$$

$$\log {}^*\beta_{22} = -5.1$$
$$\log {}^*\beta_{43} = -8.4$$

$$\log {}^*K_{s0} = 12.7$$

In air-saturated solutions Cr(III) ions should, theoretically, be oxidized to dichromate above pH 3–4. In practice, the oxidation does not proceed to a measurable extent and can be neglected.

Chromium(III)

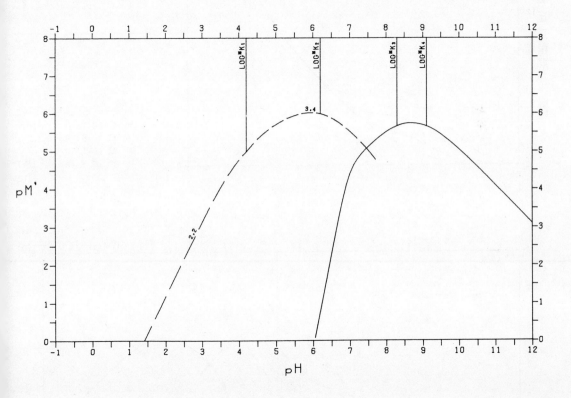

Chromium(III)-Acetate [Ch. 11]

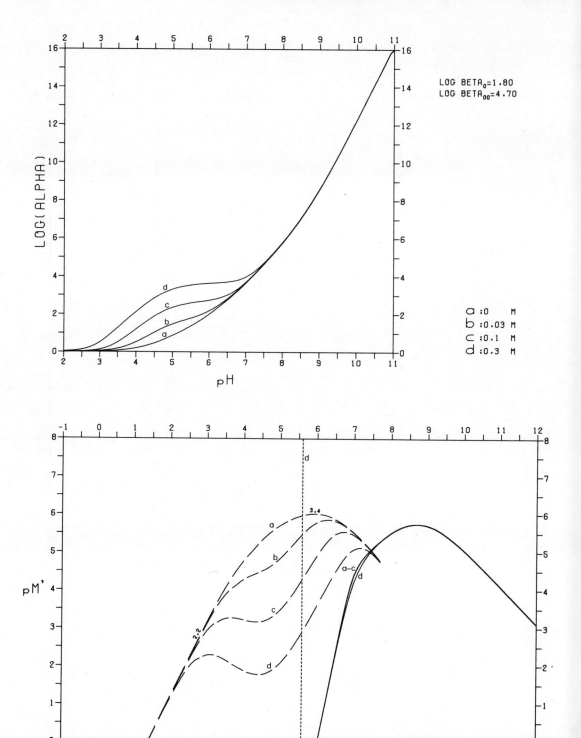

LOG BETA₀=1.80
LOG BETA₀₀=4.70

a : 0   M
b : 0.03 M
c : 0.1 M
d : 0.3 M

Chromium(III)-Citrate

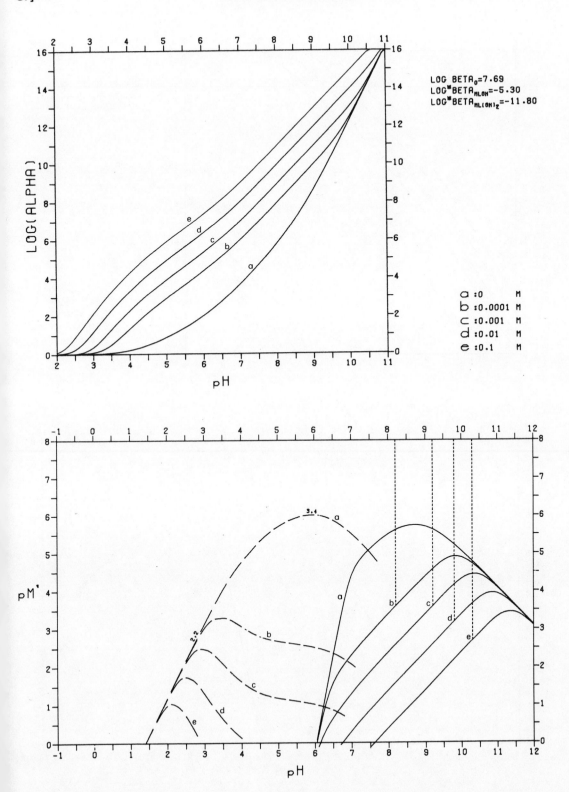

Chromium(III)–EDTA [Ch. 11

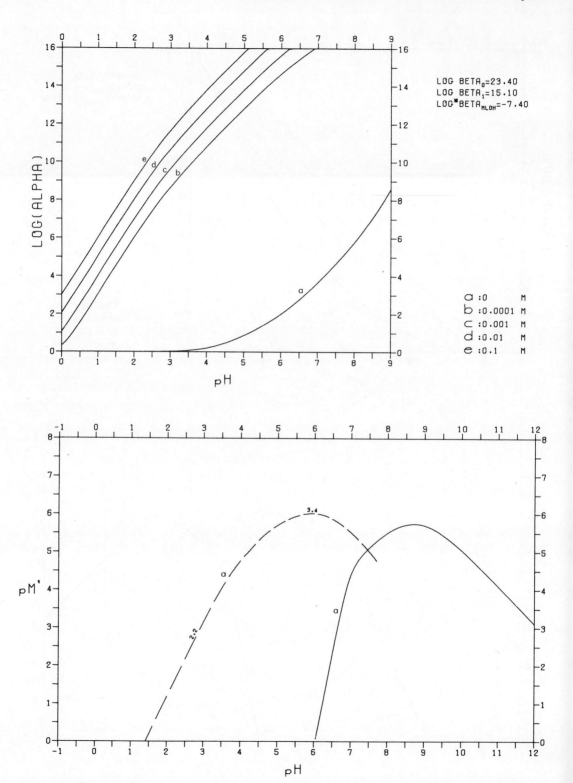

Chromium(III)-Fluoride

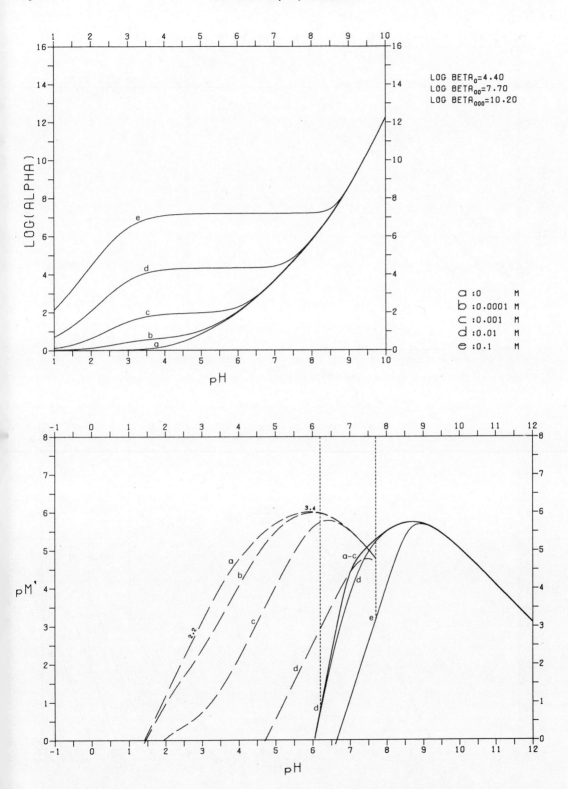

220 **Chromium(III)-Iminodiacetate** [Ch. 11

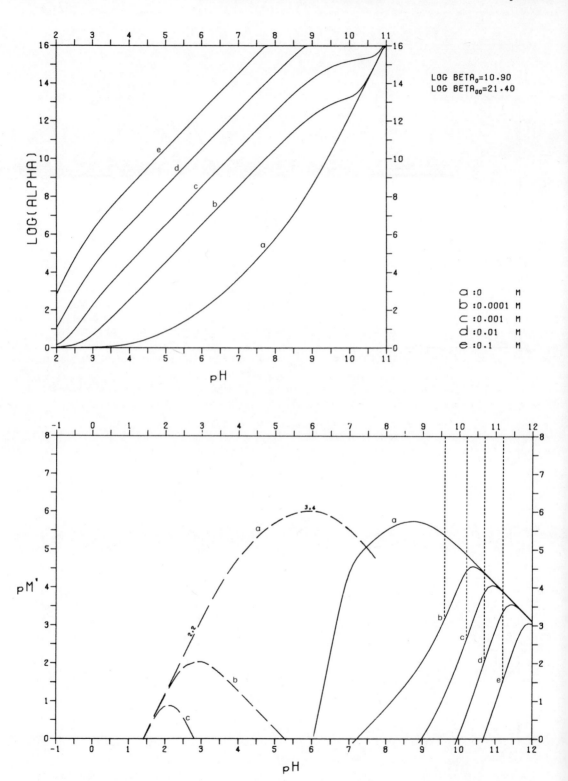

LOG BETA$_0$=10.90
LOG BETA$_{00}$=21.40

a : 0   M
b : 0.0001 M
c : 0.001 M
d : 0.01   M
e : 0.1   M

Chromium(III)-Oxalate

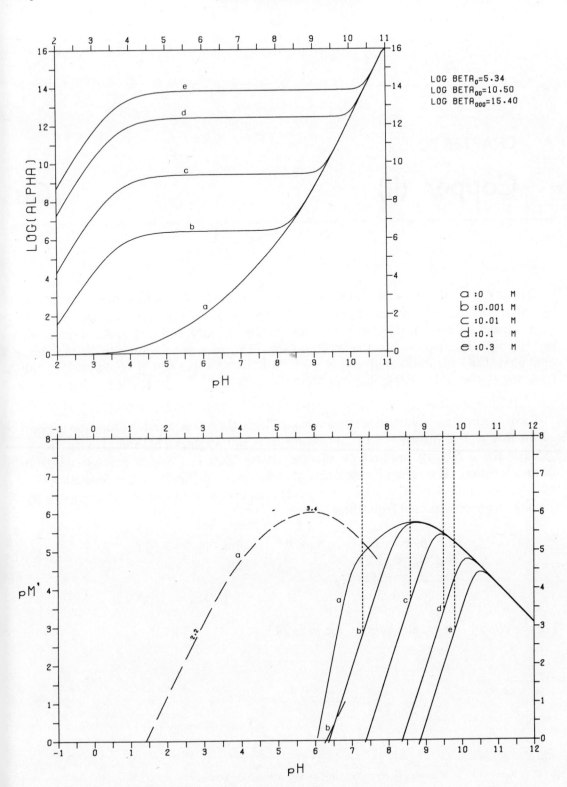

LOG BETA$_0$=5.34
LOG BETA$_{00}$=10.50
LOG BETA$_{000}$=15.40

a : 0 M
b : 0.001 M
c : 0.01 M
d : 0.1 M
e : 0.3 M

CHAPTER 12

Copper (II)　　　　　　　　　　　　　　　　　　　　　　　　　　　Cu

The stable oxidation state of Cu in aqueous medium is +II, and the principal cationic hydrolysis product of Cu^{2+} is $Cu_2(OH)_2^{2+}$ The formation of $Cu(OH)^+$ has been reported by a number of investigators, but the value of the stability constant remains uncertain; log $*\beta_1 = -8.2$ is probably the most reliable. The increasing solubility of the stable solid phase CuO in carbonate-free 0.04 - 0.08M KOH solutions can be considered to result from formation of $Cu(OH)_3^-$ and $Cu(OH)_4^{2-}$, with log $*\beta_4 = -39.1$ and a lower limit of -27.8 for log $*\beta_3$.

Some preliminary solubility experiments utilizing flameless atomic absorption supported the estimate of log $*\beta = -17.5$ made by Baes and Mesmer [1], and gave log $*K_{s0} = 8.85$ for the solubility product of the metastable $Cu(OH)_2$ which precipitates from aqueous solution. The solubility experiments were made difficult by the formation of a colloidal precipitate of $Cu(OH)_2$, but centrifuging improved the results remarkably. The plots were constructed from these values:

$$\log *\beta_1 = -8.2$$
$$\log *\beta_2 = -17.5$$
$$\log *\beta_3 = -27.8$$
$$\log *\beta_4 = -39.1$$

$$\log *\beta_{22} = -10.6$$

$$\log *K_{s0} = 8.85$$

For the copper – carbonate system, see page 245.

[1] C. F. Baes and R. E. Mesmer, *The Hydrolysis of Cations*, Wiley, New York, 1976.

Copper(II)

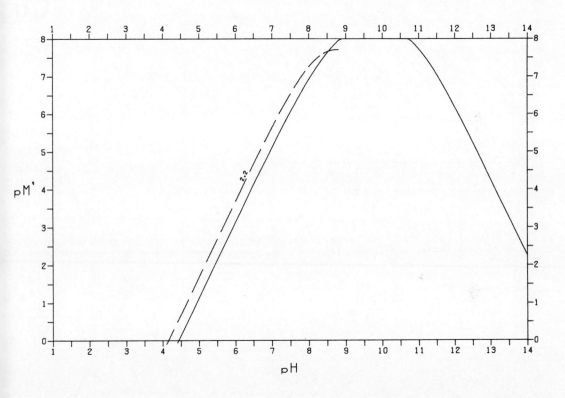

Copper(II)-Acetylacetone [Ch. 12

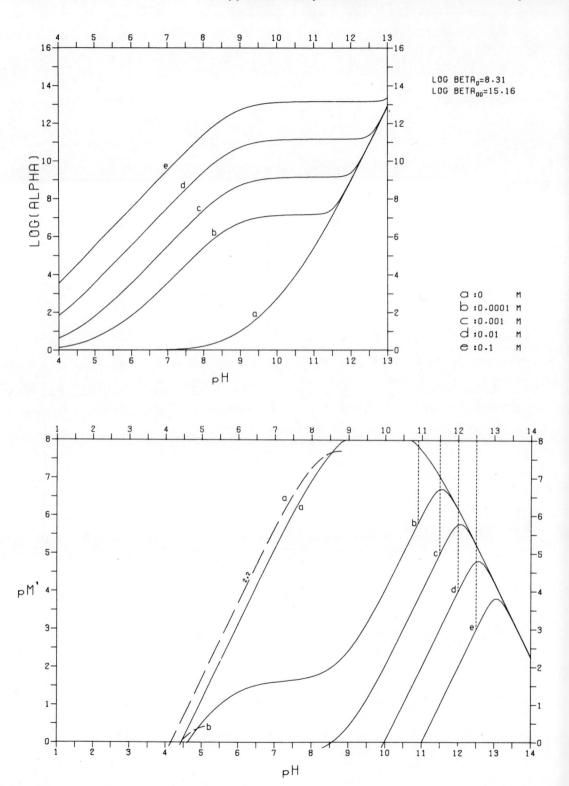

LOG BETA$_0$=8.31
LOG BETA$_{00}$=15.16

a : 0   M
b : 0.0001 M
c : 0.001   M
d : 0.01   M
e : 0.1   M

Copper(II)-Ammonia

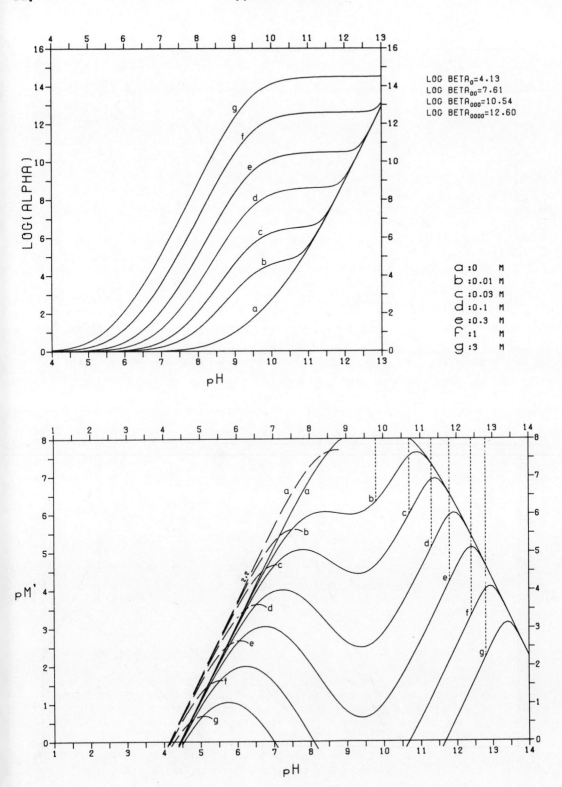

LOG BETA$_0$=4.13
LOG BETA$_{00}$=7.61
LOG BETA$_{000}$=10.54
LOG BETA$_{0000}$=12.60

a : 0   M
b : 0.01 M
c : 0.03 M
d : 0.1 M
e : 0.3 M
f : 1   M
g : 3   M

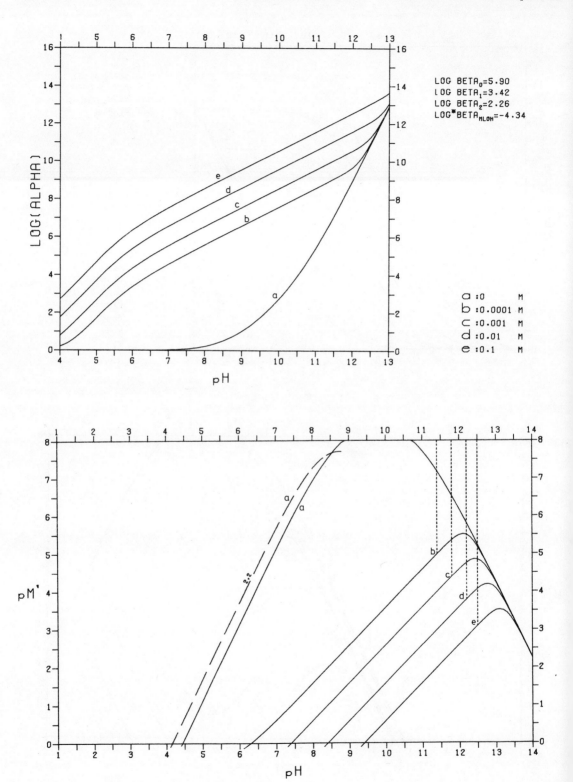

Copper(II)-Cyanide

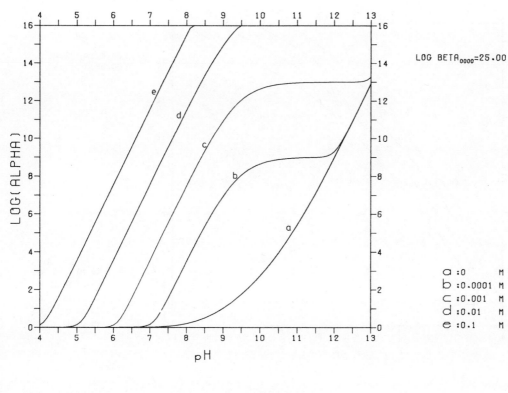

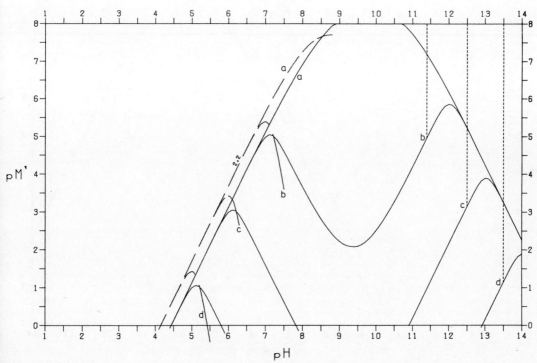

Copper(II)-DCTA

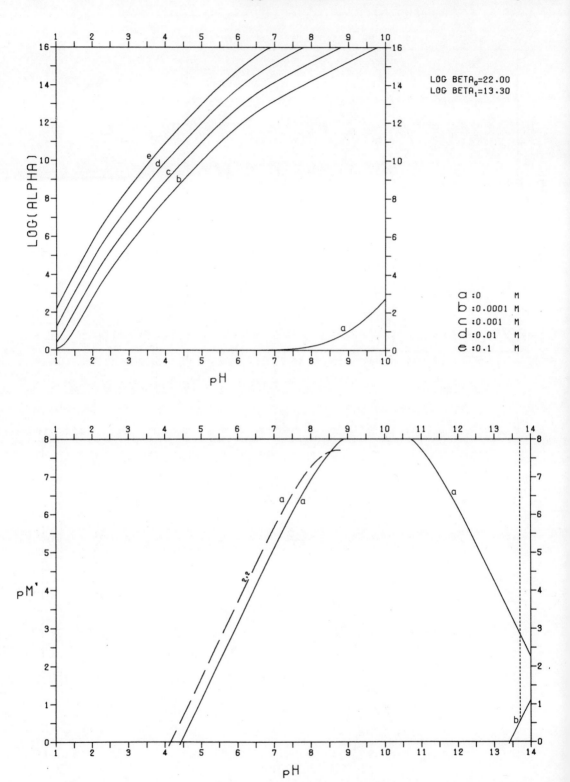

Copper(II)-DTPA

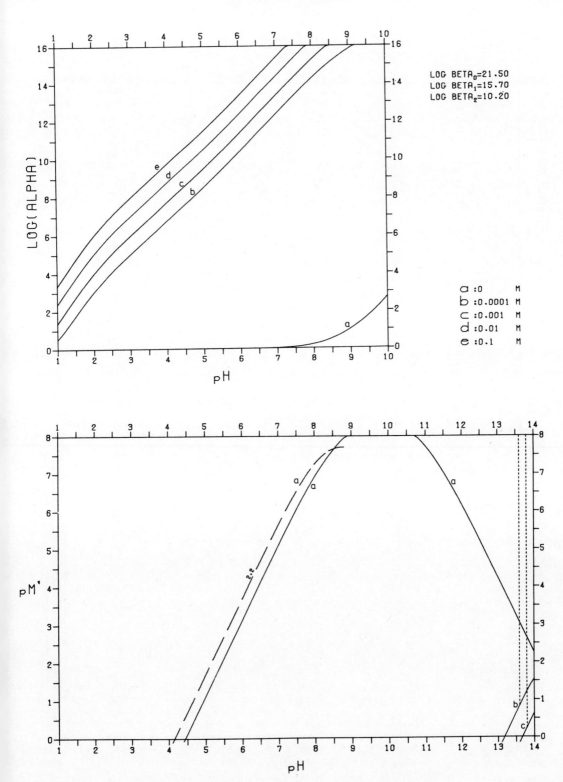

LOG BETA₀=21.50
LOG BETA₁=15.70
LOG BETA₂=10.20

a : 0 M
b : 0.0001 M
c : 0.001 M
d : 0.01 M
e : 0.1 M

Copper(II)-EDTA [Ch. 12]

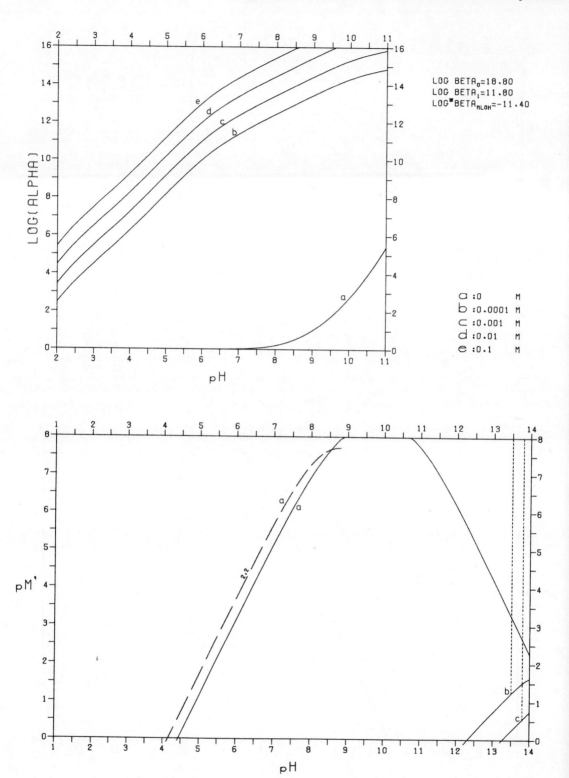

Copper(II)-Ethylenediamine

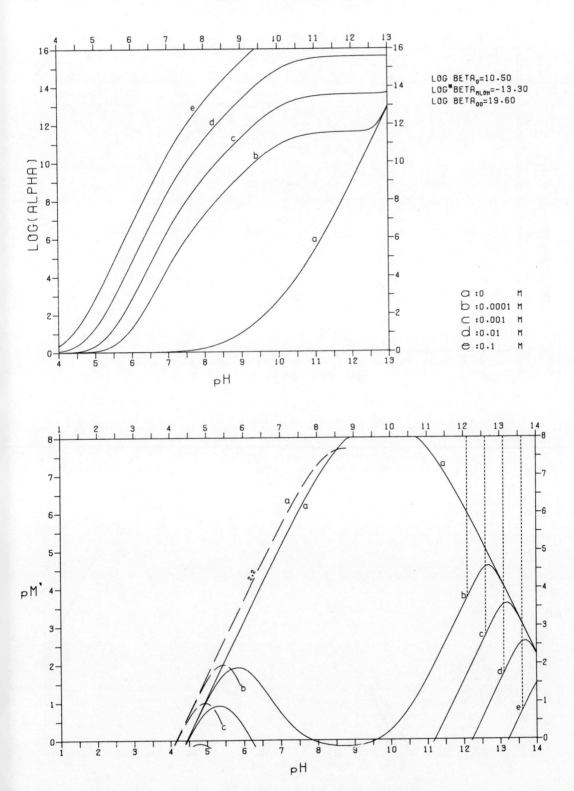

LOG BETA$_0$ = 10.50
LOG*BETA$_{MLOH}$ = -13.30
LOG BETA$_{00}$ = 19.60

a : 0 M
b : 0.0001 M
c : 0.001 M
d : 0.01 M
e : 0.1 M

Copper(II)-Glycine [Ch. 12

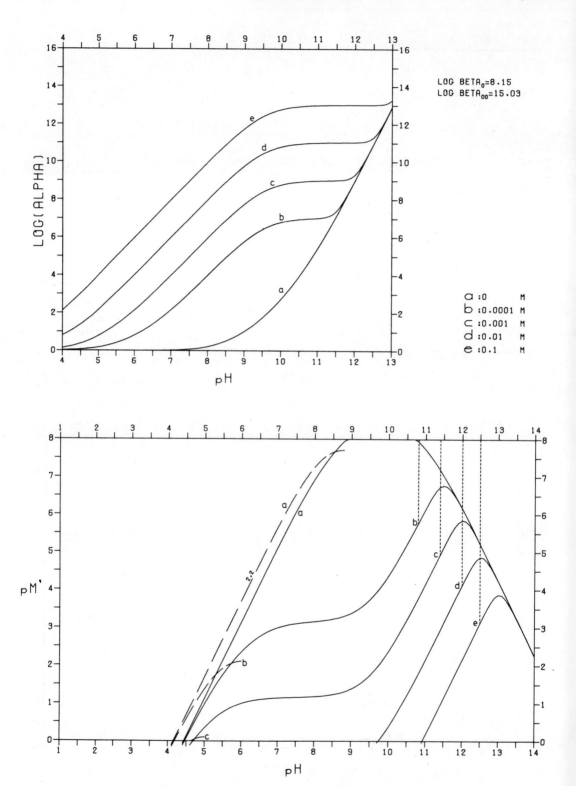

LOG BETA$_0$=8.15
LOG BETA$_{00}$=15.03

a : 0   M
b : 0.0001   M
c : 0.001   M
d : 0.01   M
e : 0.1   M

Copper(II)-Hydroxylamine

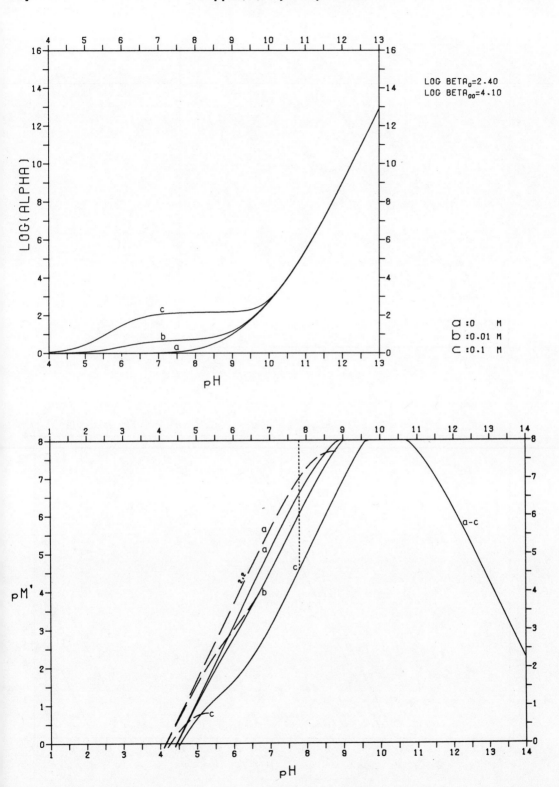

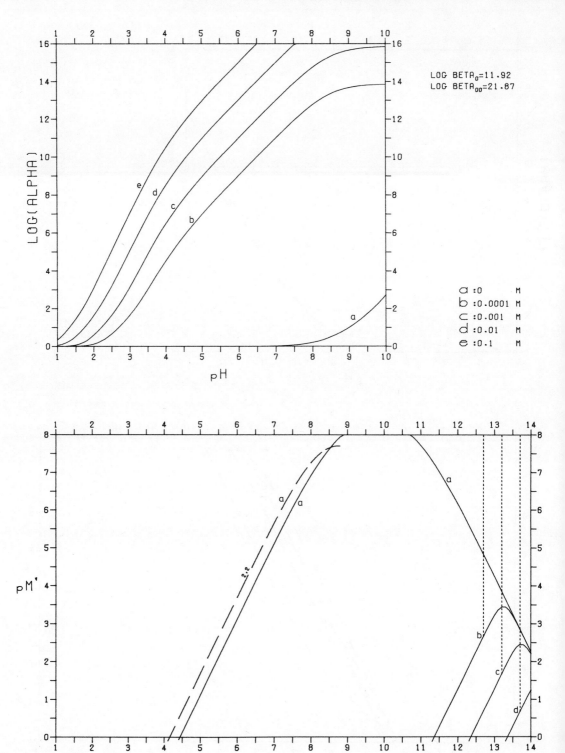

Copper(II)-Iminodiacetate

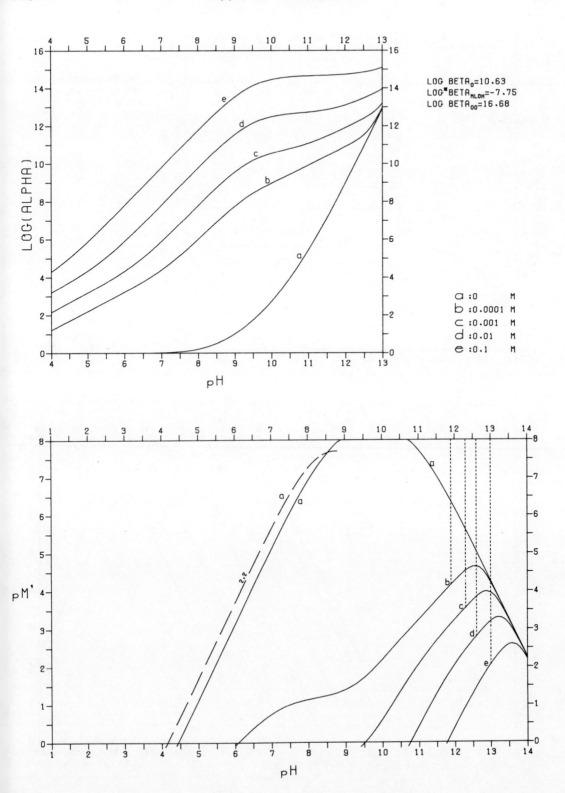

Copper(II)-Oxalate [Ch. 12]

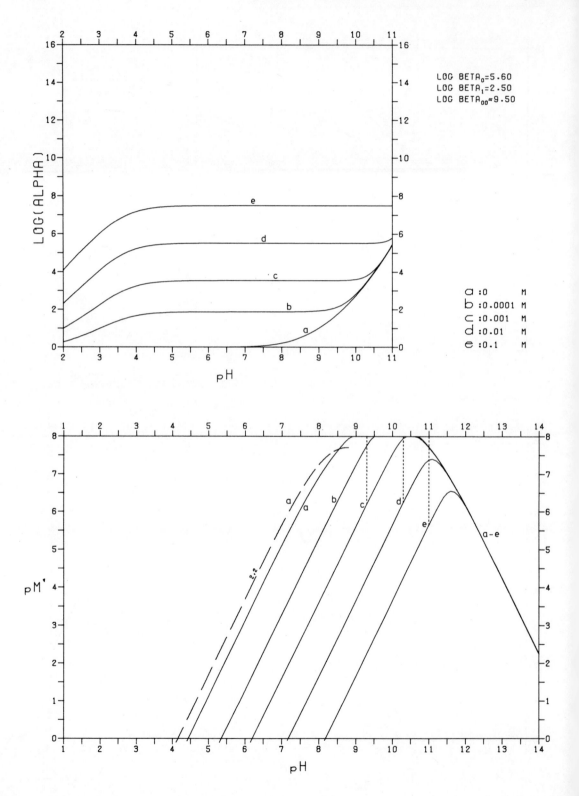

Copper(II)-Pyridine-2,6-dicarboxylate

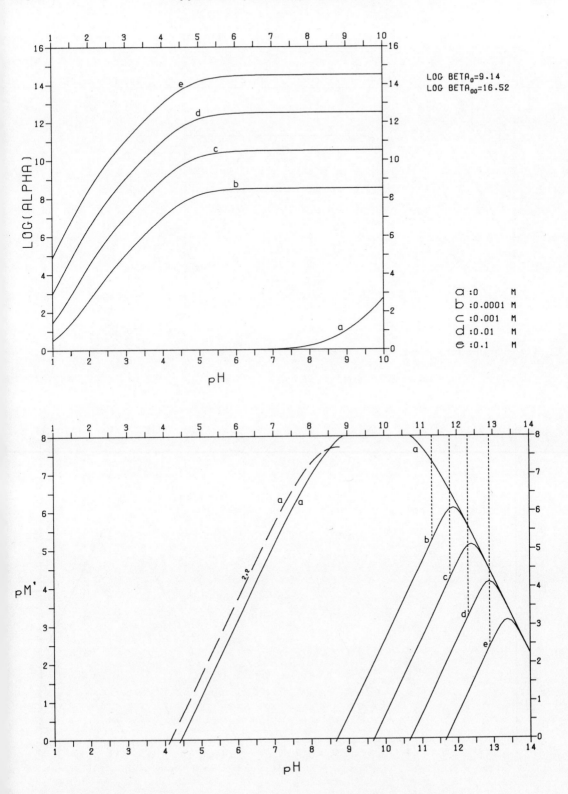

Copper(II)-1,10-Phenanthroline [Ch. 12]

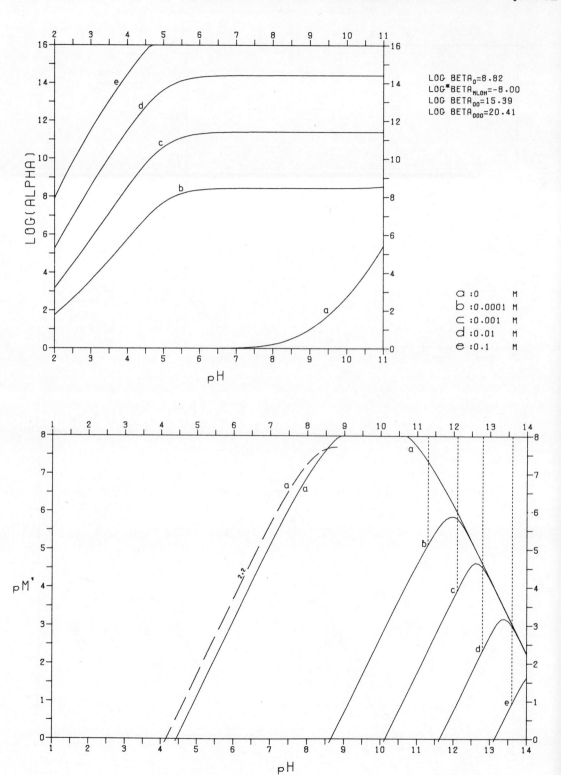

Copper(II)-Salicylate

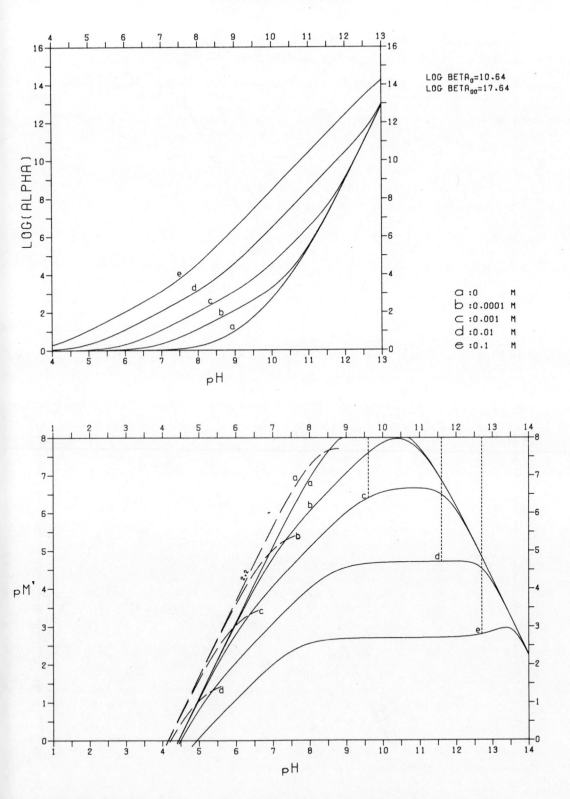

LOG BETA$_0$=10.64
LOG BETA$_{00}$=17.64

a : 0 M
b : 0.0001 M
c : 0.001 M
d : 0.01 M
e : 0.1 M

240   **Copper(II)-5-Sulphosalicylate**   [Ch. 12

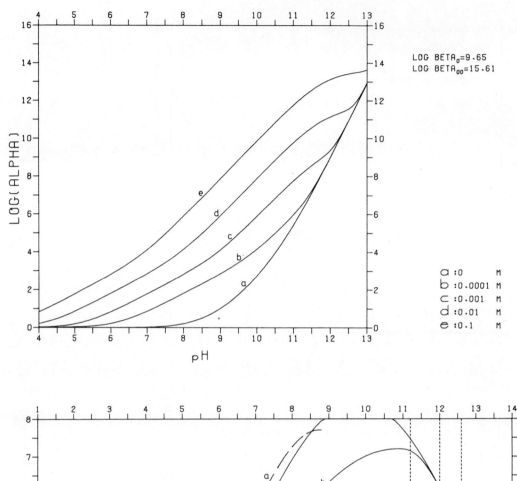

LOG BETA$_0$ = 9.65
LOG BETA$_{00}$ = 15.61

a : 0 M
b : 0.0001 M
c : 0.001 M
d : 0.01 M
e : 0.1 M

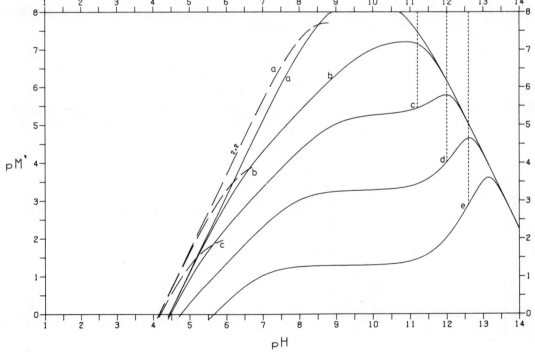

Copper(II)-Tartrate

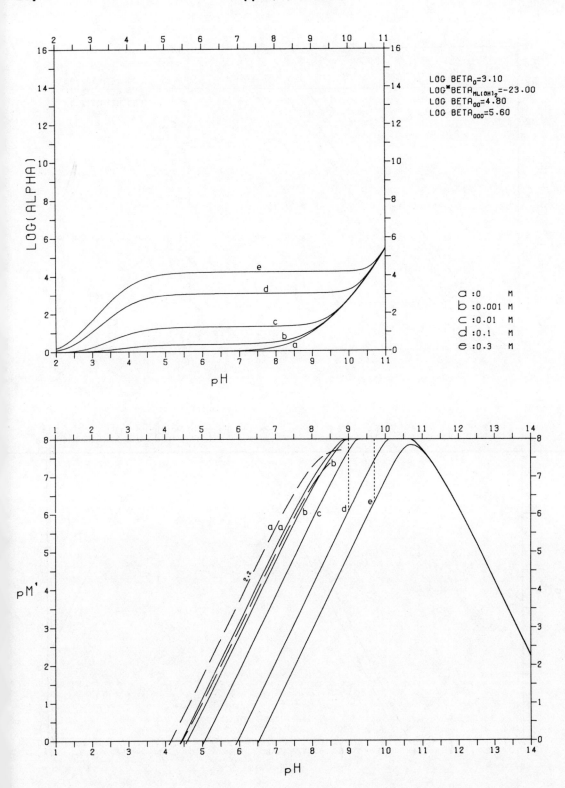

LOG BETA$_0$=3.10
LOG*BETA$_{ML(OH)_2}$=-23.00
LOG BETA$_{00}$=4.80
LOG BETA$_{000}$=5.60

a : 0 M
b : 0.001 M
c : 0.01 M
d : 0.1 M
e : 0.3 M

242   **Copper(II)-Tetren**   [Ch. 12

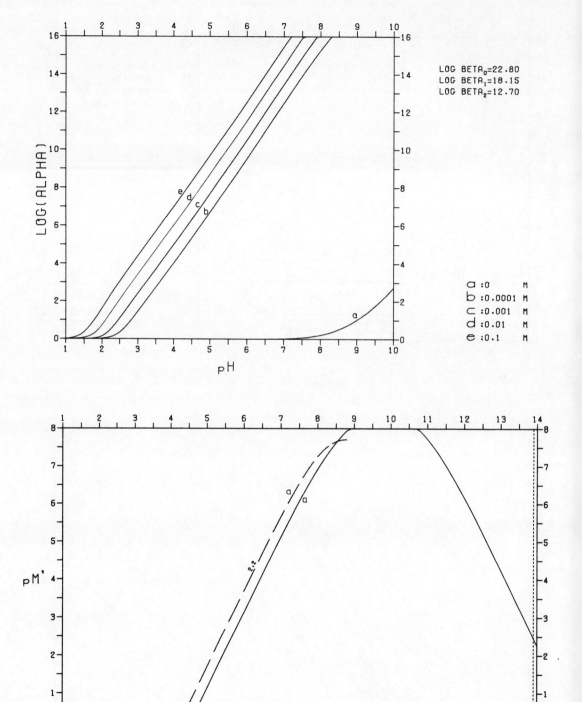

Copper(II)-Tiron

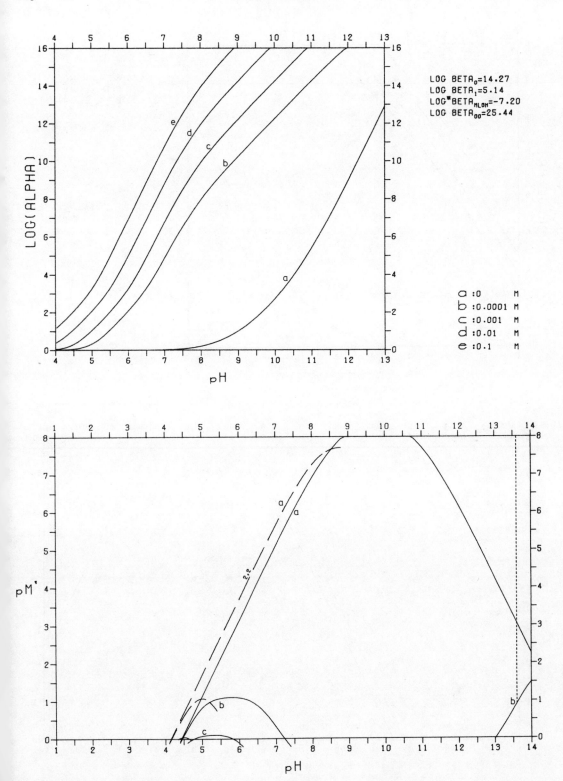

LOG BETA$_0$=14.27
LOG BETA$_1$=5.14
LOG*BETA$_{MLOH}$=-7.20
LOG BETA$_{00}$=25.44

a : 0 M
b : 0.0001 M
c : 0.001 M
d : 0.01 M
e : 0.1 M

Copper(II)–Trien [Ch. 12

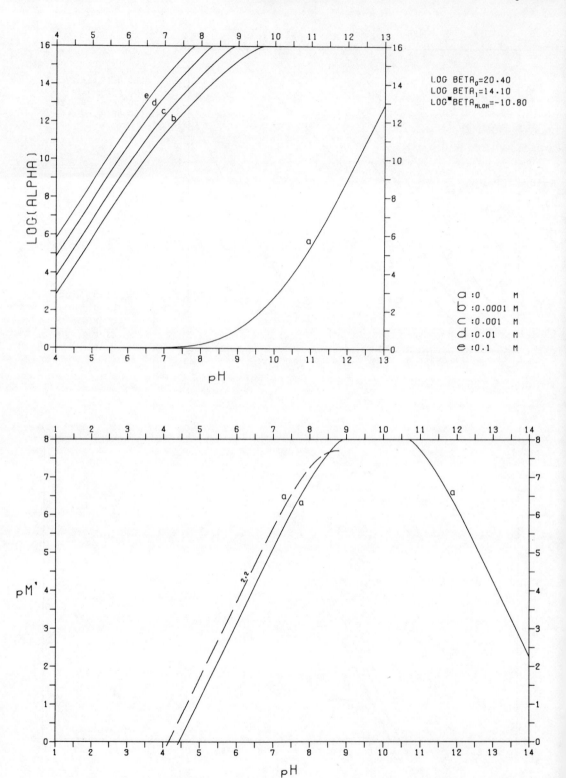

Cu — carbonate system

The occurrence of the stable, soluble carbonate complexes $CuCO_3$ and $Cu(CO_3)_2^{2-}$ has no influence on the low-pH side of the precipitation region in the pM' – pH diagram, because $Cu(OH)_2$ is still the less soluble solid in air-saturated water. At high pH, however, copper hydroxide dissolves to form $Cu(CO_3)_2^{2-}$.

The equation for the high-pH borderline of the precipitation region in air-saturated water [log P_{CO_2} = −3.52 (P in atmospheres)], derived from the stability constants log β_0 = 6.7 and log β_{00} = 9.8, is pM' = (−2 pH + 26.3). This corresponds to an apparent value of 35.1 for log β_4 (see page 26). The following set of constants have been used for the plots for air-saturated solutions.

$$\log {}^*\beta_1 = -8.2 \qquad \log {}^*\beta_{22} = -10.6$$
$$\log {}^*\beta_2 = -17.5$$
$$\log {}^*\beta_3 = -35.1 \qquad \log {}^*K_{s0} = 8.85$$

Carbonate cannot be treated as a separate ligand here, because the concentration of carbonate is not constant in air-saturated water; it increases enormously at high pH.

The pM'–pH plots for air-saturated solutions containing other ligands can be constructed from the corresponding plots for CO_2-free solutions by moving the high-pH borderline of precipitation to the left by 2 pH units.

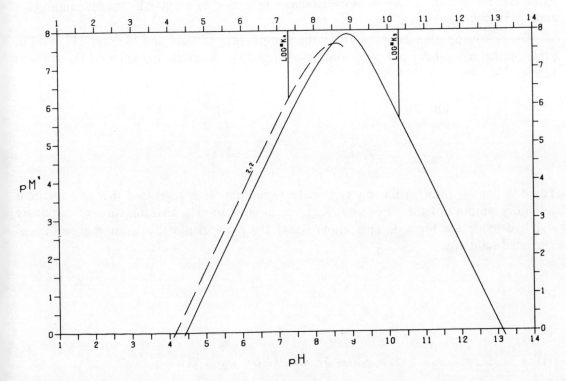

CHAPTER 13
Dysprosium (III)   Dy

Hydrolysis of the large Dy^{3+} ion does not become appreciable until fairly high pH (>7). Hydrolysis studies are quite difficult, since, at ordinary concentrations, hydroxide invariably precipitates when the average number of OH^- ions coordinated is small.

There is no evidence for polycomplexation. A limited amount of reliable information suggests that $Dy(OH)^{2+}$ is formed on the low-pH side of the precipitation region.

The study of hydroxide precipitation is difficult because of supersaturation, colloid formation and dependence of $^*K_{s0}$ on particle size, rapid partial recrystallization and carbonate formation. These phenomena make definition of a generally useful equilibrium state somewhat arbitrary.

The following data adopted from the compilation of Baes and Mesmer [1] hold in principle for a one-hour-old precipitate prepared under inert gas to exclude CO_2, at room temperature, and at ionic strength = 0.1.

$$\log {}^*\beta_1 = -8.0 \qquad \log {}^*K_{s0} = 17$$
$$\log {}^*\beta_2 = -16.2$$
$$\log {}^*\beta_3 = -24.7$$
$$\log {}^*\beta_4 = -35$$

There is limited information on carbonate formation. It is suggested that the apparent solubility product of the carbonate, K_{carb}, is so close to $^*K_{s0}$ that mixtures of carbonate and hydroxide are formed. This implies that the pM'-pH plot does not change for air-saturated solutions.

[1] C. F. Baes and R. E. Mesmer, *The Hydrolysis of Cations*, Wiley, New York, 1976.

Dysprosium(III)

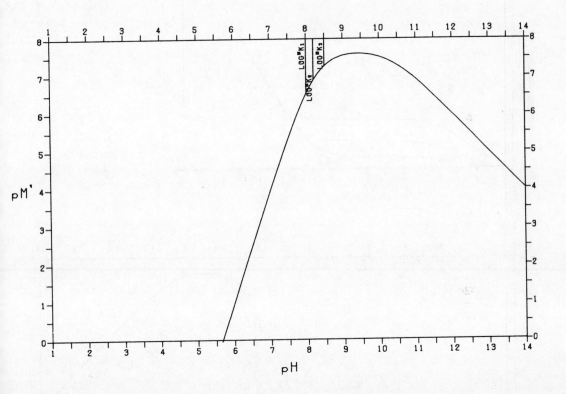

Dysprosium(III)-Acetate

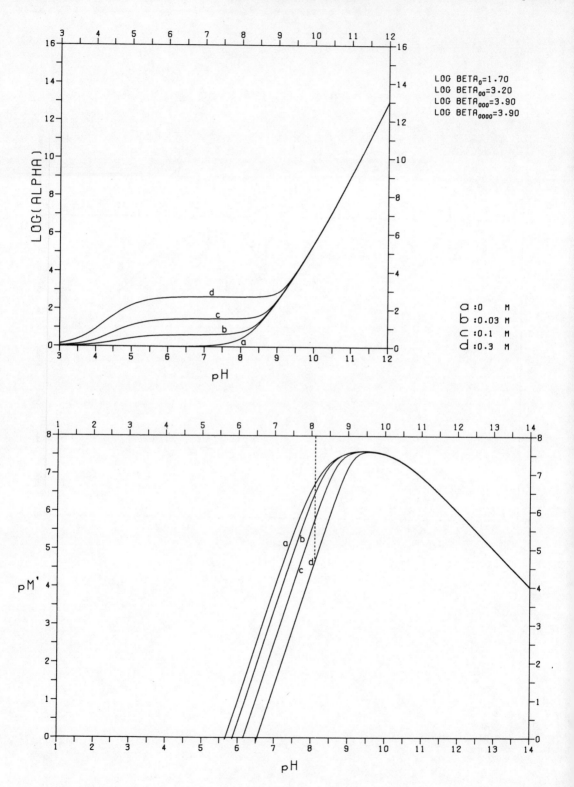

Dysprosium(III)-Acetylacetone

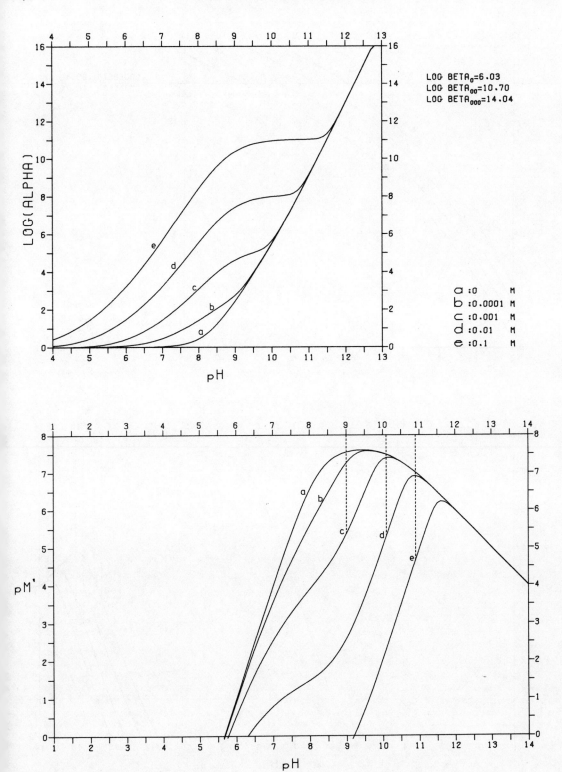

LOG BETA$_0$=6.03
LOG BETA$_{00}$=10.70
LOG BETA$_{000}$=14.04

a : 0 M
b : 0.0001 M
c : 0.001 M
d : 0.01 M
e : 0.1 M

Dysprosium(III)-DCTA [Ch. 13

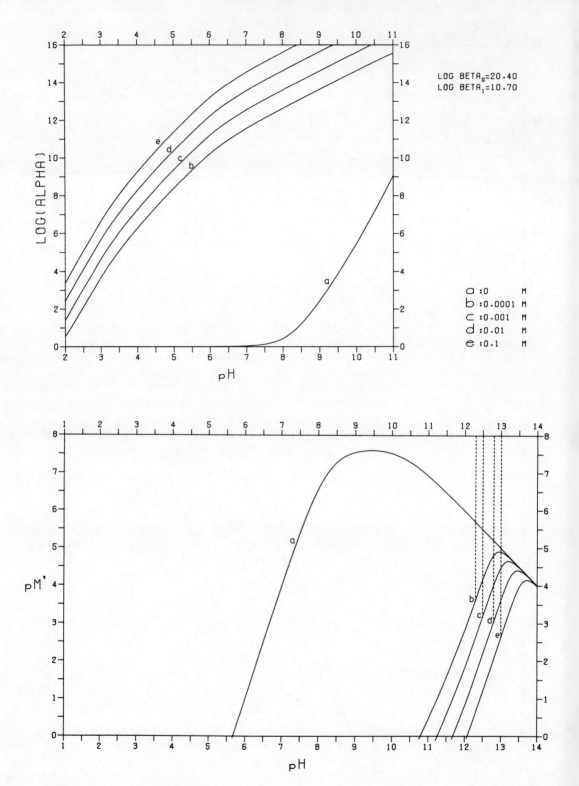

Dysprosium(III)-DTPA

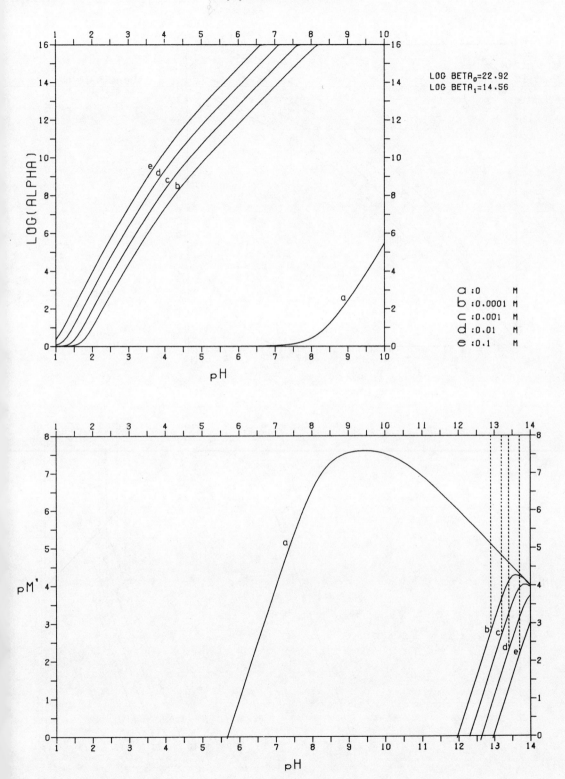

LOG BETA$_0$=22.92
LOG BETA$_1$=14.56

a :0 M
b :0.0001 M
c :0.001 M
d :0.01 M
e :0.1 M

Dysprosium(III)-EDTA

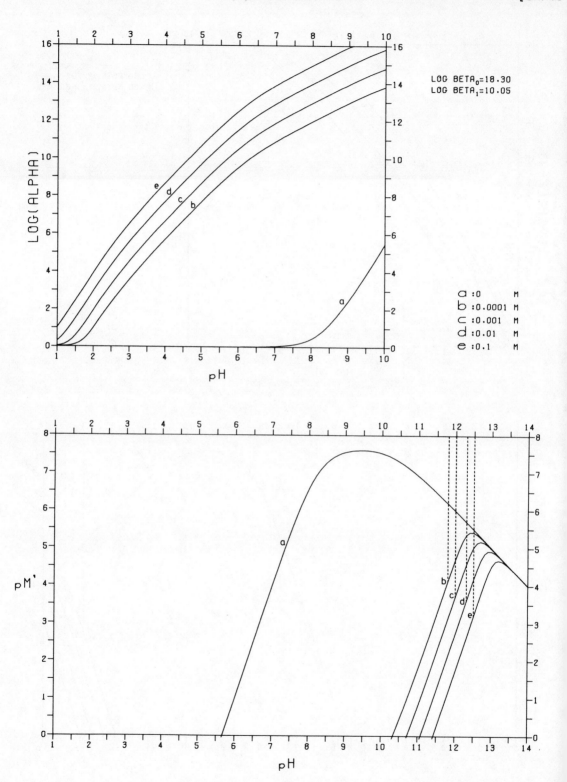

Dysprosium(III)-Iminodiacetate

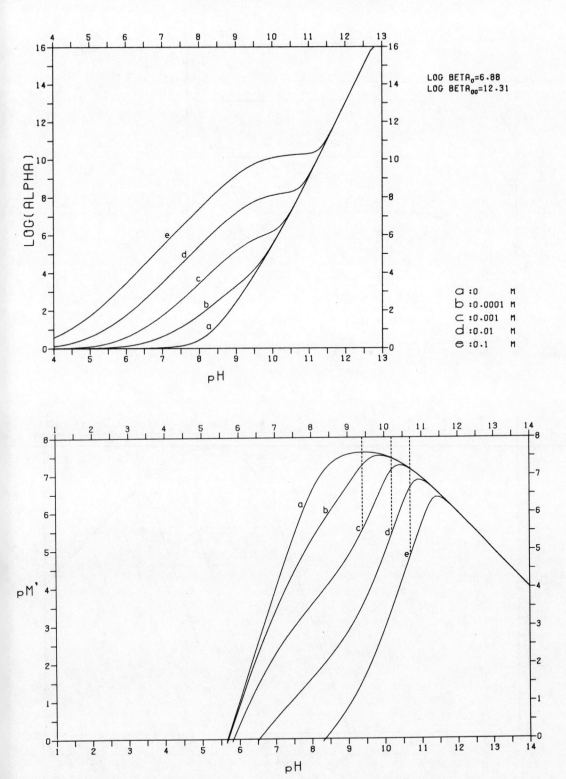

LOG BETA$_0$=6.88
LOG BETA$_{00}$=12.31

a : 0 M
b : 0.0001 M
c : 0.001 M
d : 0.01 M
e : 0.1 M

254 Dysprosium(III)-Pyridine-2,6-dicarboxylate [Ch. 13

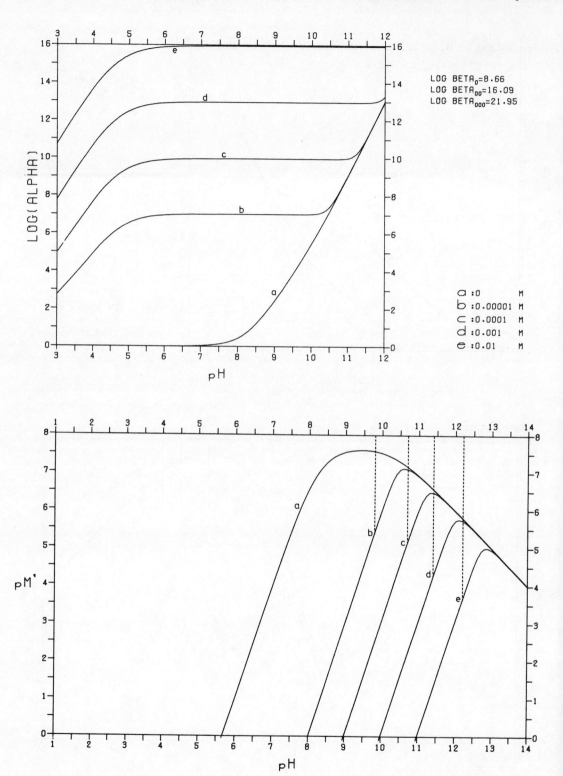

Dysprosium(III)-5-Sulphosalicylate

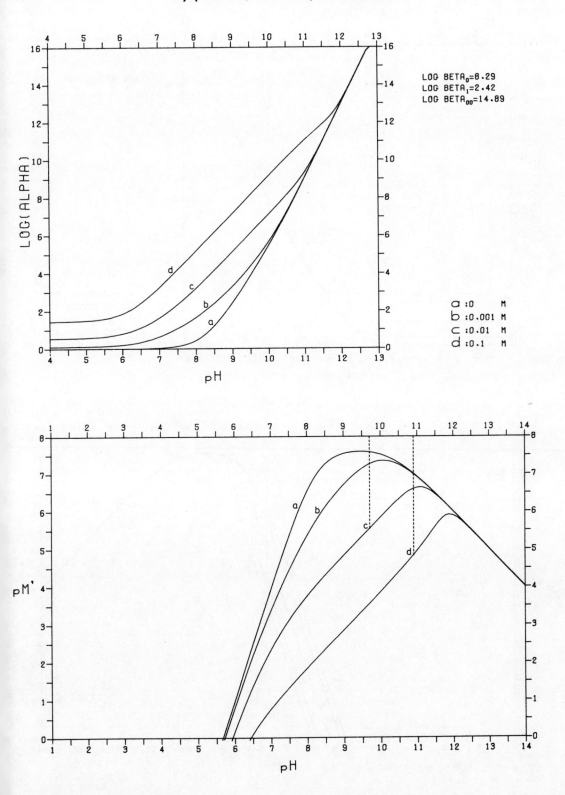

LOG BETA$_0$=8.29
LOG BETA$_1$=2.42
LOG BETA$_{00}$=14.89

a : 0   M
b : 0.001 M
c : 0.01 M
d : 0.1   M

Dysprosium(III)-Tartrate [Ch. 13]

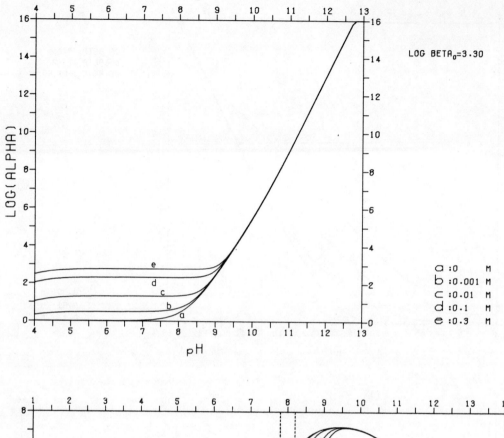

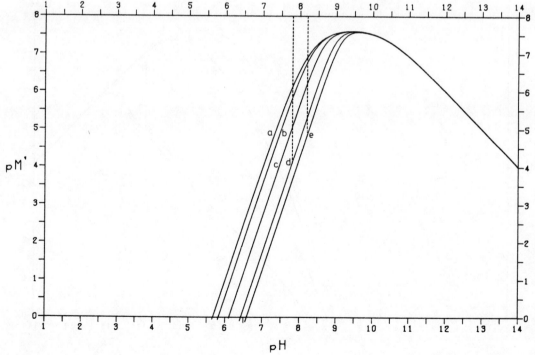

CHAPTER 14
Erbium (III) Er

Hydrolysis of the rather large Er^{3+} ion does not start until fairly high pH. At ordinary concentrations, hydroxide is always precipitated when the hydrolysis is being studied; precipitation starts even when only small amounts of the ion are hydrolysed. There is conclusive evidence, however, from the limited amount of reliable information available, that the polycomplex $Er_2(OH)_2^{4+}$ is formed.

As with most rare earths, the study of hydroxide precipitation is difficult because of supersaturation, colloid formation and the dependence of $*K_{s0}$ on particle size, and rapid partial recrystallization. These phenomena make the definition of a generally useful equilibrium situation somewhat arbitrary.

The following data adopted from the compilation of Baes and Mesmer [1] hold for a one-hour-old precipitate.

$$\log *\beta_1 = -7.9$$
$$\log *\beta_2 = -16.0$$
$$\log *\beta_3 = -24.2$$
$$\log *\beta_4 = -33.0$$

$$\log *\beta_{22} = -13.7$$

$$\log *K_{s0} = 16.0$$

The information available about carbonate formation suggests that it should not occur, and that CO_2 will not be absorbed.

[1] C. F. Baes and R. E. Mesmer, *The Hydrolysis of Cations*, Wiley, New York, 1976.

Erbium(III)

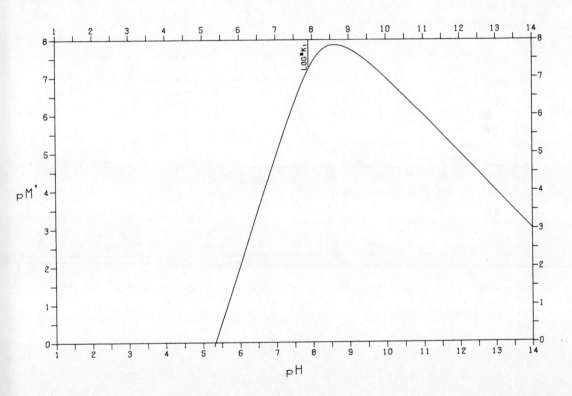

Erbium(III)-Acetate [Ch. 14]

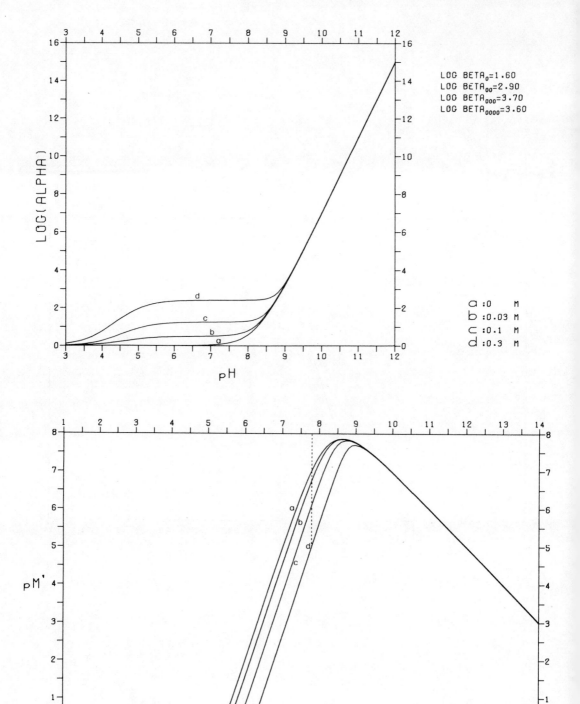

Erbium(III)-Acetylacetone

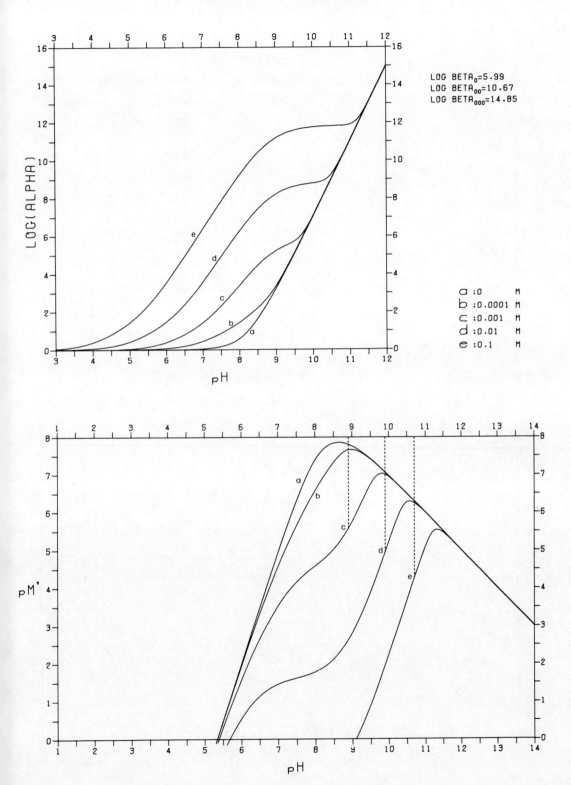

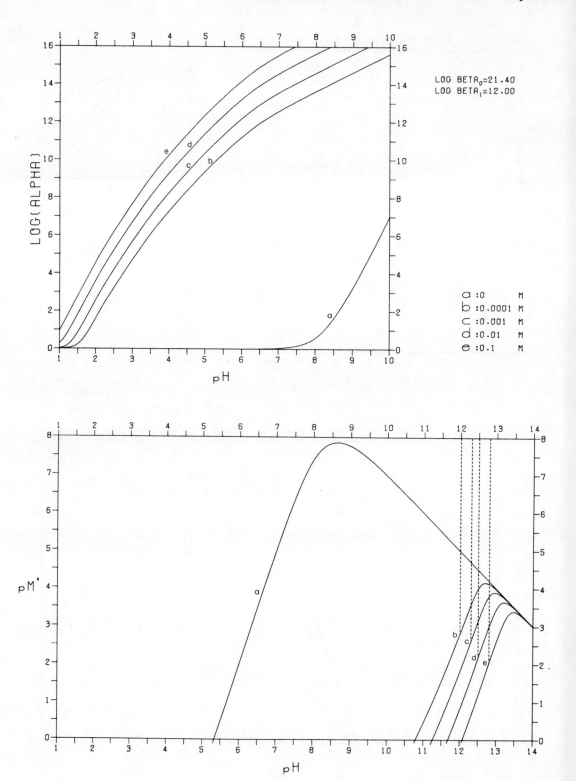

Erbium(III)–DTPA

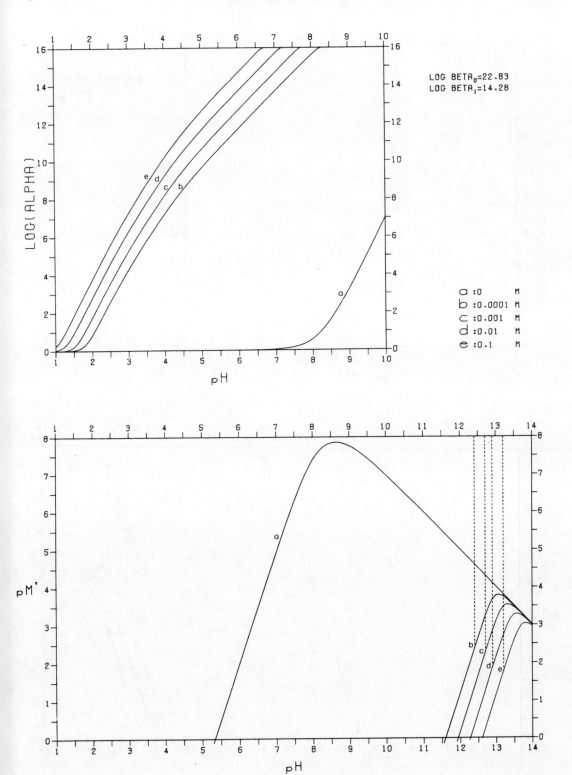

LOG BETA$_0$ = 22.83
LOG BETA$_1$ = 14.28

a : 0 M
b : 0.0001 M
c : 0.001 M
d : 0.01 M
e : 0.1 M

264   Erbium(III)-EDTA   [Ch. 14]

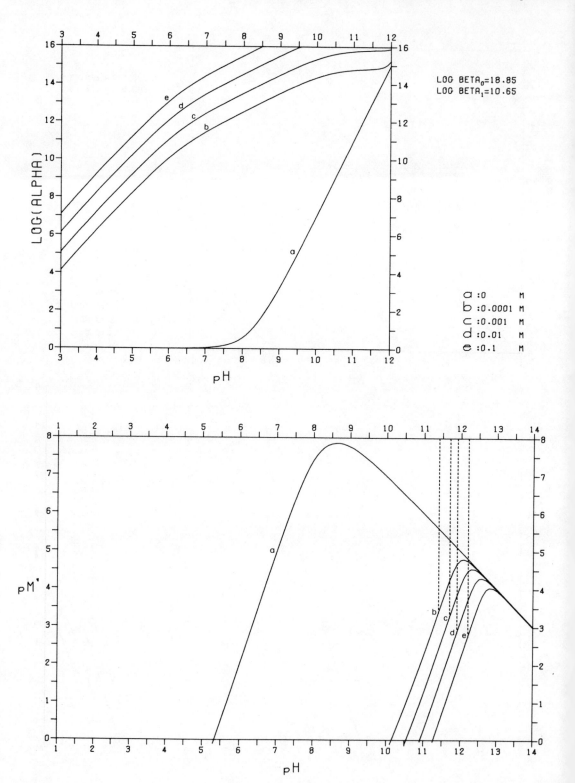

Erbium(III)-(OH)-Quinolinesulphonate

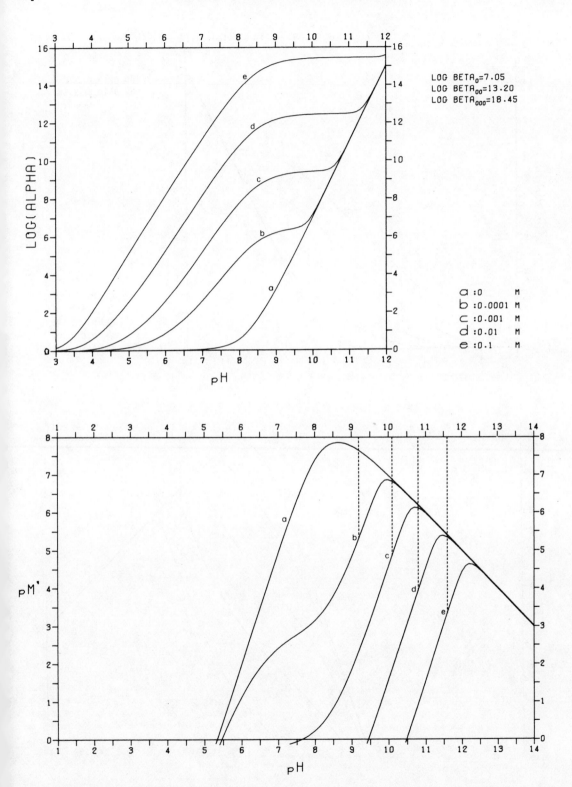

Erbium(III)-Iminodiacetate

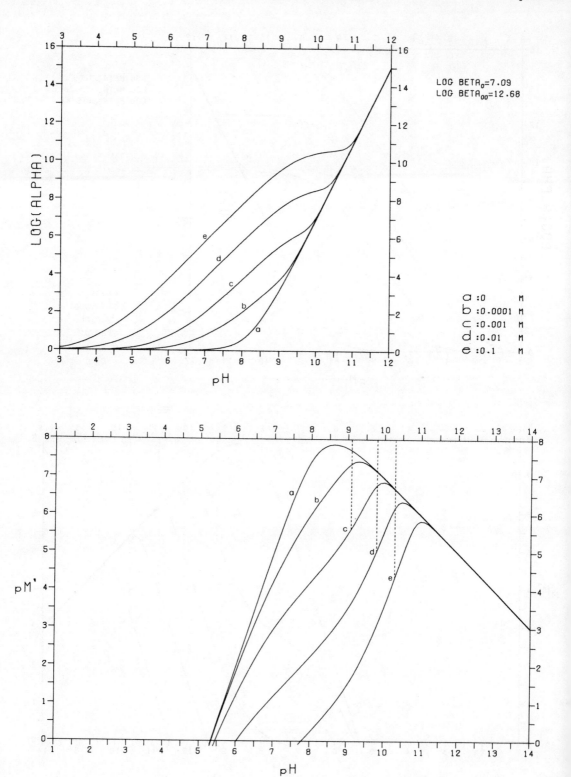

Erbium(III)-Oxalate

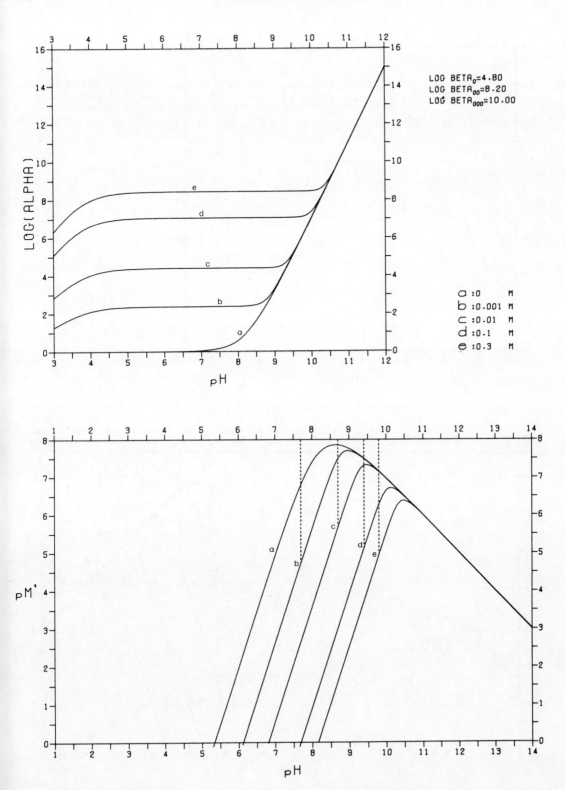

268 **Erbium(III)–Pyridine-2,6-dicarboxylate** [Ch. 14

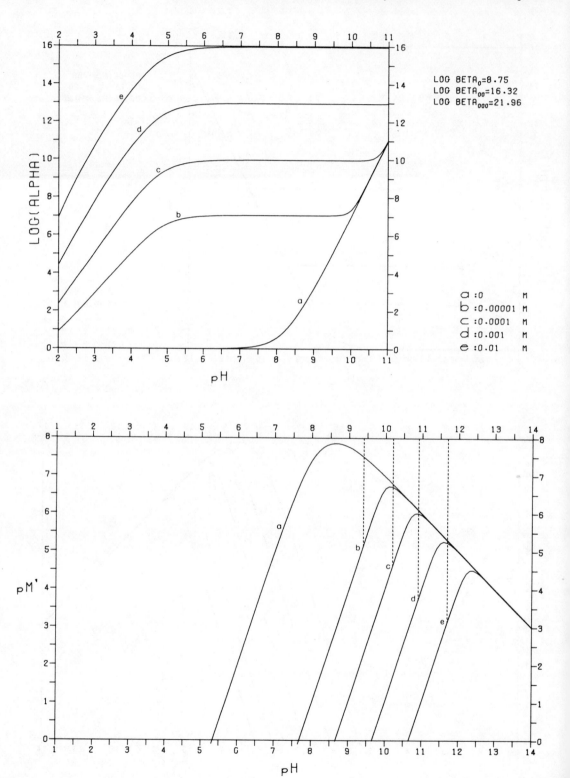

Erbium(III)-5-Sulphosalicylate

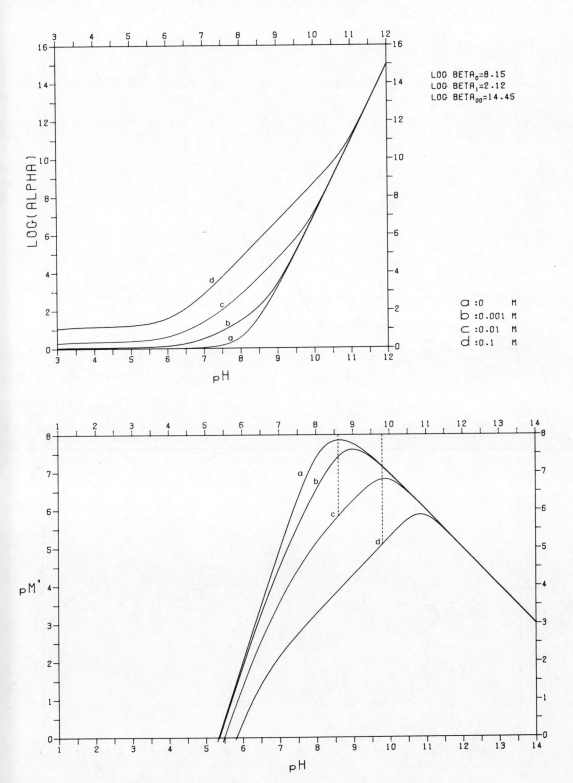

LOG BETA$_0$=8.15
LOG BETA$_1$=2.12
LOG BETA$_{00}$=14.45

a : 0 M
b : 0.001 M
c : 0.01 M
d : 0.1 M

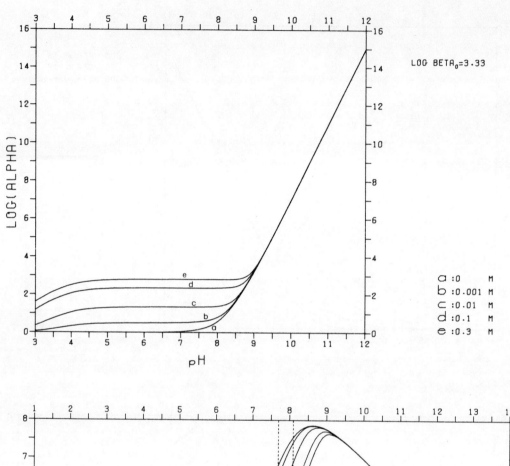
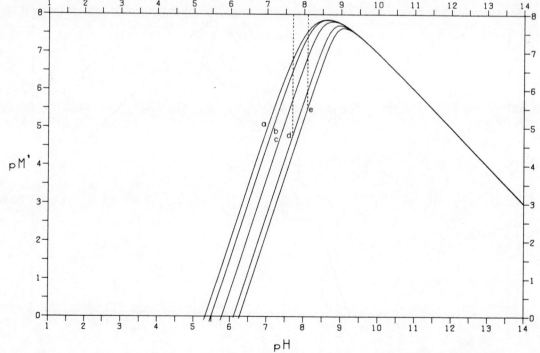

CHAPTER 15

Europium (III)   Eu

Only a little information is available about the hydrolysis of Eu^{3+} ions; no polycomplexes are known.

The following data used for the construction of the plots are estimates by comparison with other rare earths, from data given by Baes and Mesmer [1].

$$\log *\beta_1 = -7.8 \qquad \log *K_{s0} = 18.0$$
$$\log *\beta_2 = -16.2$$
$$\log *\beta_3 = -25$$
$$\log *\beta_4 = -37$$

It is unlikely that the carbonate is formed in air-saturated solutions.

[1] C. F. Baes and R. E. Mesmer, *The Hydrolysis of Cations,* Wiley, New York, 1976.

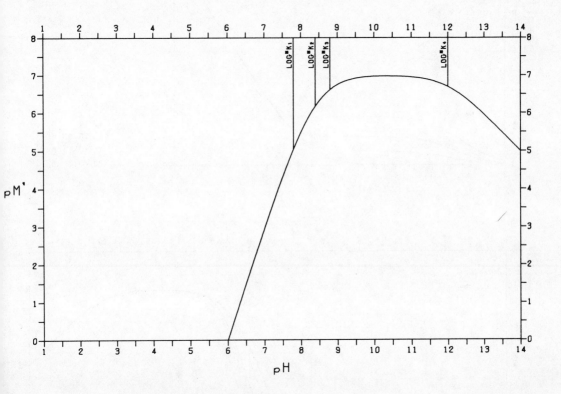

Europium(III)-Acetate

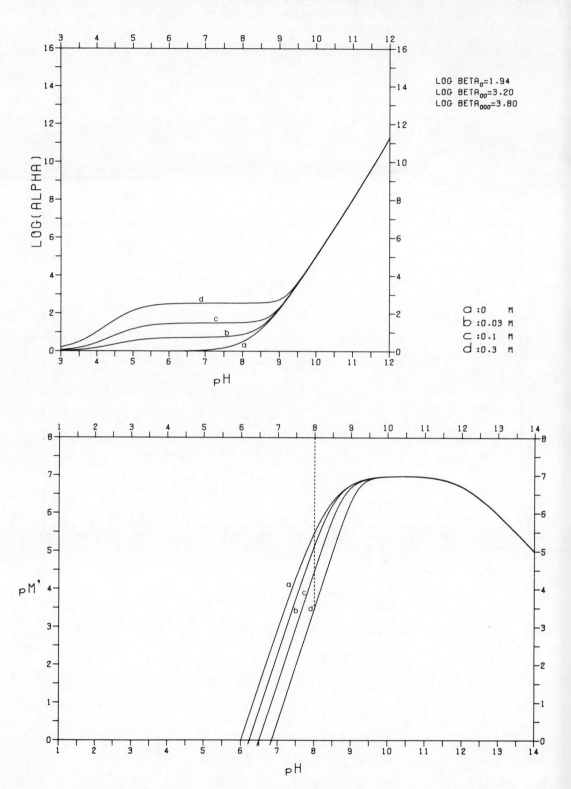

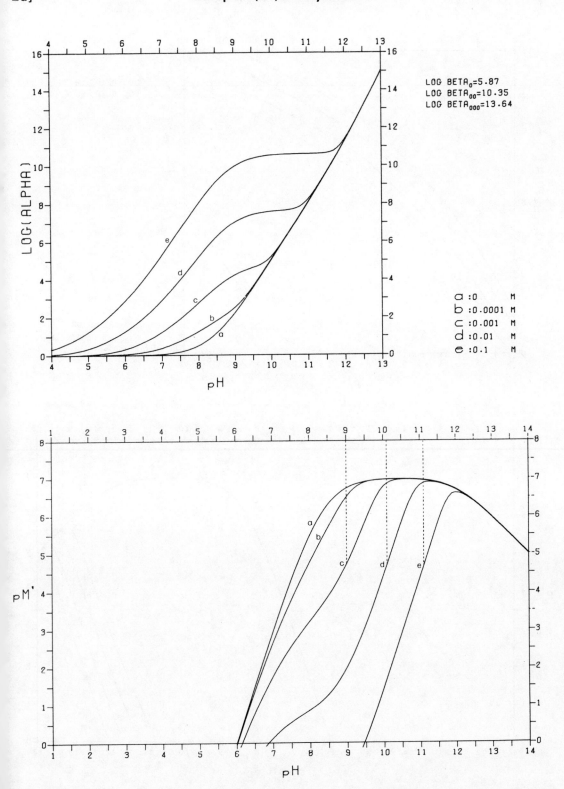

Europium(III)-DCTA

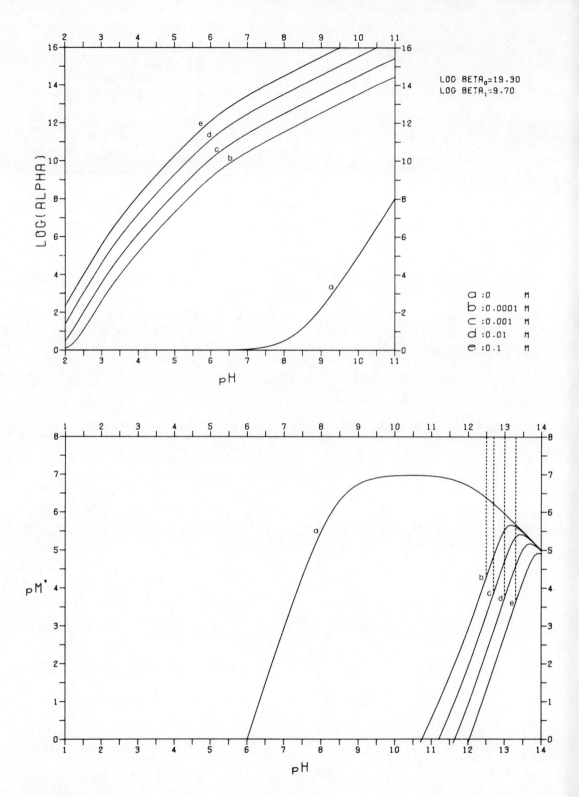

Europium(III)-DTPA

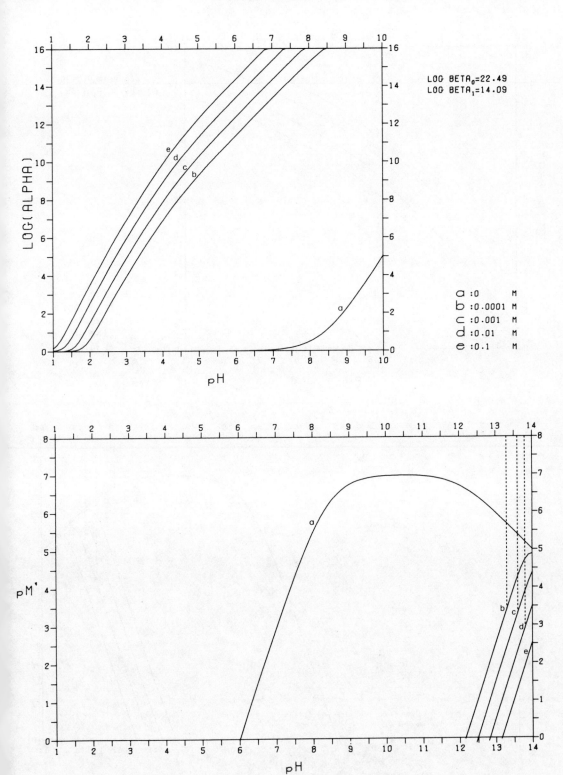

LOG BETA$_0$=22.49
LOG BETA$_1$=14.09

a : 0   M
b : 0.0001 M
c : 0.001 M
d : 0.01   M
e : 0.1   M

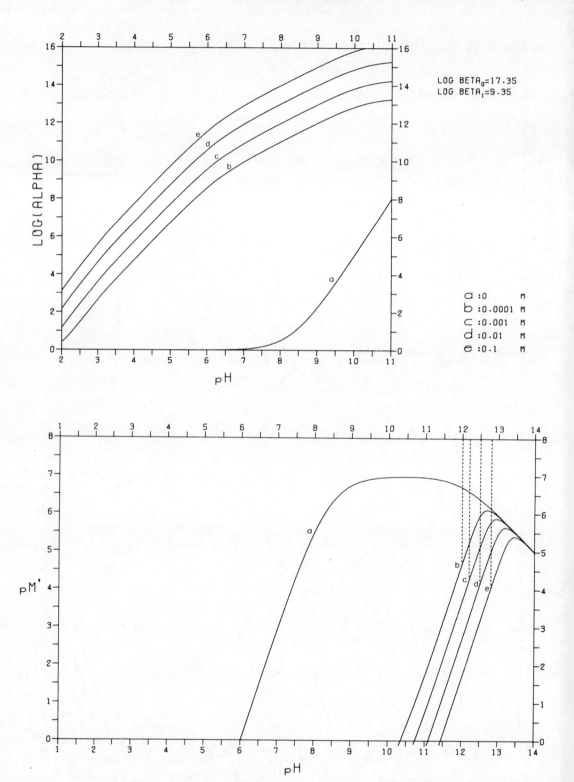

Europium(III)-Iminodiacetate

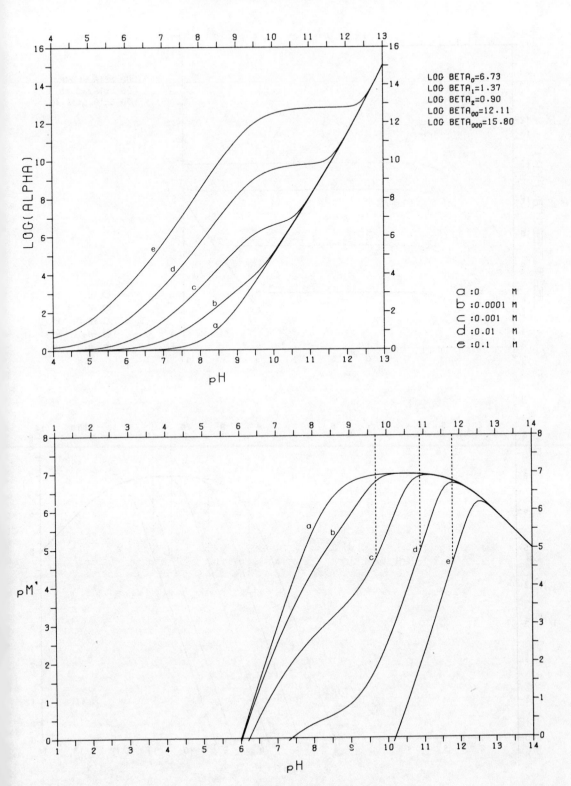

Europium(III)-Oxalate [Ch. 15]

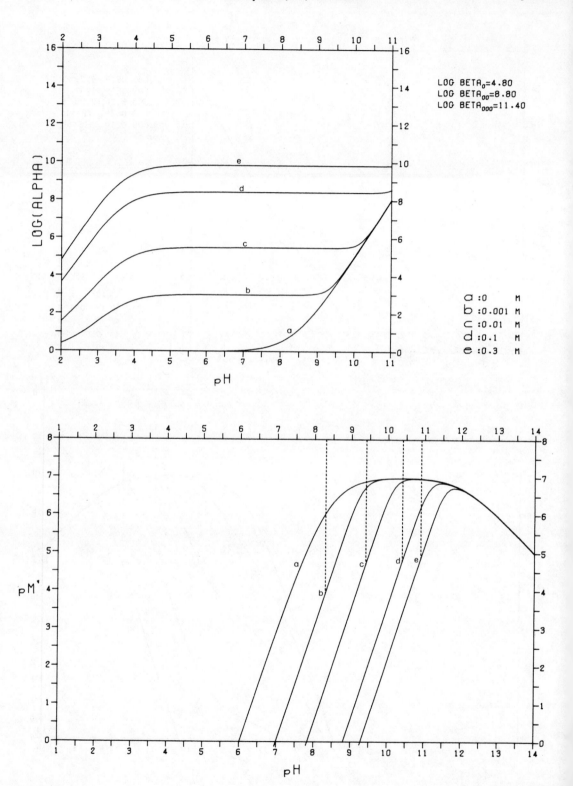

Europium(III)-Pyridine-2,6-dicarboxylate

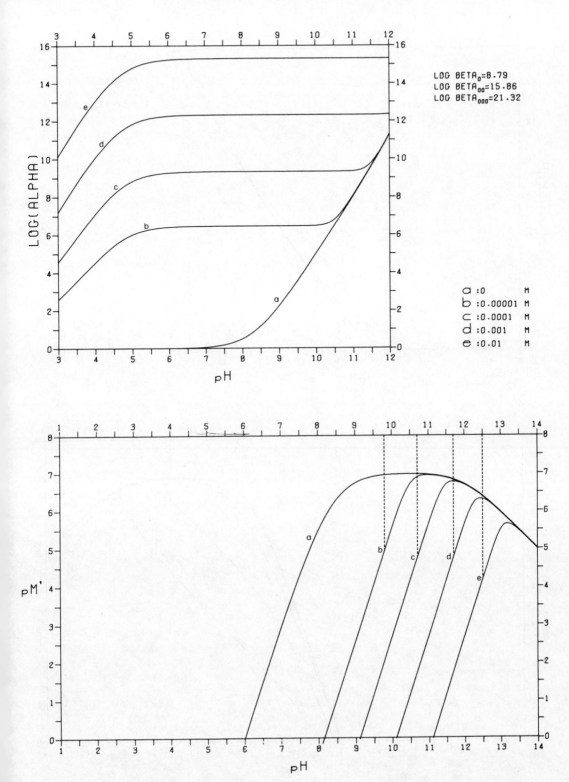

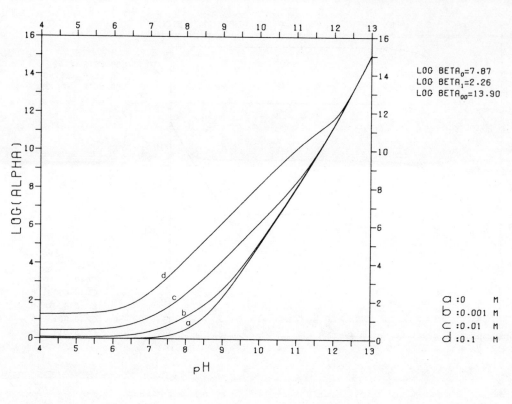
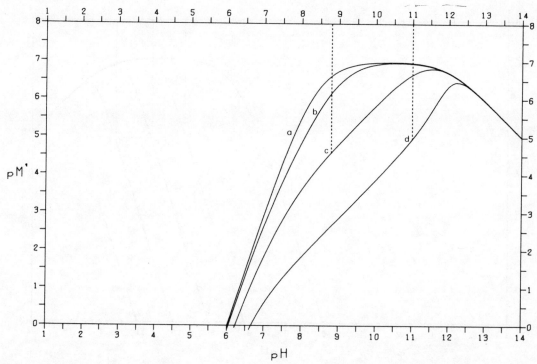

Europium(III)-Tartrate

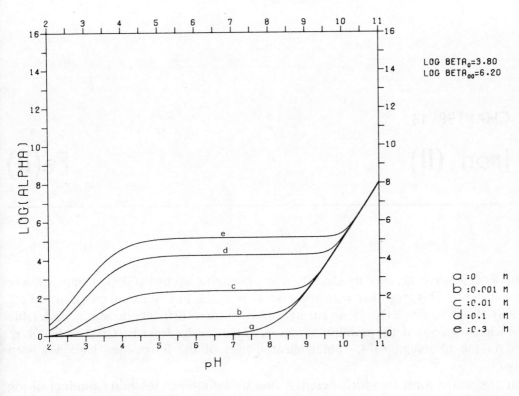

LOG BETA$_0$=3.80
LOG BETA$_{00}$=6.20

a : 0   M
b : 0.001   M
c : 0.01   M
d : 0.1   M
e : 0.3   M

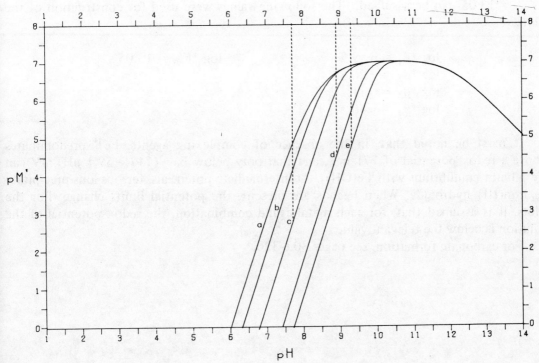

CHAPTER 16

Iron (II)   Fe(II)

The Fe^{2+} ion hydrolyses only slightly before precipitation of $Fe(OH)_2$ begins, even in dilute solutions. This, together with the ease with which Fe^{2+} ions are oxidized to Fe^{3+} [and $Fe(OH)_2(s)$ to $Fe(OH)_3(s)$] makes hydrolysis studies difficult; there is considerable lack of agreement about the stability constants of the species from $Fe(OH)^+$ to $Fe(OH)_4^{2-}$ formed in the pH range 7–14. Polynuclear species of the ferrous ion have not been reported.

For the active form in which $Fe(OH)_2$ first precipitates, a solubility product of $\log {}^*K_{s0} = 13.05$ can be assumed. The following values were used for construction of the plots.

$$\log {}^*\beta_1 = -9.7 \qquad \log {}^*K_{s0} = 13.05$$
$$\log {}^*\beta_2 = -20.8$$
$$\log {}^*\beta_3 = -31$$
$$\log {}^*\beta_4 = -46$$

It must be noted that, in the absence of complexing agents, Fe^{3+} predominates above a redox potential of 771 mV, and that only below $E = (147 - 59.1 \text{ pH})$ mV can Fe^{2+} be in equilibrium with $Fe(OH)_2$. At intermediate potentials, ferrous ions precipitate as iron(III) hydroxide. When ligands are present, the potential limits change. For the plots, it is assumed that, for each metal-ligand combination, the redox potential of the solution is below the relevant value.

For carbonate formation, see pages 303–320.

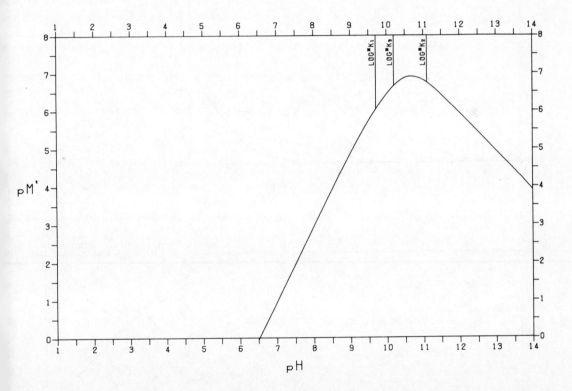

Iron(II)-Acetylacetone [Ch. 16

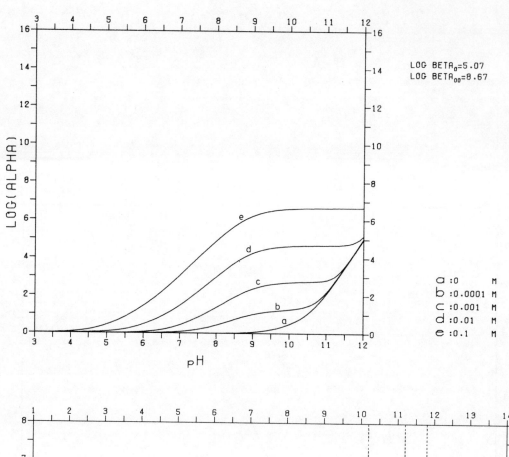

LOG BETA$_0$=5.07
LOG BETA$_{00}$=8.67

a : 0   M
b : 0.0001 M
c : 0.001 M
d : 0.01   M
e : 0.1   M

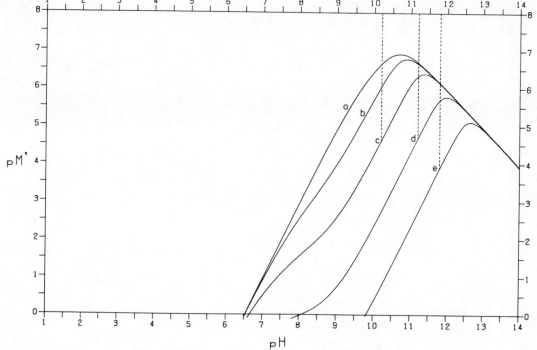

Iron(II)-Citrate

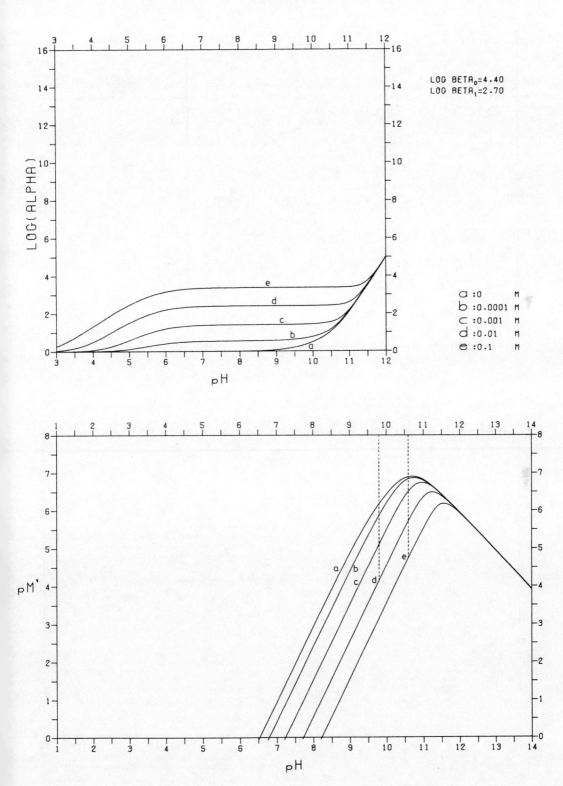

288   Iron(II)-Cyanide   [Ch. 16

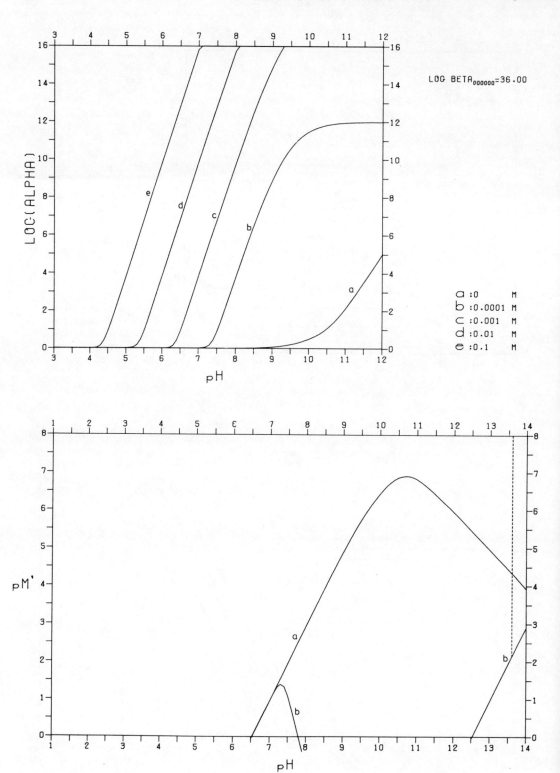

Iron(II)-DCTA

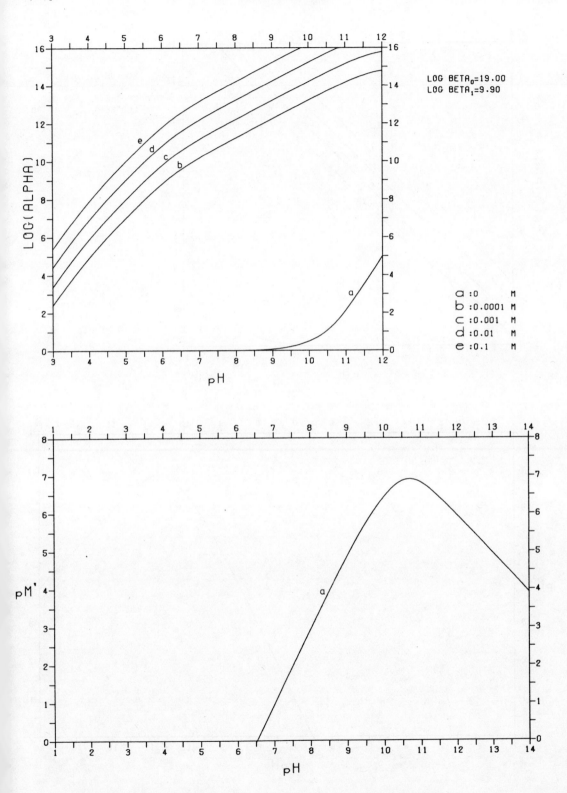

Iron(II)-DTPA [Ch. 16]

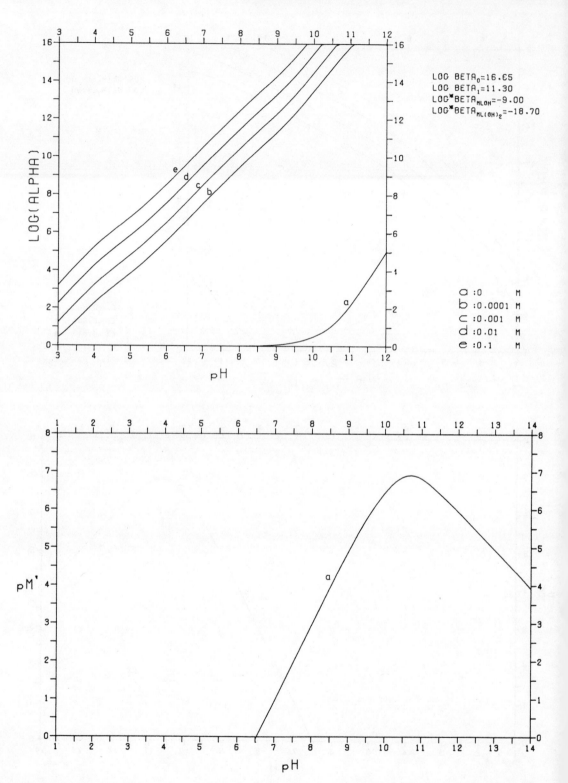

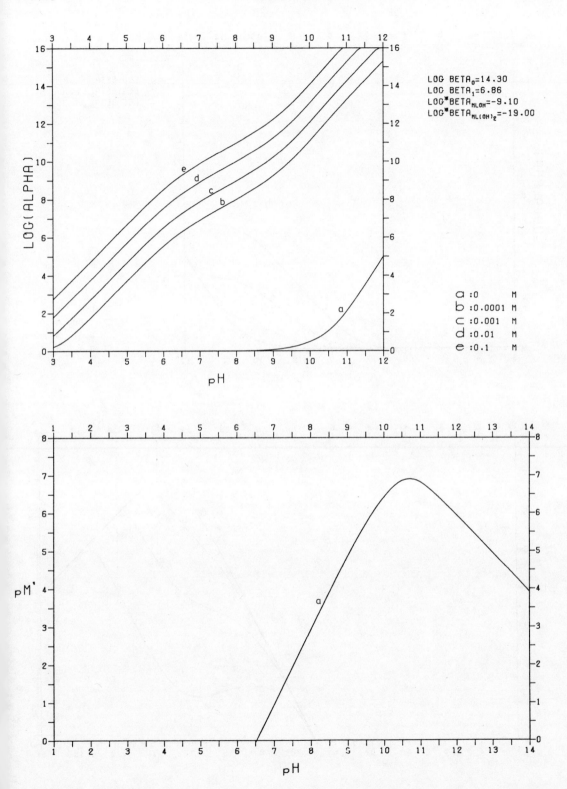

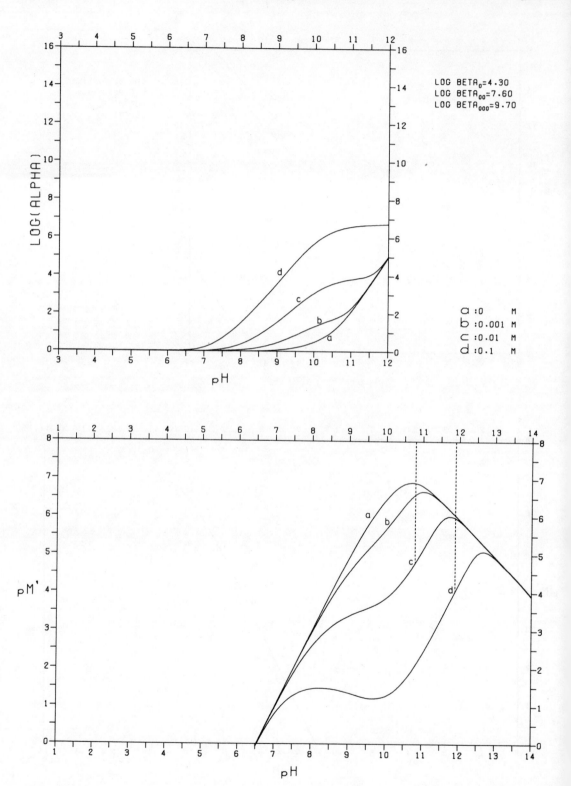

Iron(II)–Glycine

LOG BETA$_0$=4.13
LOG BETA$_{00}$=7.65

a : 0 M
b : 0.001 M
c : 0.01 M
d : 0.1 M

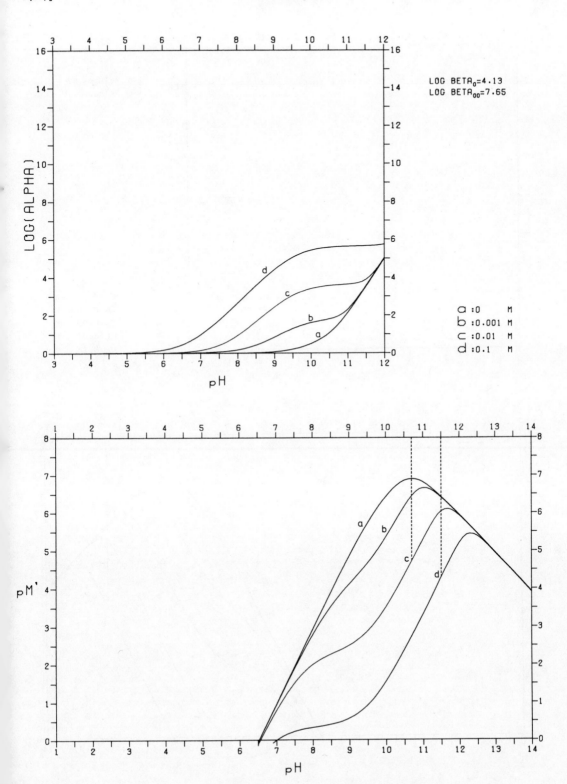

Iron(II)-Iminodiacetate

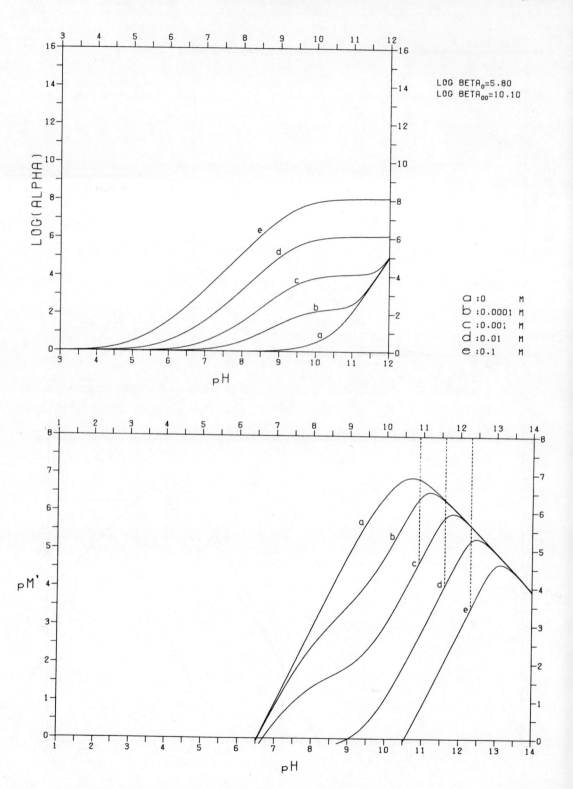

Iron(II)–Oxalate

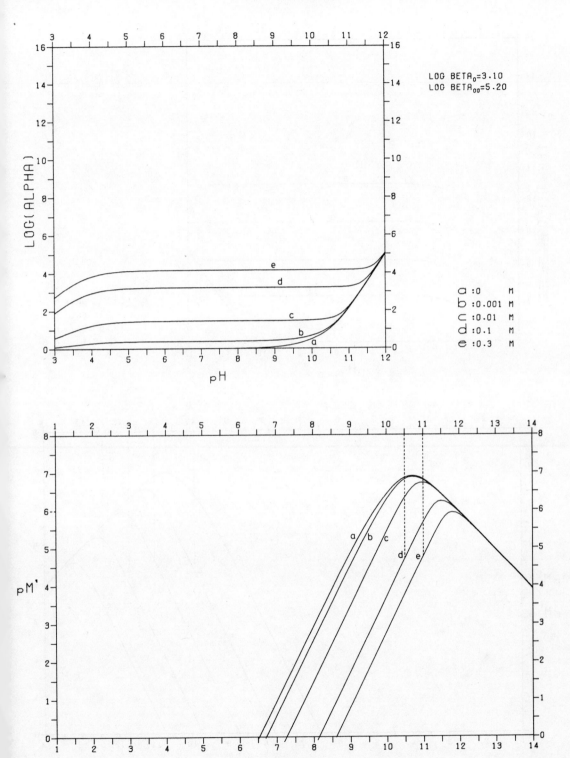

Iron(II)-Pyridine-2,6-dicarboxylate [Ch. 16

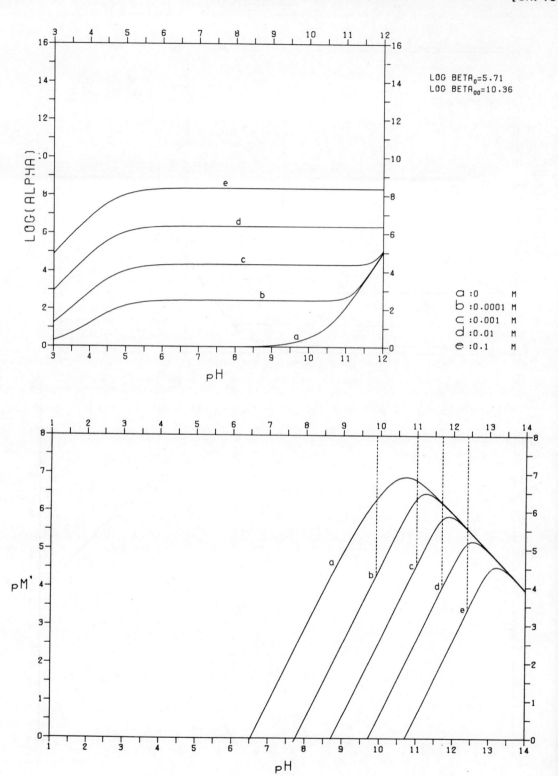

LOG BETA$_0$=5.71
LOG BETA$_{00}$=10.36

a : 0 M
b : 0.0001 M
c : 0.001 M
d : 0.01 M
e : 0.1 M

Iron(II)-1,10-Phenanthroline

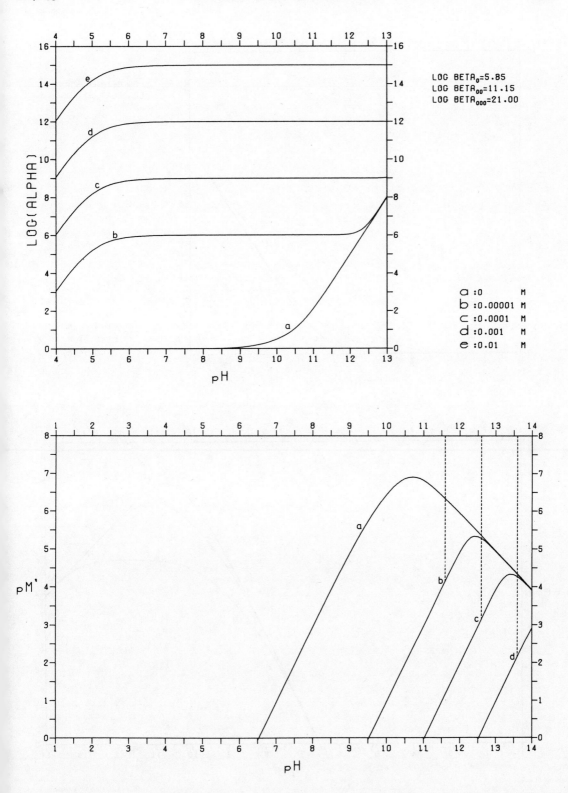

Iron(II)-Salicylate

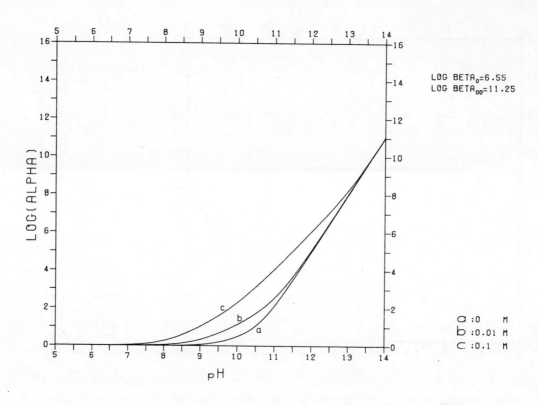

LOG BETA$_0$=6.55
LOG BETA$_{00}$=11.25

a : 0   M
b : 0.01 M
c : 0.1 M

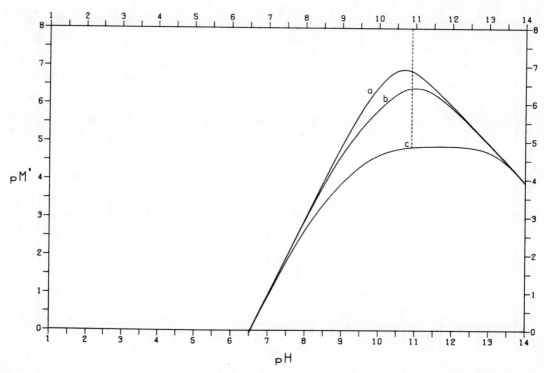

Iron(II)-5-Sulphosalicylate

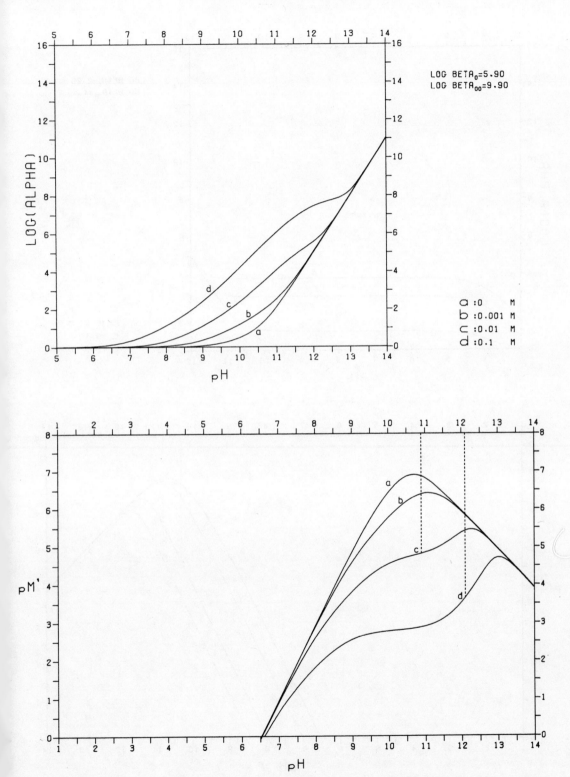

Iron(II)-Tartrate [Ch. 16

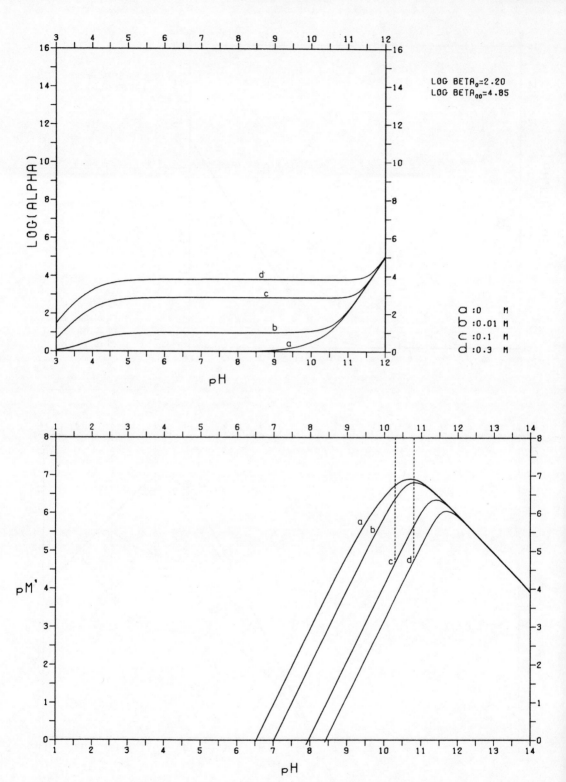

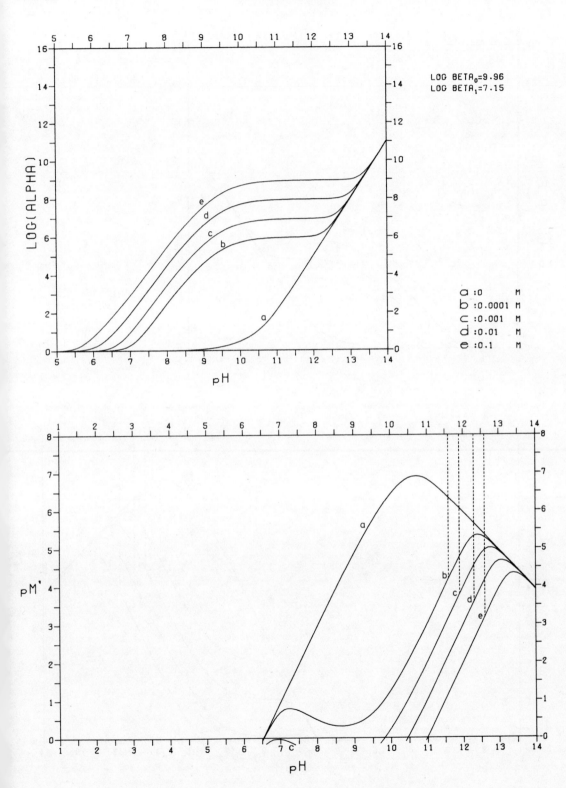

Iron(II)-Trien [Ch. 16

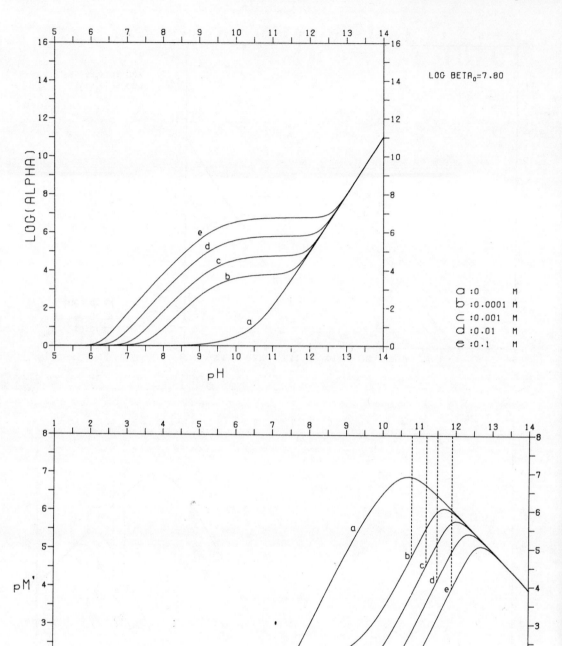

Fe (II) — carbonate system

Fe^{2+} will precipitate mainly as the carbonate from air-saturated solutions. The solubility product for iron(II) carbonate, log K_{s0} = −10.7 leads to an apparent constant of K_{carb} = 11.0. This value has been used for construction of the plots which follow.

$$\log {}^*\beta_1 = -9.7$$
$$\log {}^*\beta_2 = -20.8$$
$$\log {}^*\beta_3 = -31$$
$$\log {}^*\beta_4 = -46$$

$$\log K_{carb} = 11.0$$

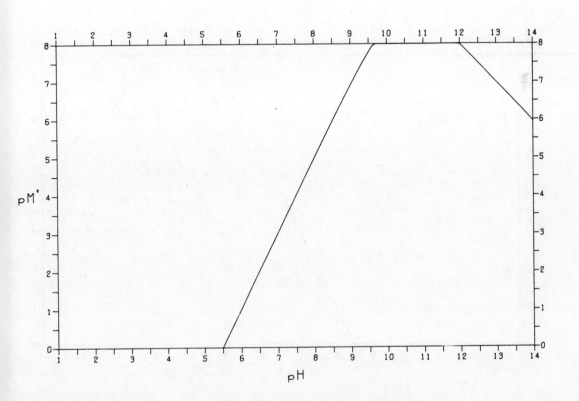

Fe(II)–Carbonate–Acetylacetone

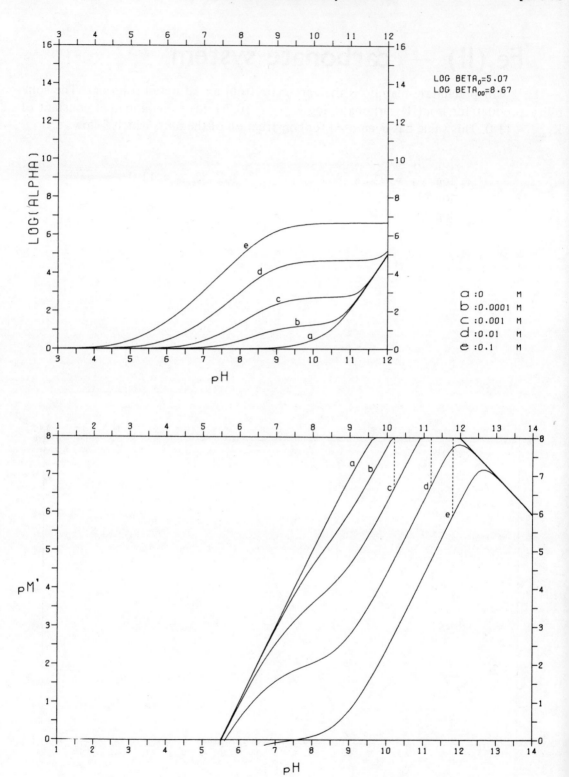

LOG BETA$_0$=5.07
LOG BETA$_{00}$=8.67

a : 0 M
b : 0.0001 M
c : 0.001 M
d : 0.01 M
e : 0.1 M

Fe(II)-Carbonate-Citrate

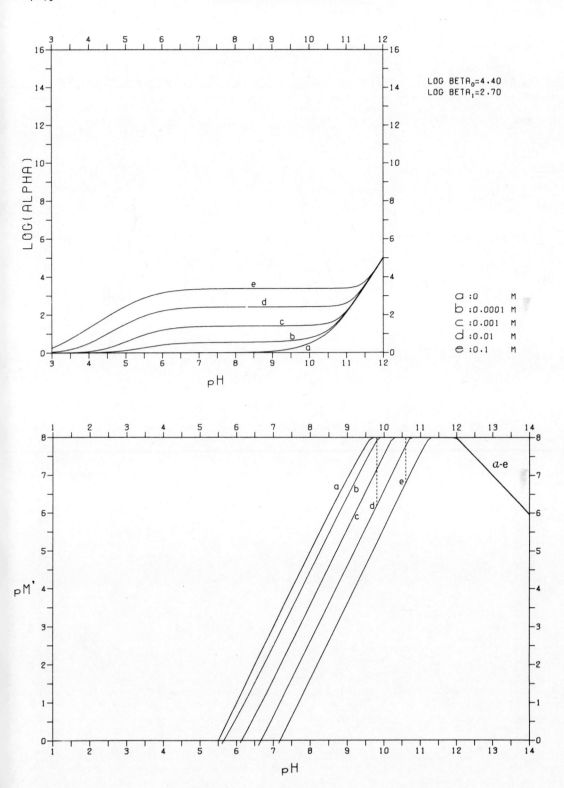

LOG BETA$_0$=4.40
LOG BETA$_1$=2.70

a : 0 M
b : 0.0001 M
c : 0.001 M
d : 0.01 M
e : 0.1 M

Fe(II)-Carbonate-Cyanide [Ch. 16]

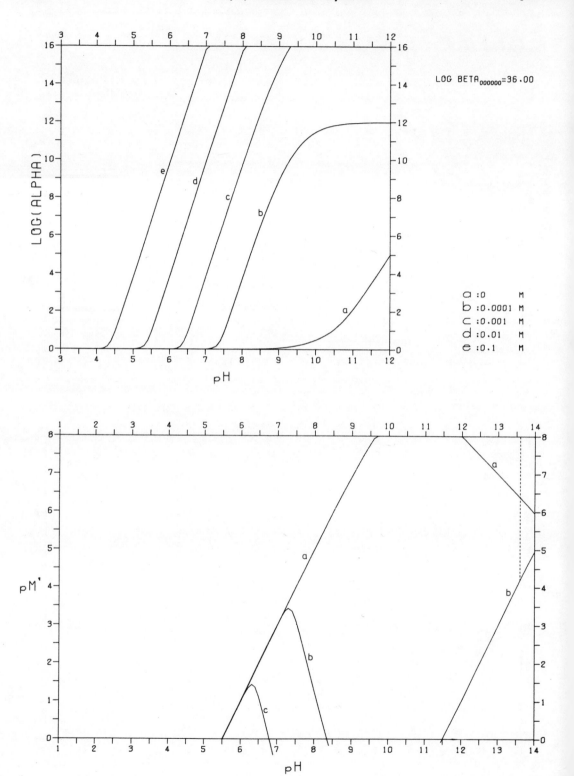

Fe(II)-Carbonate-DCTA

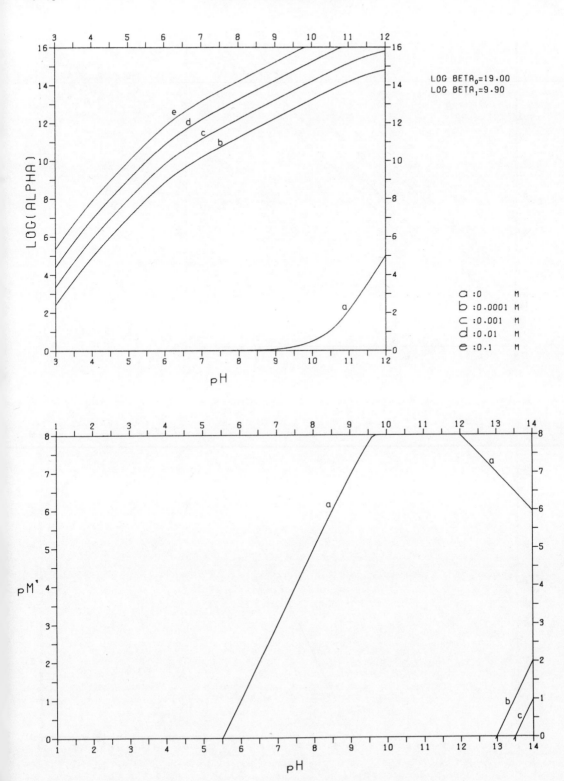

Fe(II)-Carbonate-DTPA [Ch. 16

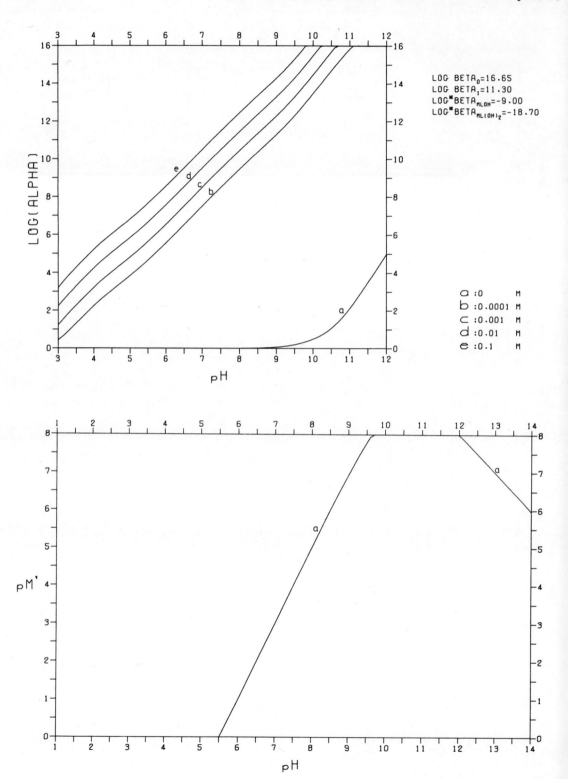

Fe(II)-Carbonate-EDTA

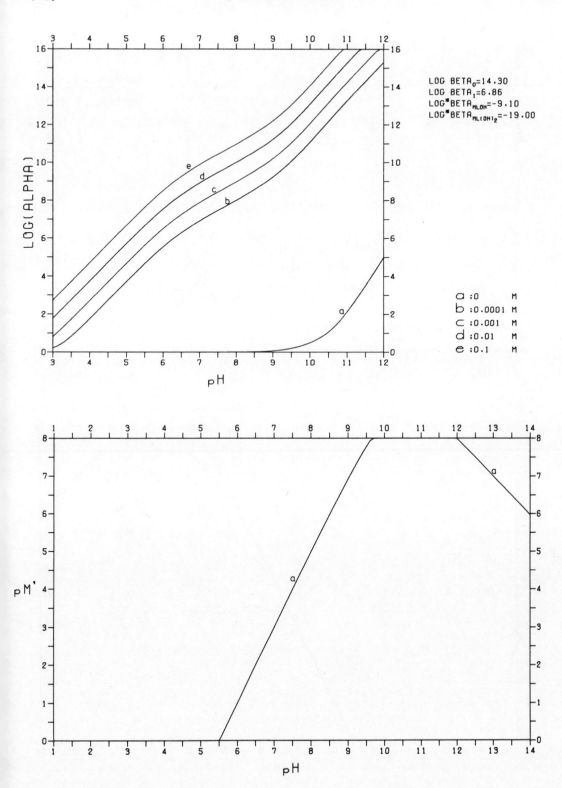

LOG BETA$_0$=14.30
LOG BETA$_1$=6.86
LOG*BETA$_{MLOH}$=-9.10
LOG*BETA$_{ML(OH)_2}$=-19.00

a : 0   M
b : 0.0001 M
c : 0.001 M
d : 0.01   M
e : 0.1   M

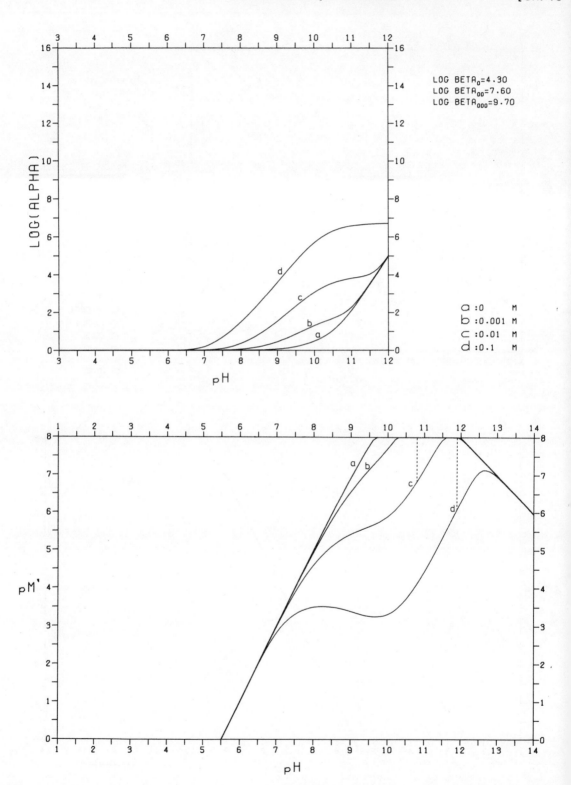

Fe(II)-Carbonate-Glycine

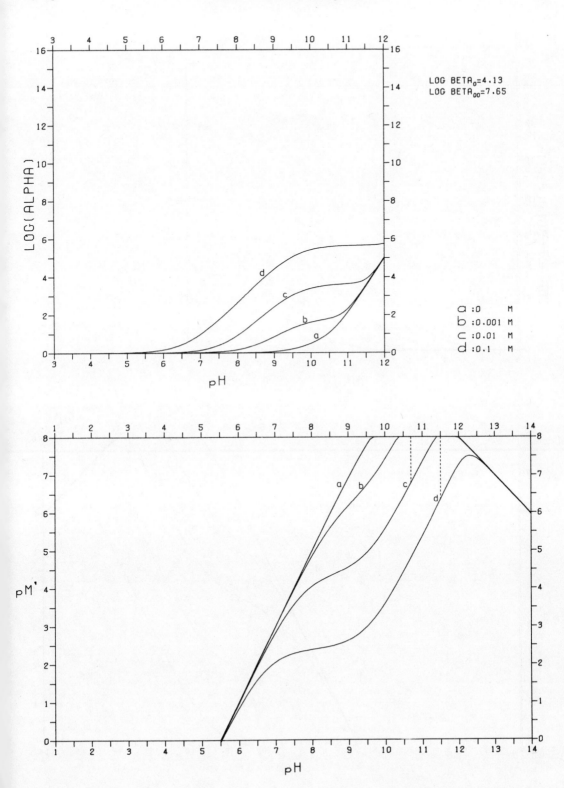

Fe(II)–Carbonate–Iminodiacetate [Ch. 16

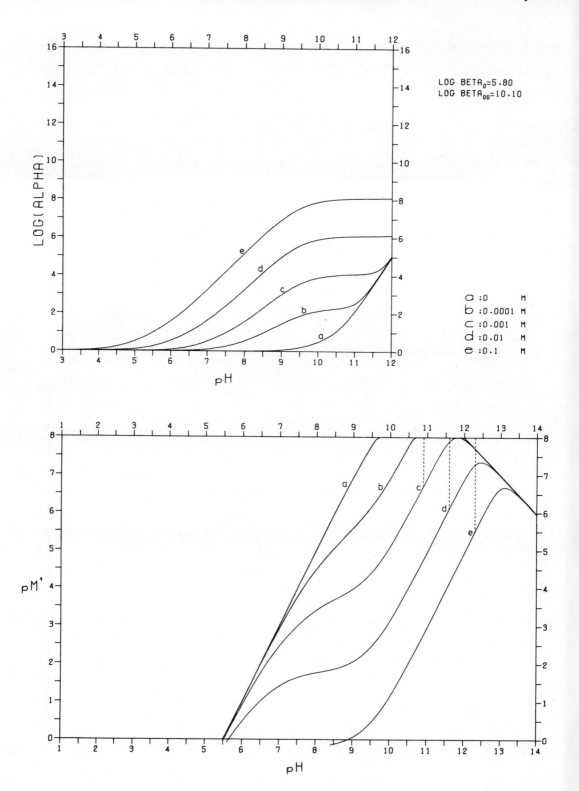

Fe(II)-Carbonate-Oxalate

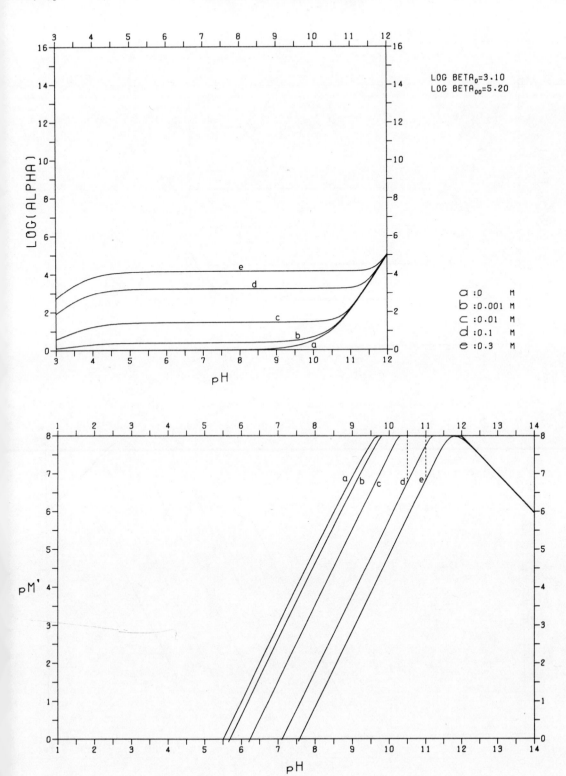

LOG BETA$_0$=3.10
LOG BETA$_{00}$=5.20

a : 0   M
b : 0.001 M
c : 0.01   M
d : 0.1   M
e : 0.3   M

Fe(II)-Carbonate-Pyridine-2,6-dicarboxylate [Ch. 16

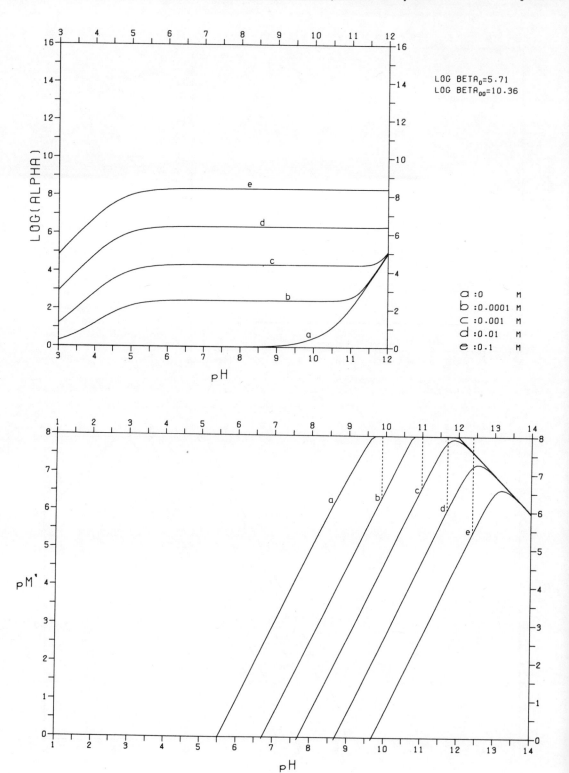

Fe(II)–Carbonate–1,10–Phenanthroline

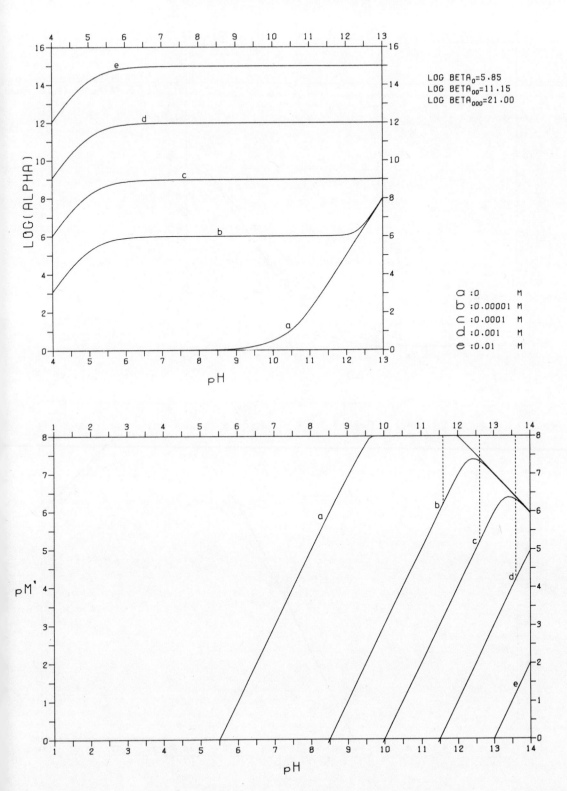

LOG BETA$_0$ = 5.85
LOG BETA$_{00}$ = 11.15
LOG BETA$_{000}$ = 21.00

a : 0   M
b : 0.00001 M
c : 0.0001 M
d : 0.001   M
e : 0.01   M

316 Fe(II)-Carbonate-Salicylate [Ch. 16]

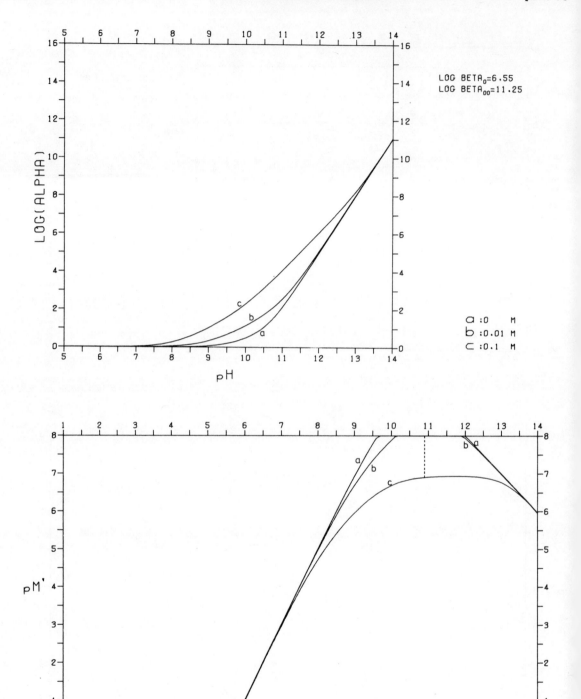

Fe(II)-Carbonate-5-Sulphosalicylate

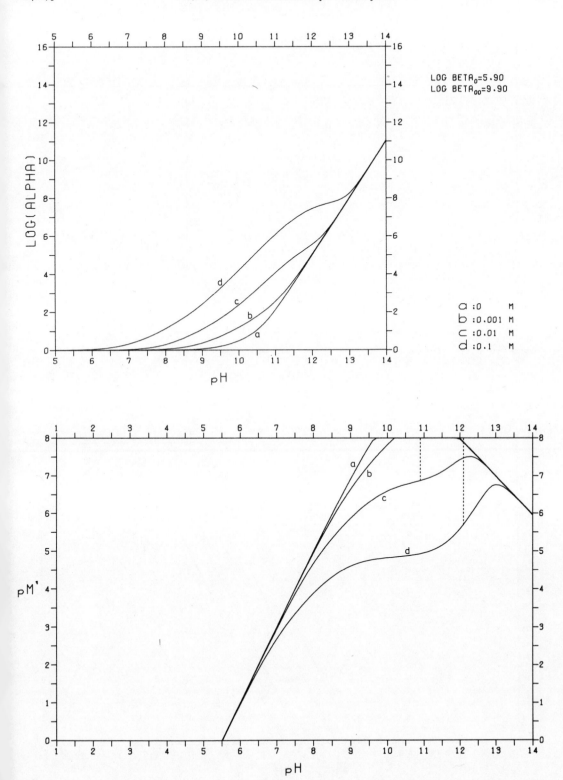

LOG BETA$_0$=5.90
LOG BETA$_{00}$=9.90

a : 0   M
b : 0.001   M
c : 0.01   M
d : 0.1   M

Fe(II)-Carbonate-Tartrate [Ch. 16]

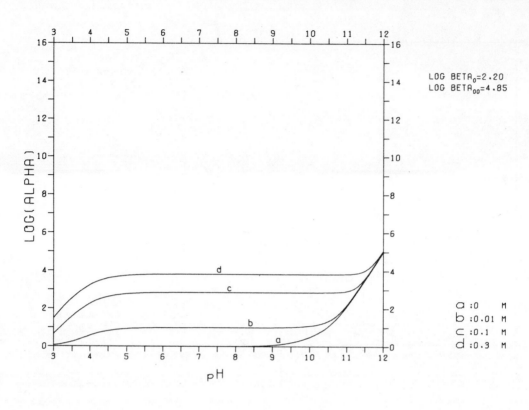

LOG BETA₀=2.20
LOG BETA₀₀=4.85

a : 0   M
b : 0.01 M
c : 0.1 M
d : 0.3 M

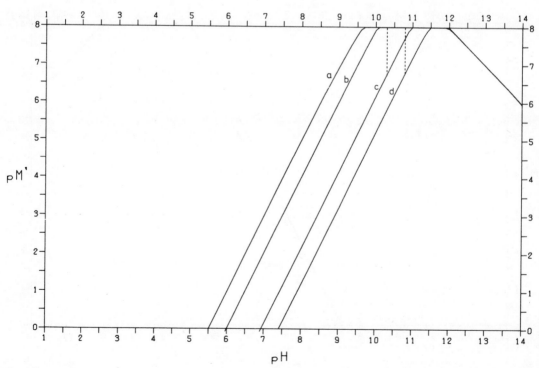

Fe(II)-Carbonate-Tetren

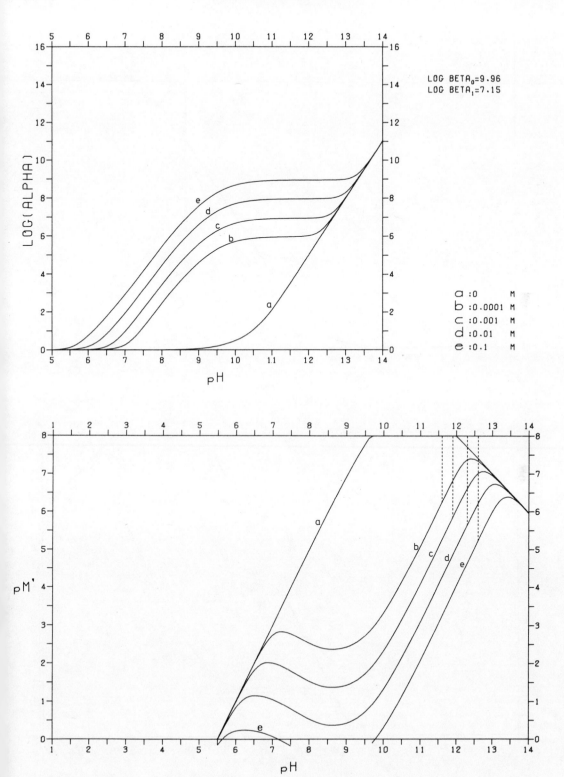

LOG BETA$_0$=9.96
LOG BETA$_1$=7.15

a : 0 M
b : 0.0001 M
c : 0.001 M
d : 0.01 M
e : 0.1 M

320 Fe(II)-Carbonate-Trien [Ch. 16

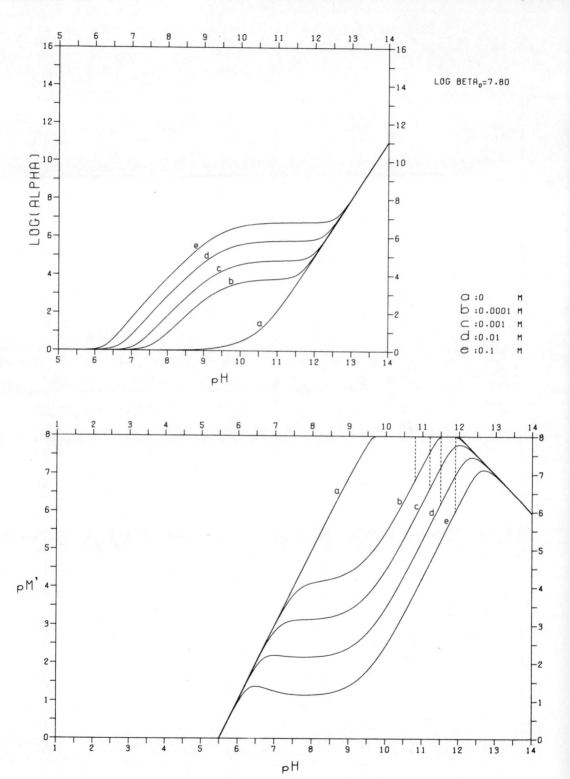

CHAPTER 17

Iron (III)   Fe(III)

Studies of the hydrolysis of Fe^{3+} ions have established that $Fe(OH)^{2+}$, $Fe(OH)_2^+$ and the dimer $Fe_2(OH)_2^{4+}$ are formed. There is also evidence that a small amount of $Fe_3(OH)_4^{5+}$ is formed before precipitation begins, but this complex does not contribute to the polycomplex line.

Several solid phases are known: α-, β- and τ-FeO(OH), α-Fe_2O_3, and the active and inactive forms of amorphous $Fe(OH)_3$. The last of these has an unknown content of hydroxide and oxide. The form in which Fe(III) hydroxide precipitates depends on conditions such as the pH, the temperature, and the anions present in solution. Usually the amorphous form transforms so slowly into α-FeO(OH) that aging for one hour generally leads to an acceptably defined steady state, for which the following constants can be adopted.

$$\log {}^*\beta_1 = -2.56 \qquad \log {}^*\beta_{22} = -2.85$$
$$\log {}^*\beta_2 = -6.19 \qquad \log {}^*\beta_{43} = -6.1$$
$$\log {}^*\beta_3 = -10$$
$$\log {}^*\beta_4 = -21.9 \qquad \log {}^*K_{s0} = 2.5$$

It must be noted that, in the absence of complexing agents, the +III oxidation state is predominant above a redox potential of 771 mV. Fe^{3+} usually forms stronger complexes than Fe^{2+}, so this redox limit tends to decrease when complexation occurs. Thus, these plots are applicable to air-saturated solutions.

Iron(III)

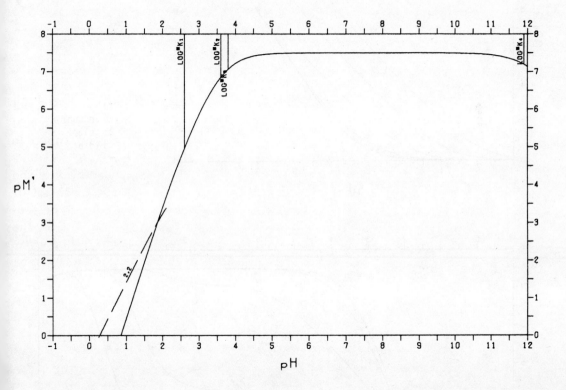

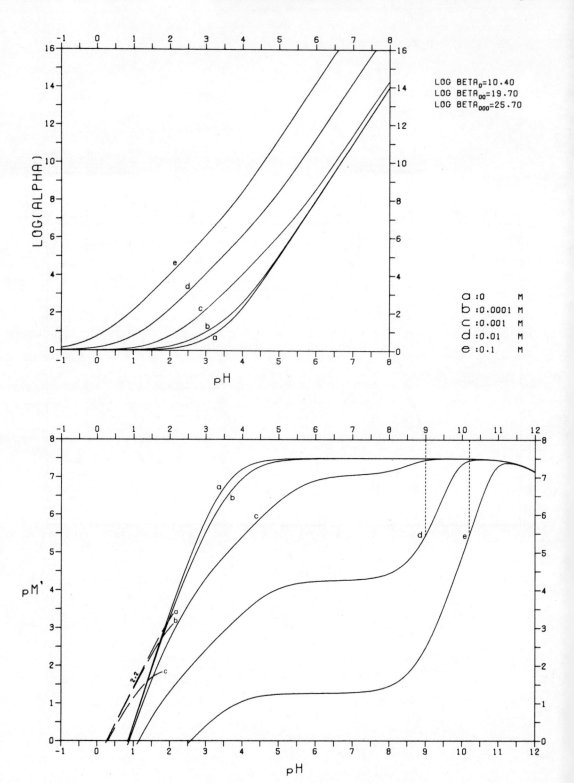

Iron(III)-Citrate

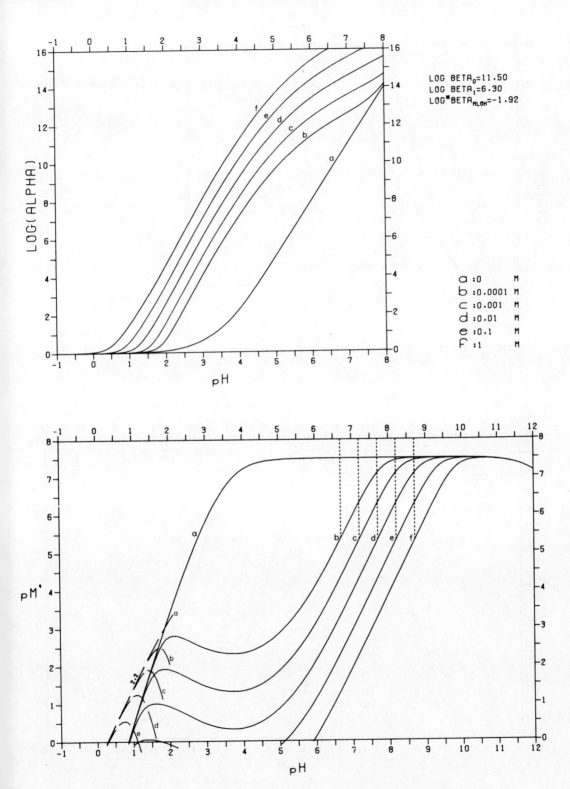

LOG BETA$_0$=11.50
LOG BETA$_1$=6.30
LOG*BETA$_{MLOH}$=-1.92

a : 0 M
b : 0.0001 M
c : 0.001 M
d : 0.01 M
e : 0.1 M
f : 1 M

326 Iron(III)-Cyanide [Ch. 17

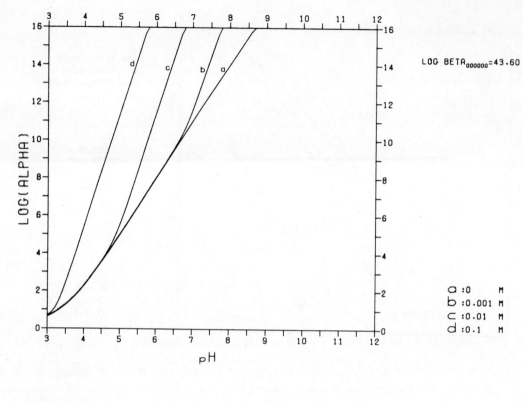

LOG BETA₀₀₀₀₀₀ = 43.60

a : 0   M
b : 0.001 M
c : 0.01   M
d : 0.1   M

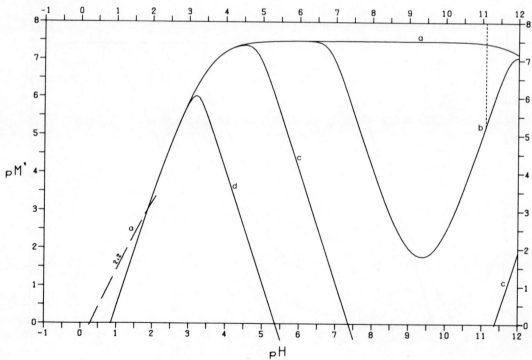

Iron(III)-DCTA

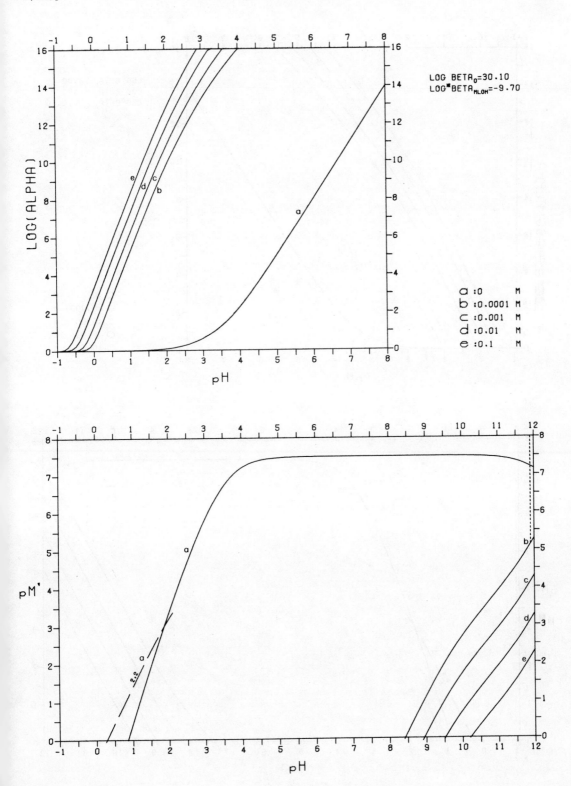

LOG BETA₀=30.10
LOG*BETA_MLOH=-9.70

a : 0   M
b : 0.0001 M
c : 0.001   M
d : 0.01   M
e : 0.1   M

Iron(III)-DTPA [Ch. 17]

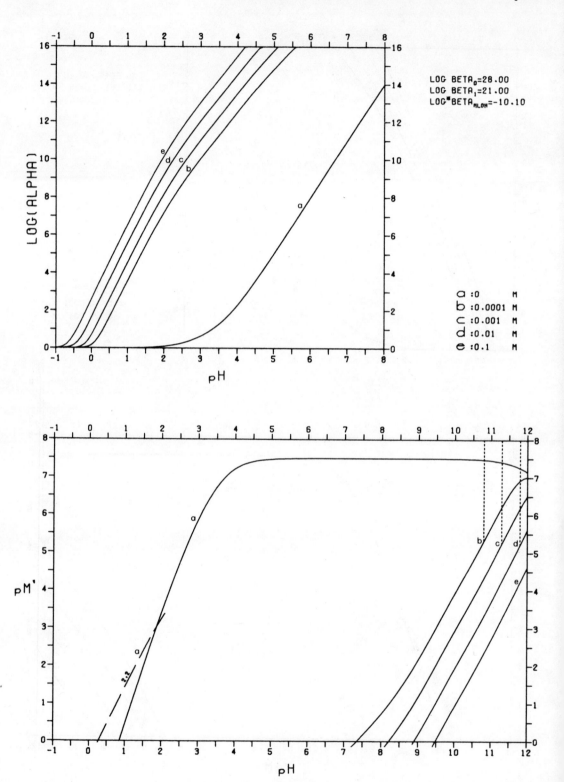

Iron(III)-EDTA

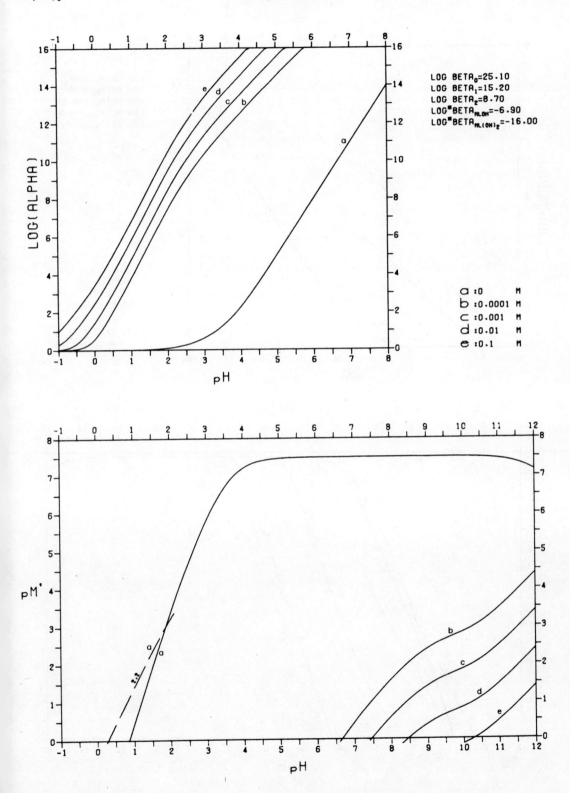

330   Iron(III)-Fluoride   [Ch. 17

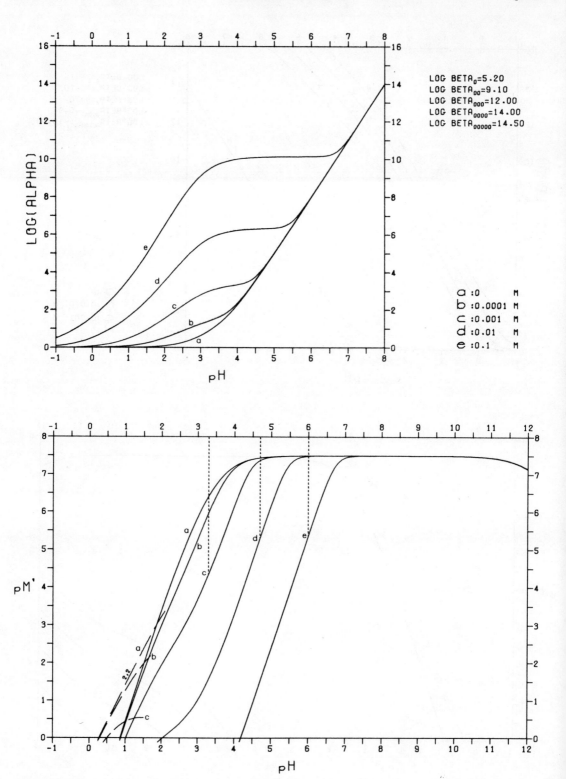

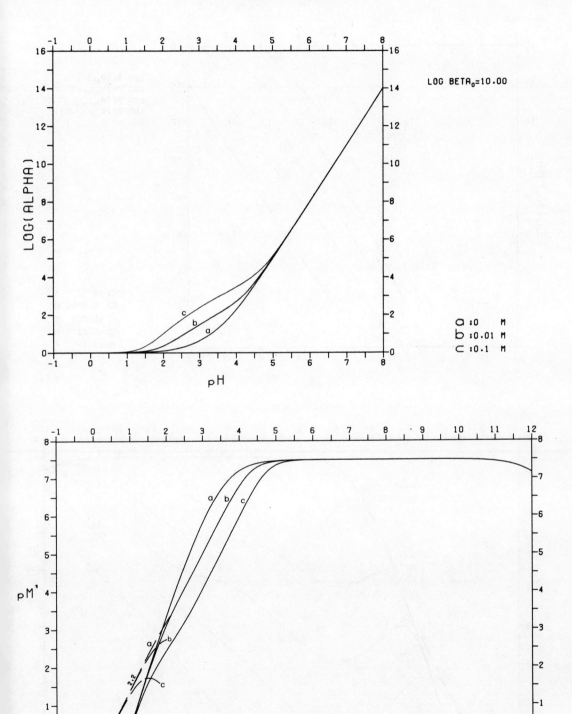

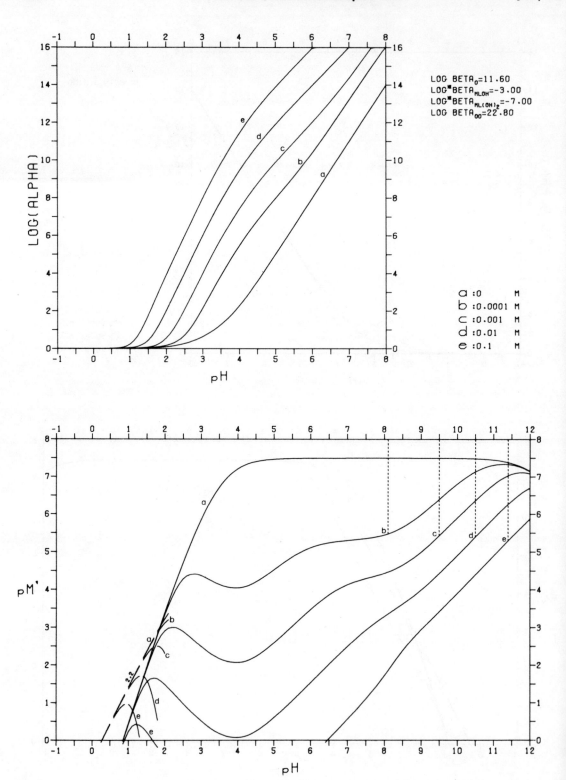

Iron(III)-Iminodiacetate

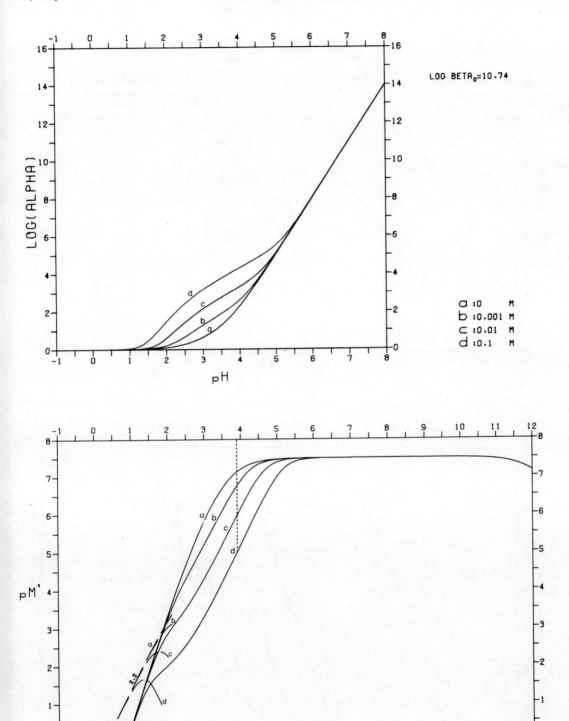

LOG BETA$_0$ = 10.74

a : 0 M
b : 0.001 M
c : 0.01 M
d : 0.1 M

334 **Iron(III)-Oxalate** [Ch. 17

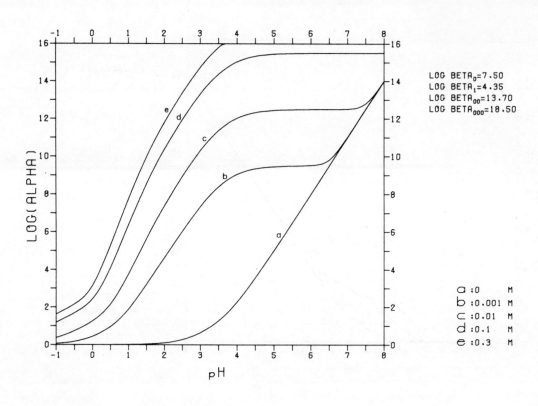

LOG BETA$_0$=7.50
LOG BETA$_1$=4.35
LOG BETA$_{00}$=13.70
LOG BETA$_{000}$=18.50

a : 0   M
b : 0.001 M
c : 0.01   M
d : 0.1   M
e : 0.3   M

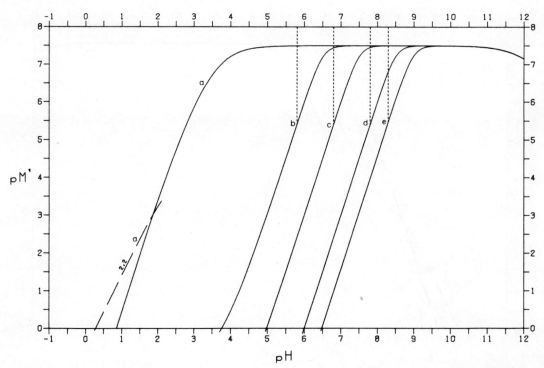

Iron(III)-Pyridine-2,6-dicarboxylate

335

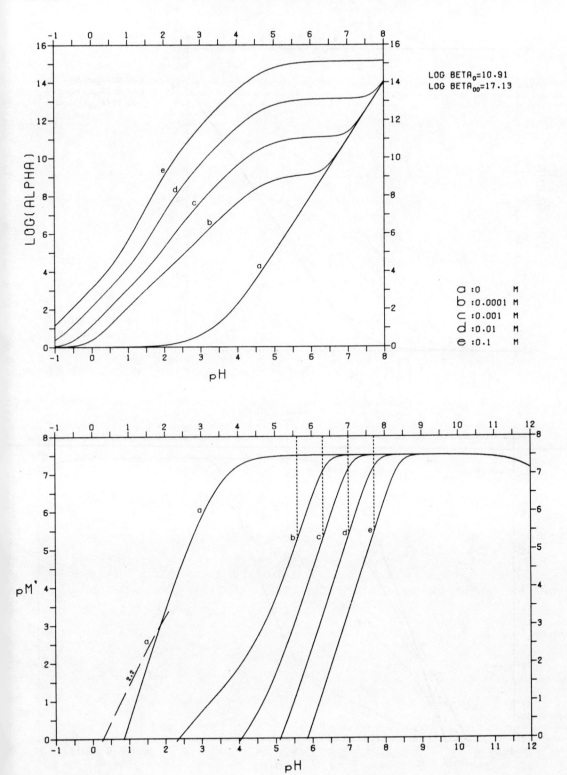

LOG BETA$_0$=10.91
LOG BETA$_{00}$=17.13

a : 0   M
b : 0.0001 M
c : 0.001   M
d : 0.01   M
e : 0.1   M

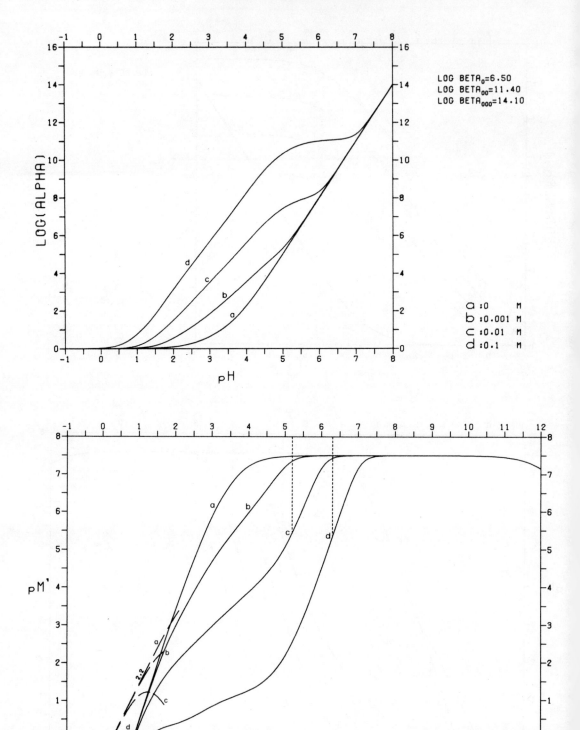

Iron(III)-Salicylate

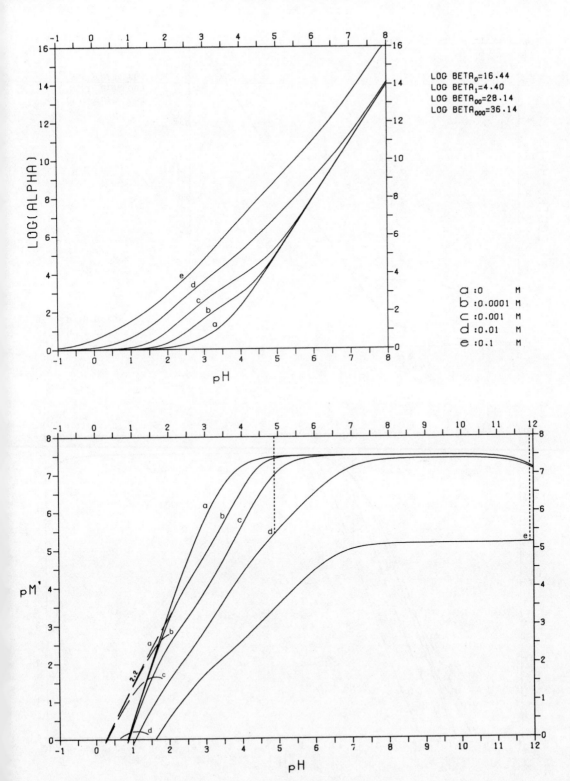

Iron(III)- Sulphate [Ch. 17

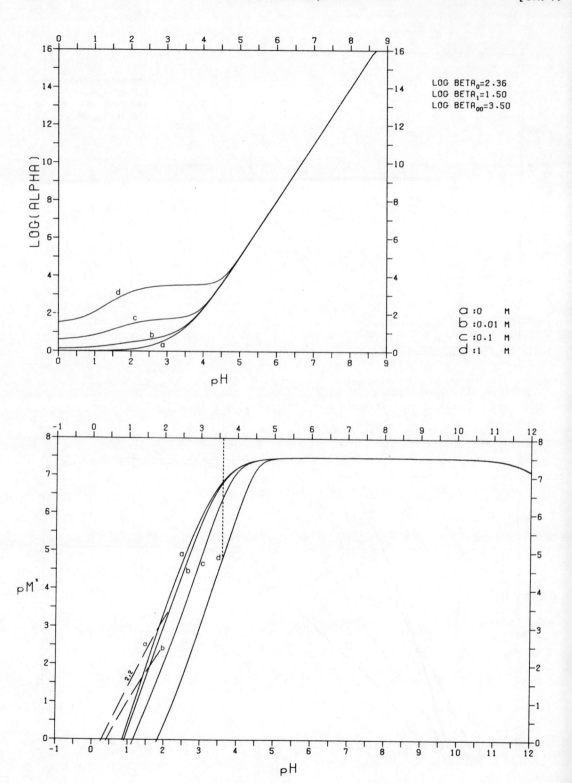

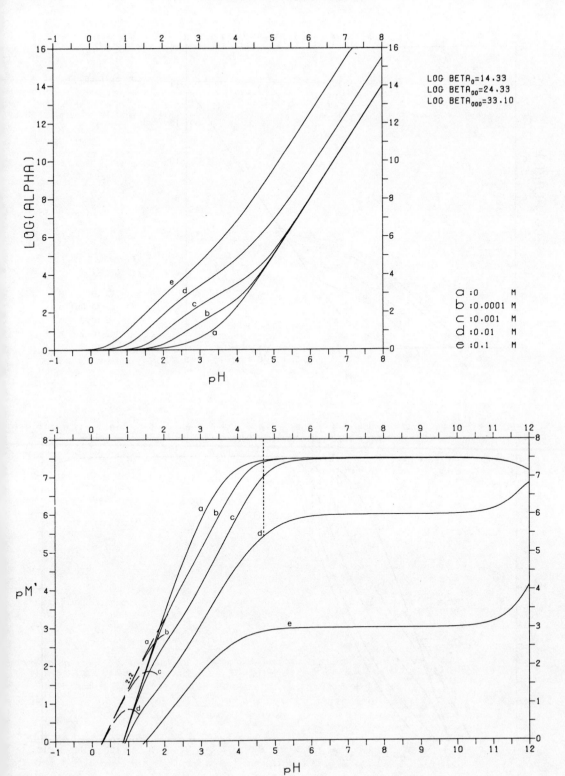

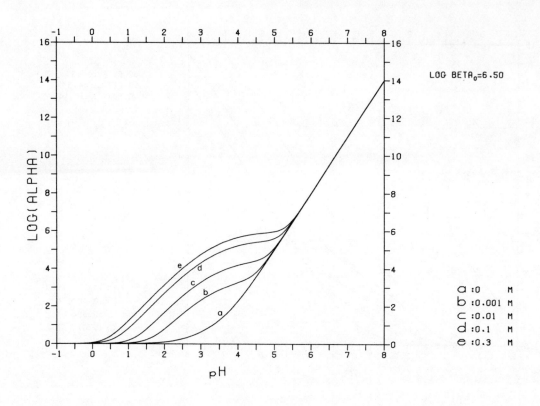
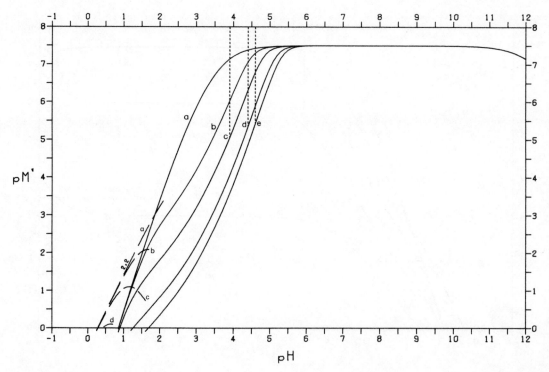

Iron(III)-Tiron

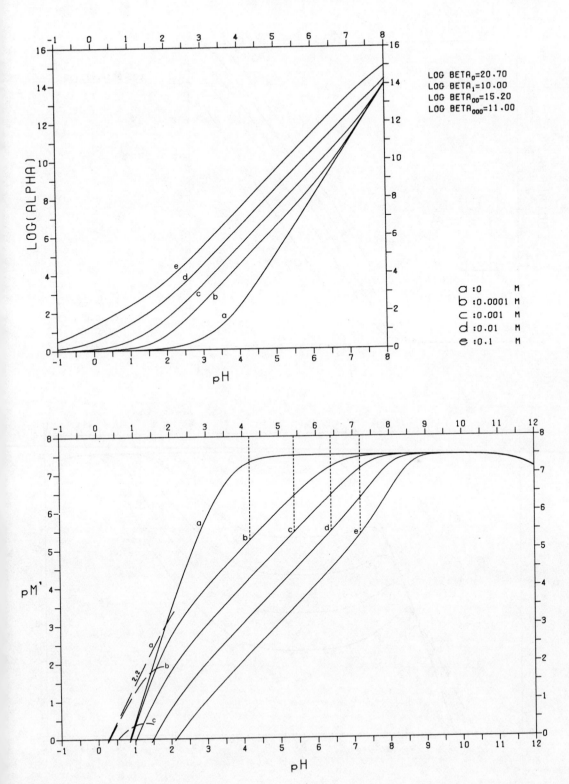

LOG BETA$_0$=20.70
LOG BETA$_1$=10.00
LOG BETA$_{00}$=15.20
LOG BETA$_{000}$=11.00

a : 0   M
b : 0.0001 M
c : 0.001   M
d : 0.01   M
e : 0.1   M

Iron(III)–Trien

[Ch. 17

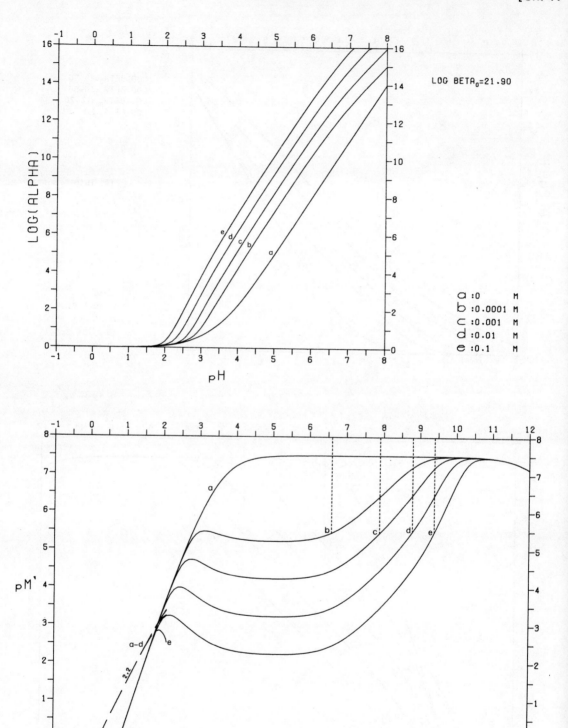

CHAPTER 18

Gallium (III)   Ga

The mononuclear species from $Ga(OH)^{2+}$ to $Ga(OH)_4^-$ are formed in dilute solutions. At higher concentrations, the slow hydrolysis reactions which occur can produce colloidal species and large polymers before precipitation begins. Supersaturation has been observed under certain conditions. The behaviour in any particular situation depends on the gallium concentration, on the sequence in which reagents are mixed, on the amount of agitation and particularly on the nature of the surface of the container.

Amorphous gallium hydroxide normally precipitates from aqueous solutions on addition of base. This amorphous hydroxide is converted into the more stable oxyhydroxide $GaO(OH)$ over several hours. Log $*K_{s0} = 3.7$ can be taken as a good estimate of log $K_{carb} = 8.6$ for air-saturated solutions. According to literature, basic carbonates are $Ga_n(OH)_{2.5n}$ have been reported to exist in supersaturated solutions, but the value given by Baes and Mesmer [1], log $*\beta_{65,26} = -139.1$, does not lead to a meaningful polycomplex line.

The values used for the plots are:

$$\log *\beta_1 = -2.6 \qquad \log *K_{s0} = 3.7$$
$$\log *\beta_2 = -6.1$$
$$\log *\beta_3 = -10.5$$
$$\log *\beta_4 = -17.4$$

---

[1] C. F. Baes and R. E. Mesmer, *The Hydrolysis of Cations*, Wiley, New York, 1976.

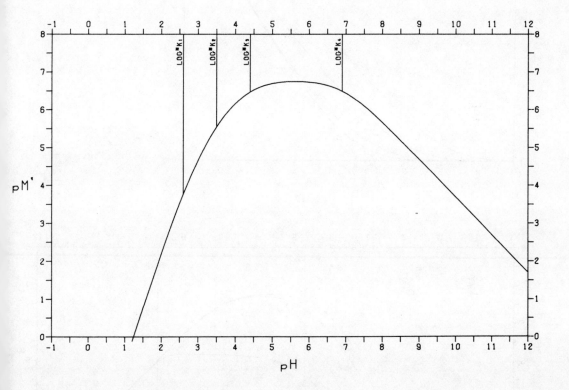

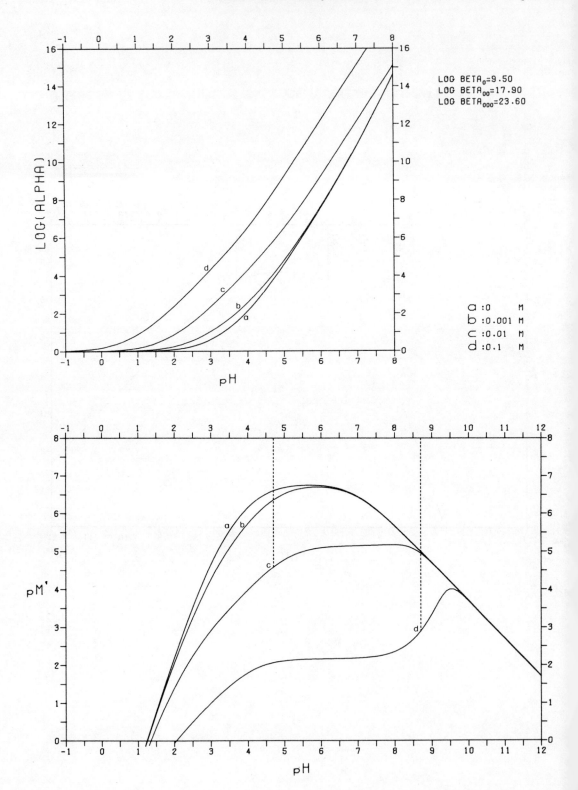

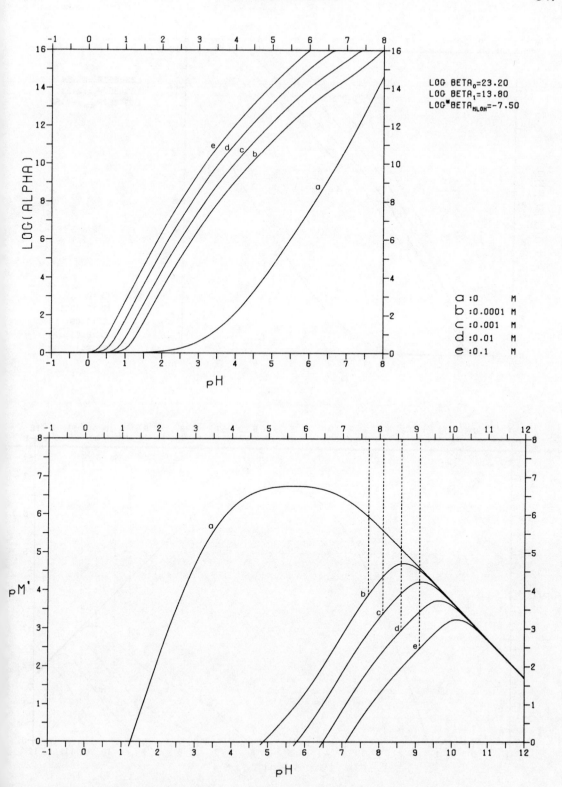

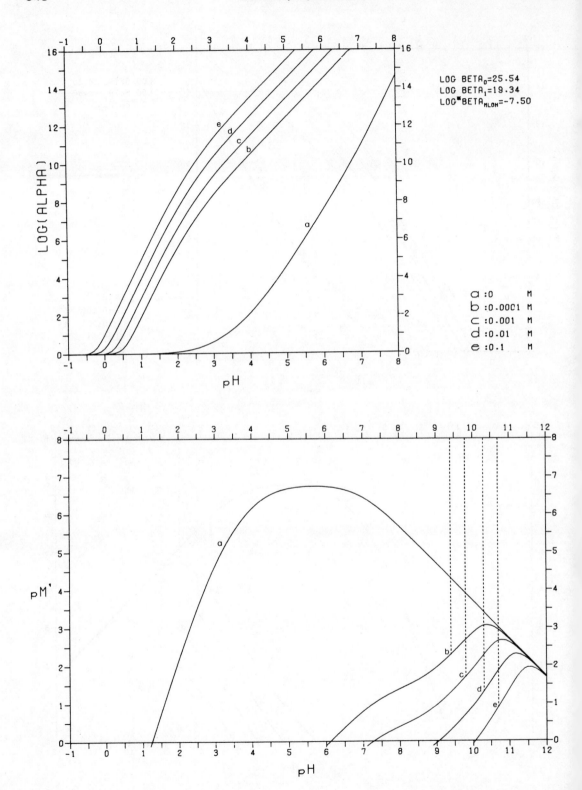

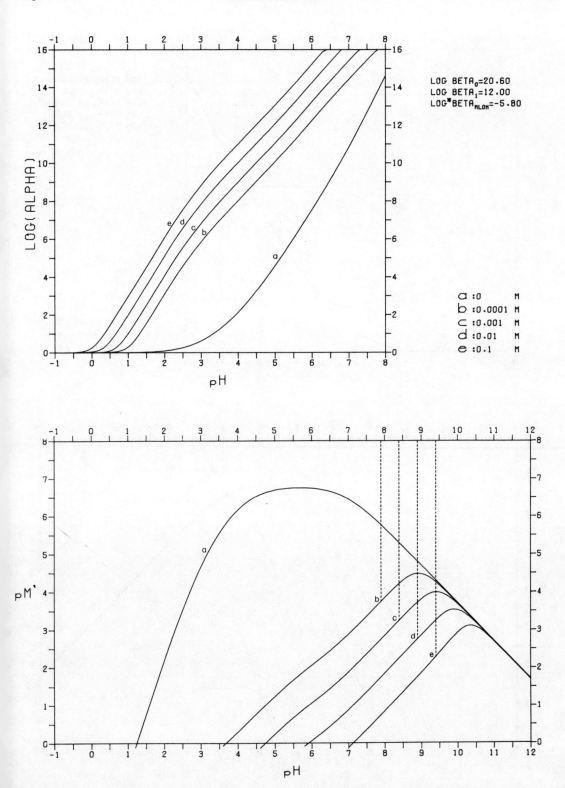

Gallium(III)-Fluoride [Ch. 18]

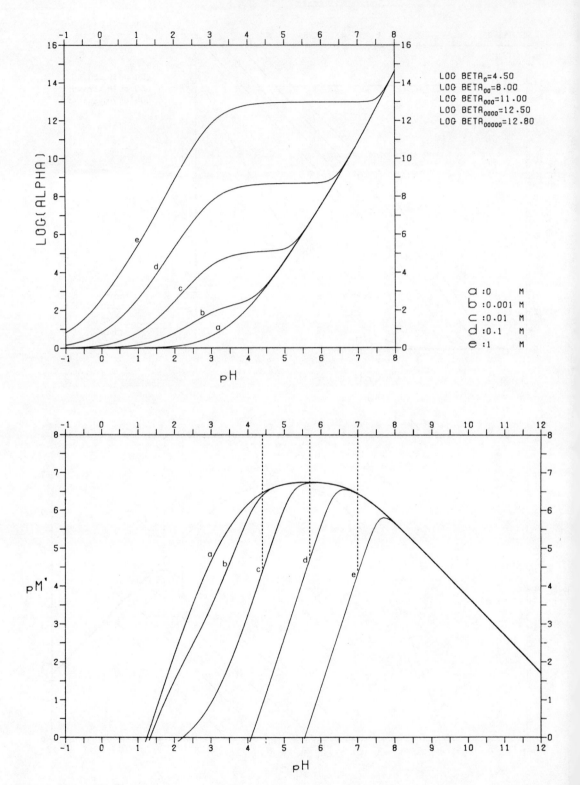

Gallium(III)-Oxalate

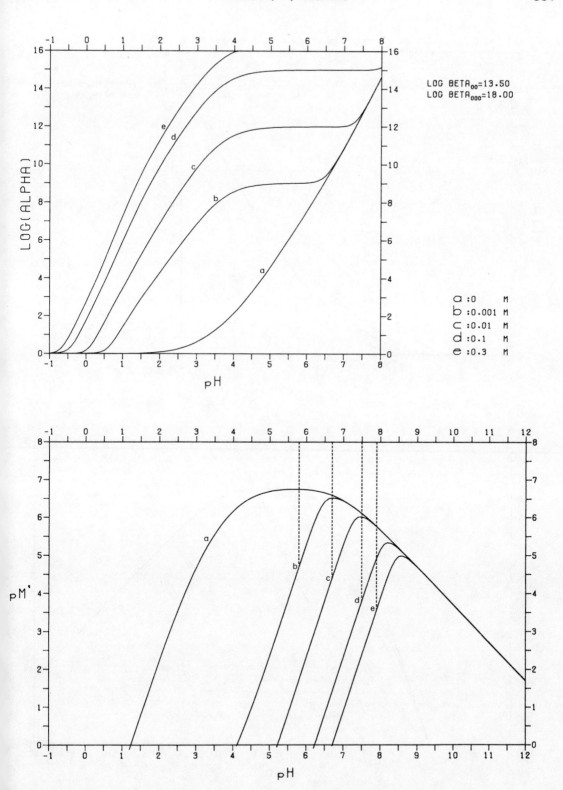

352 Gallium(III)-1,10-Phenanthroline [Ch. 18

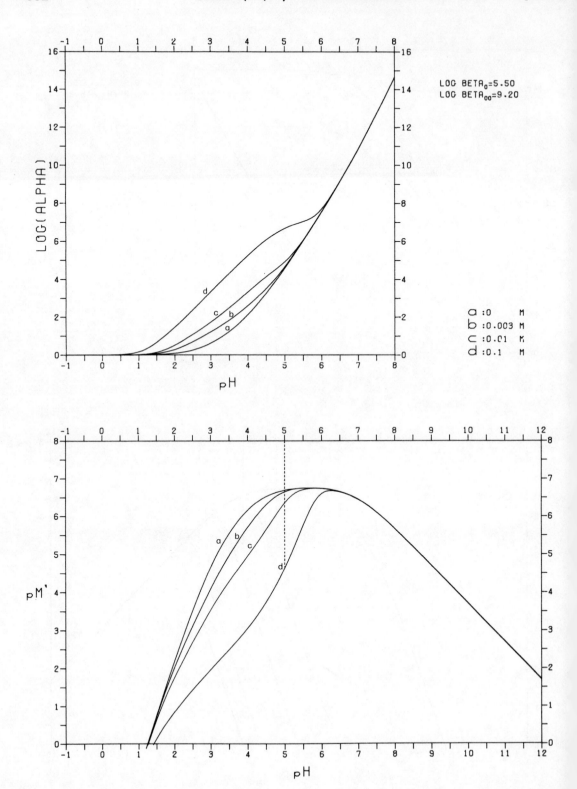

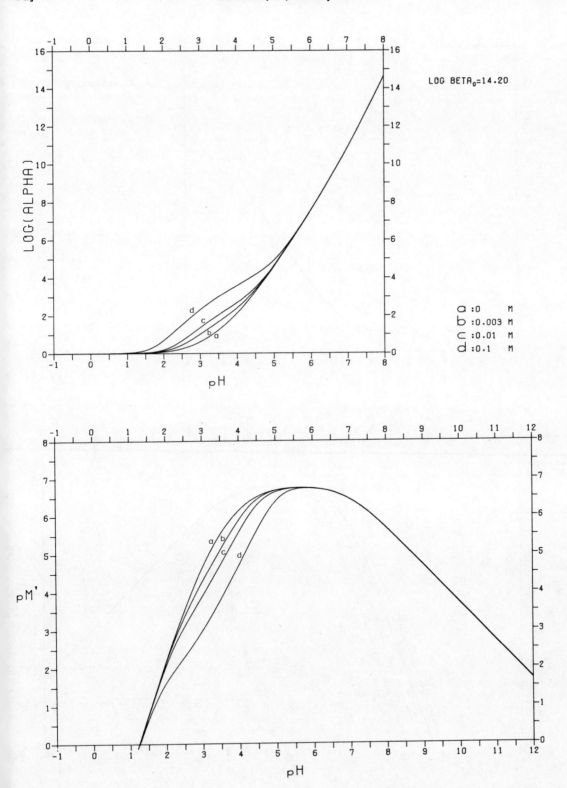

Gallium(III)-Sulphate [Ch. 18

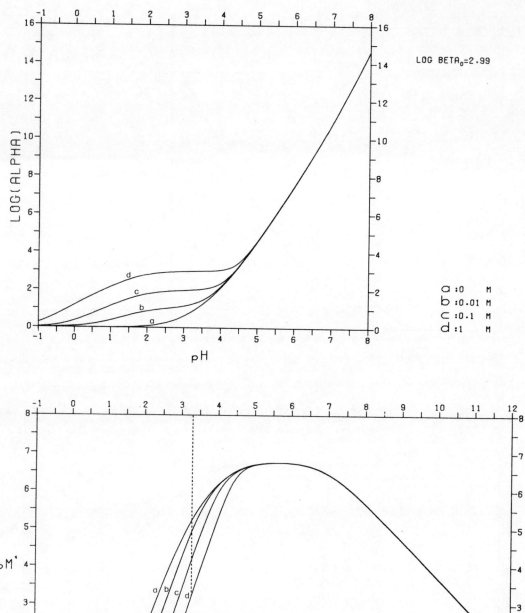

LOG BETA$_0$=2.99

a : 0 M
b : 0.01 M
c : 0.1 M
d : 1 M

Gallium(III)–5-Sulphosalicylate

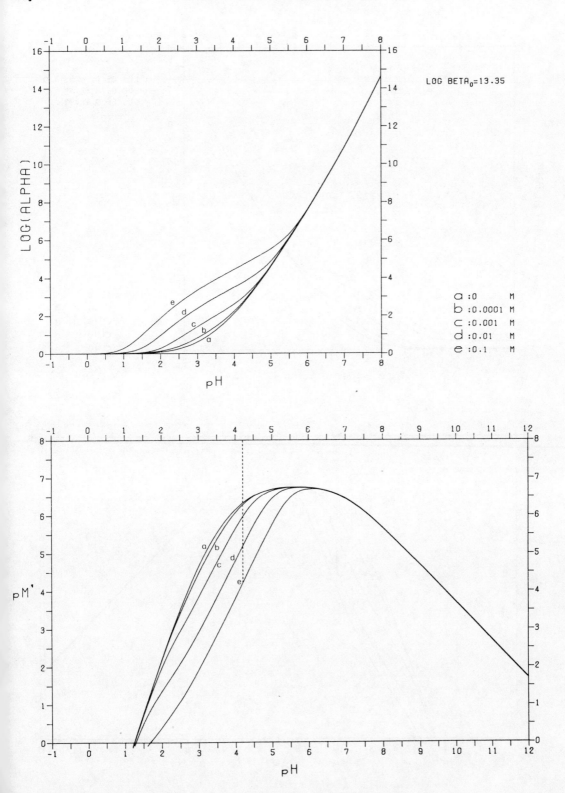

Gallium(III)-Tartrate [Ch. 18]

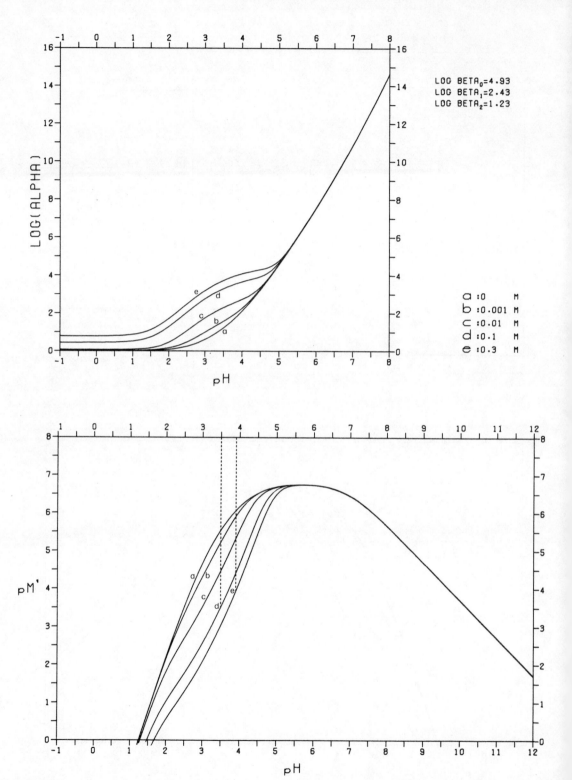

CHAPTER 19

Gadolinium (III)   Gd

At ordinary concentrations, hydrolysis of Gd^{3+} ions does not occur until precipitation has begun. There is some evidence for formation of $Gd(OH)^{2+}$, but only a limited amount of reliable information is available.

Studies of hydroxide-precipitate formation are difficult because of supersaturation, colloid formation, and changes in pH caused by recrystallization. These phenomena also make it difficult to define an equilibrium situation which is of practical utility.

The following data adopted from Baes and Mesmer [1] hold for a one-hour-old precipitate prepared under inert gas to exclude CO_2.

$$\log *\beta_1 = -8.0 \qquad \log *K_{s0} = 17.2$$
$$\log *\beta_2 = -16.4$$
$$\log *\beta_3 = -25.2$$
$$\log *\beta_4 = -35.5$$

There is little information about the formation of a carbonate precipitate. The solubility product seems to be close to that of $Gd(OH)_3$ in air-saturated solutions. Absorption of CO_2 from the air has been observed, and it seems likely that a mixture of carbonate and hydroxide is precipitated from such a solution. This has no effect on the shapes of the curves in the plots.

[1] C. F. Baes and R. E. Mesmer, *The Hydrolysis of Cations,* Wiley, New York, 1976.

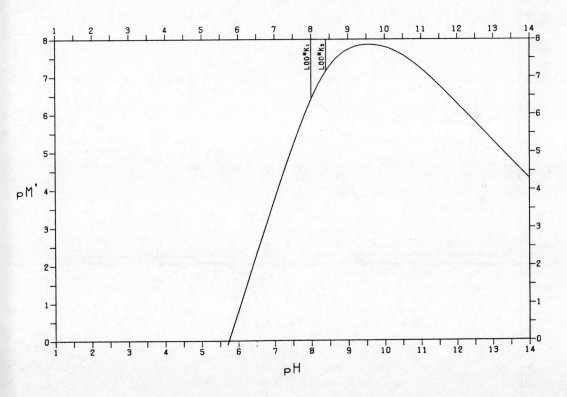

Gadolinium(III)-Acetate [Ch. 19]

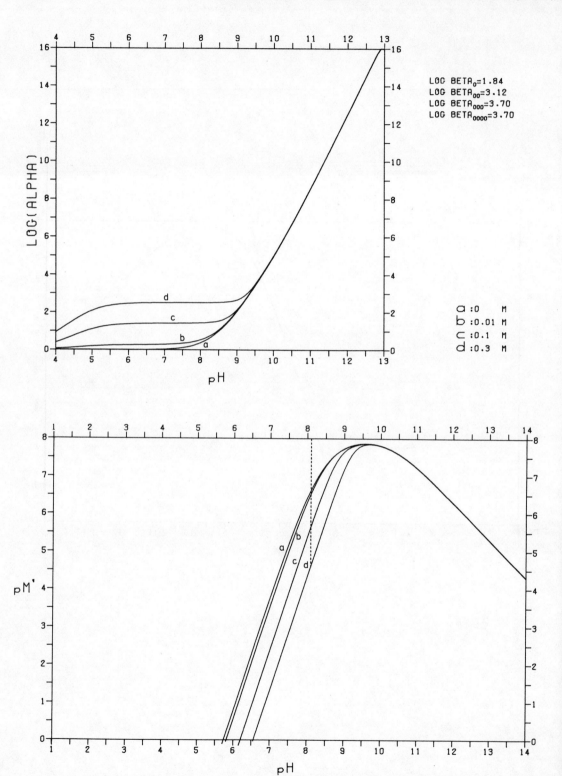

Gadolinium(III)-Acetylacetone

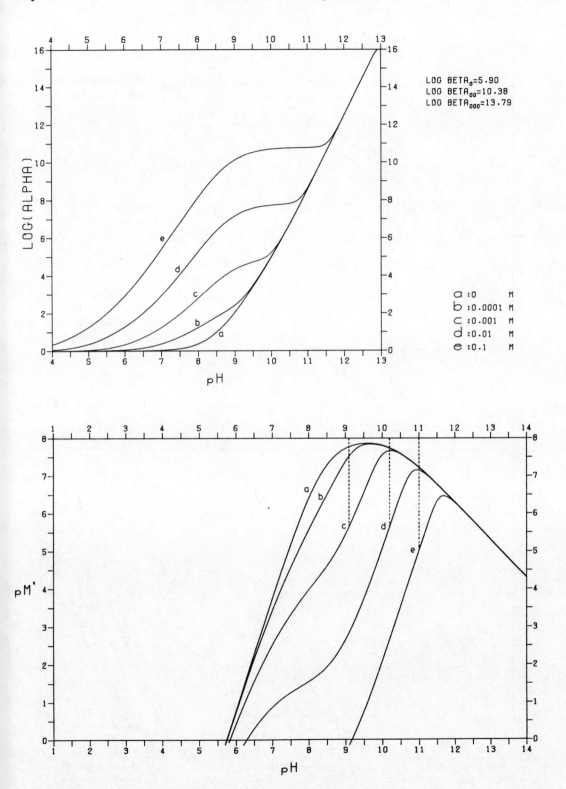

Gadolinium(III)-DCTA

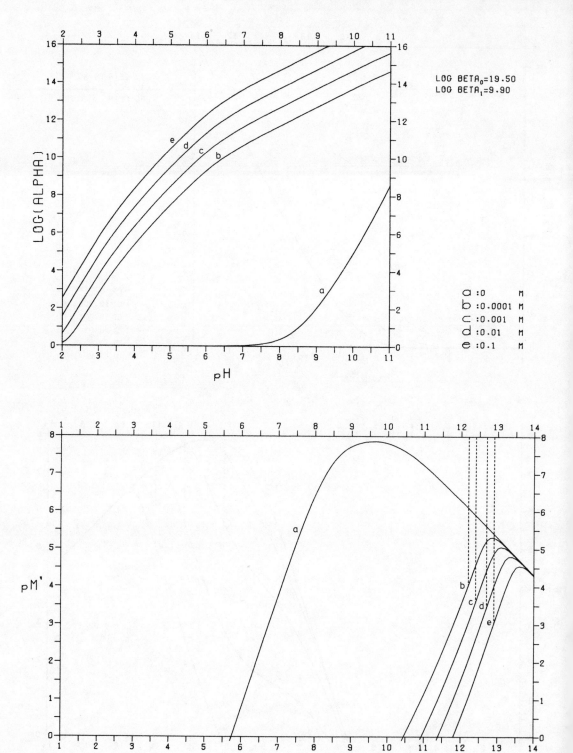

Gadolinium(III)-DTPA

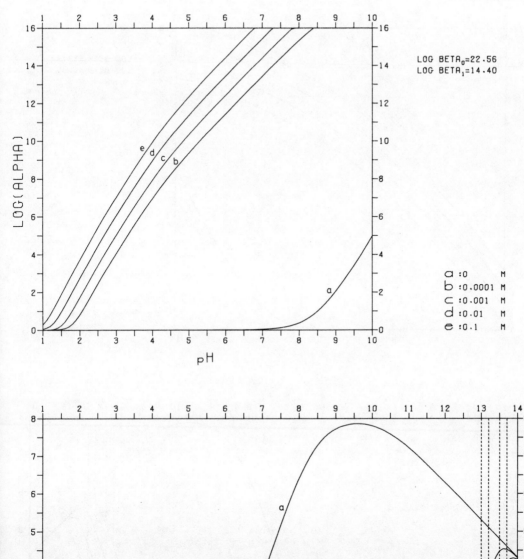

Gadolinium(III)-EDTA

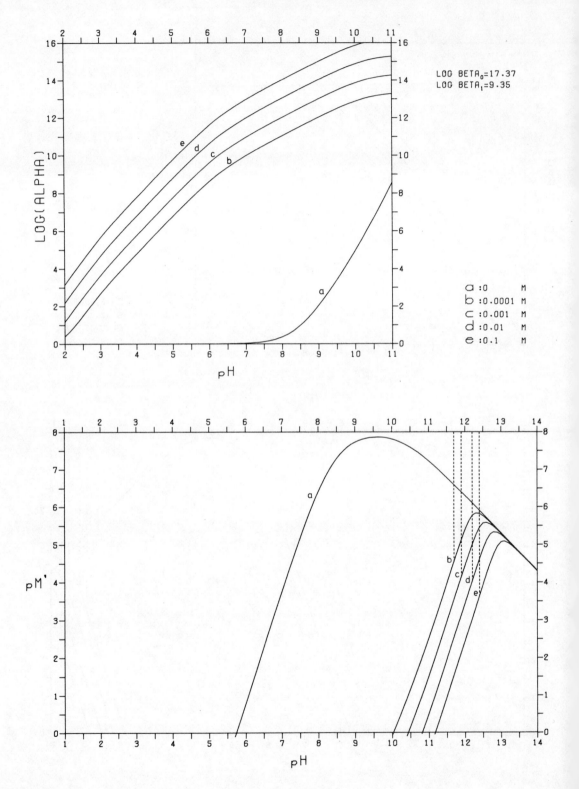

Gadolinium(III)-(OH)-Quinolinesulphonate

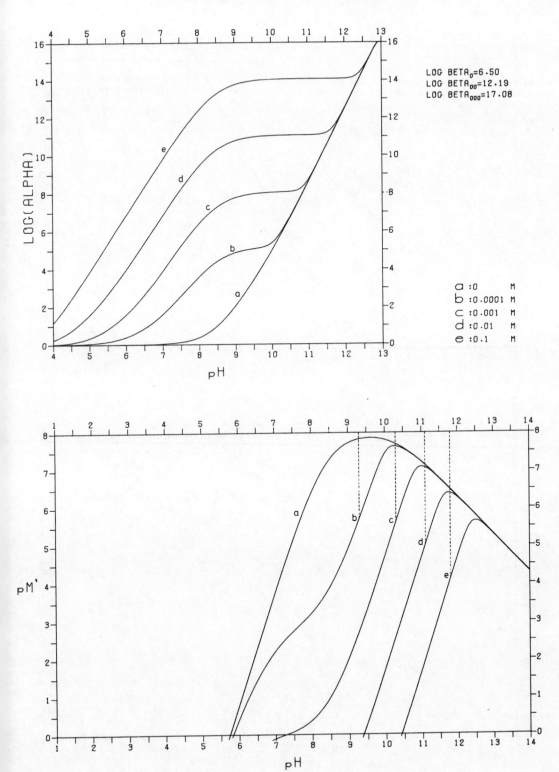

366 Gadolinium(III)-Iminodiacetate [Ch. 19

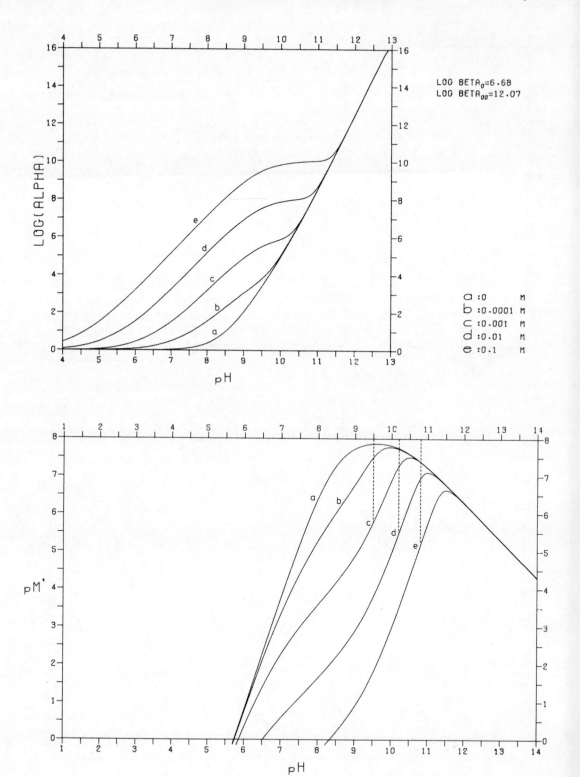

Gadolinium(III)-Oxalate

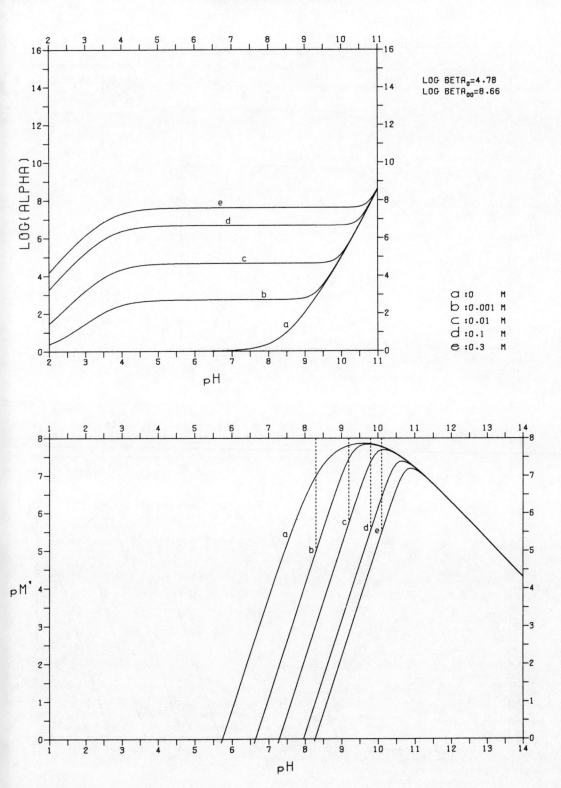

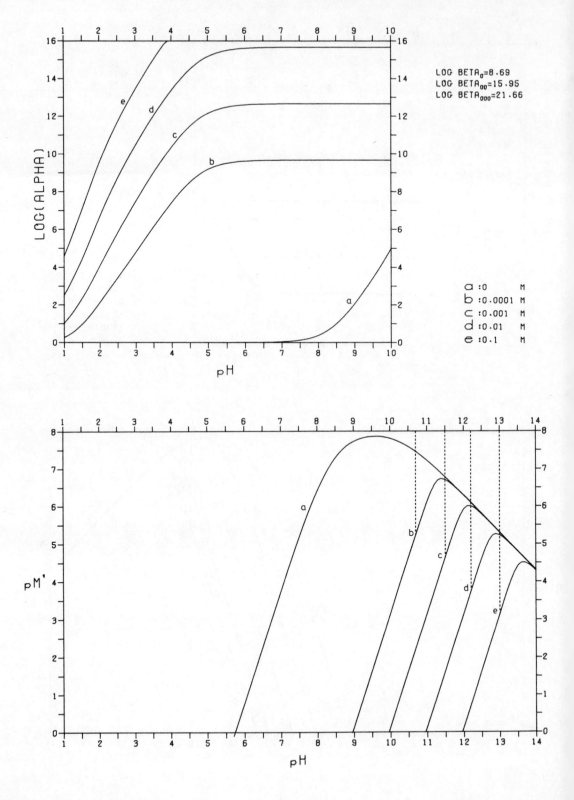

Gadolinium(III)-5-Sulphosalicylate

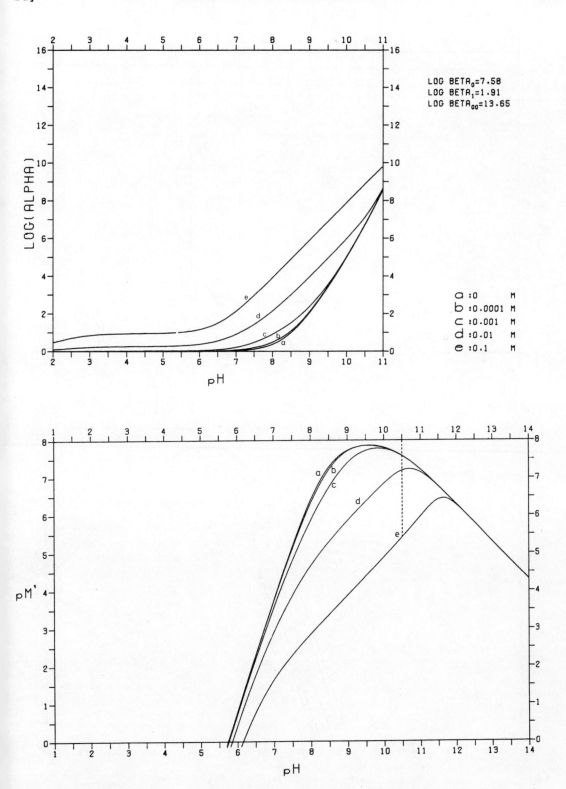

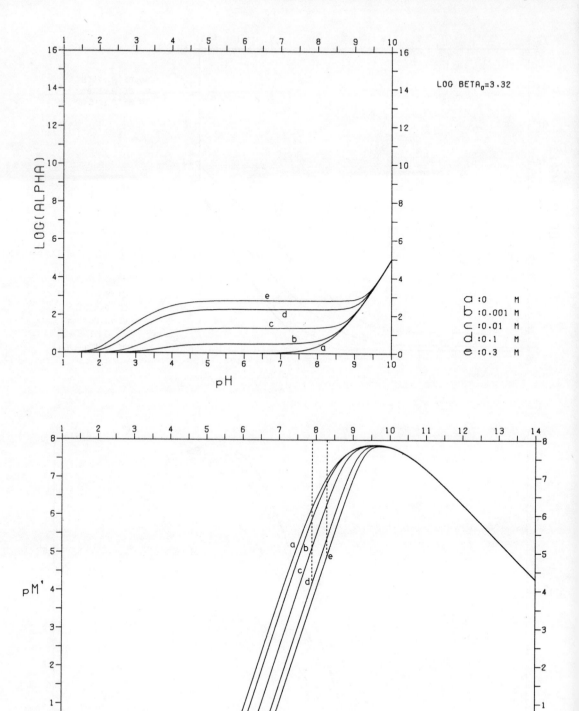

CHAPTER 20

Hafnium (IV)   Hf

Reliable information about hafnium ions is available only for strongly acidic solutions. The stability constant of $Hf(OH)^{3+}$ has been determined with reasonable accuracy; the formation constants of the species from $Hf(OH)_2^{2+}$ to $Hf(OH)_5^-$ are rough estimates based on measured solubilities of ill-defined forms of the oxide. Although there is some evidence for the formation of the polynuclear complexes $Hf_4(OH)_8^{8+}$ and $Hf_3(OH)_4^{8+}$, no useful estimates of the stability constants are available. The composition of freshly precipitated hydrated active HfO_2 depends on the anions present, the temperature, the ionic strength, and on the age, so it cannot be characterized adequately. The value goven for $^*K_{s0}$ is an estimate for ionic strength $\cong 2$. The other constants for the hafnium plots also refer to ionic strength $\cong 2$, which corresponds to the strongly acidic solutions employed for Hf chemistry.

$$\log {}^*\beta_1 = -1.2 \qquad \log {}^*\beta_4 = -12.2$$
$$\log {}^*\beta_2 = -2.8 \qquad \log {}^*\beta_5 = -18.2$$
$$\log {}^*\beta_3 = -7.5 \qquad \log {}^*K_{s0} = 0.3$$

Hafnium(IV)

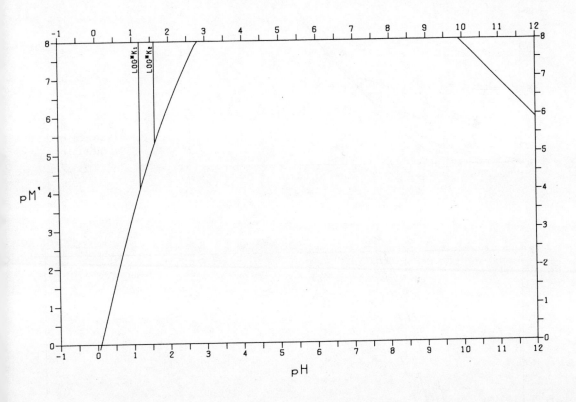

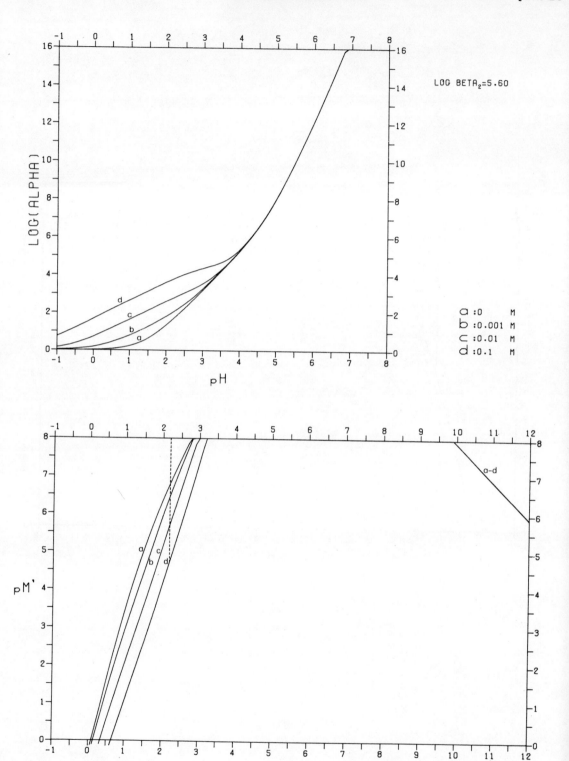

Hafnium(IV)-DTPA

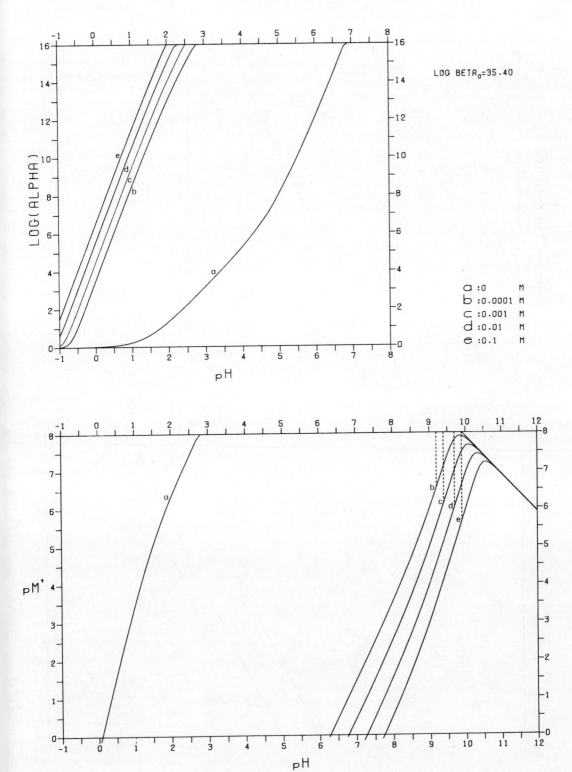

Hafnium(IV)-EDTA

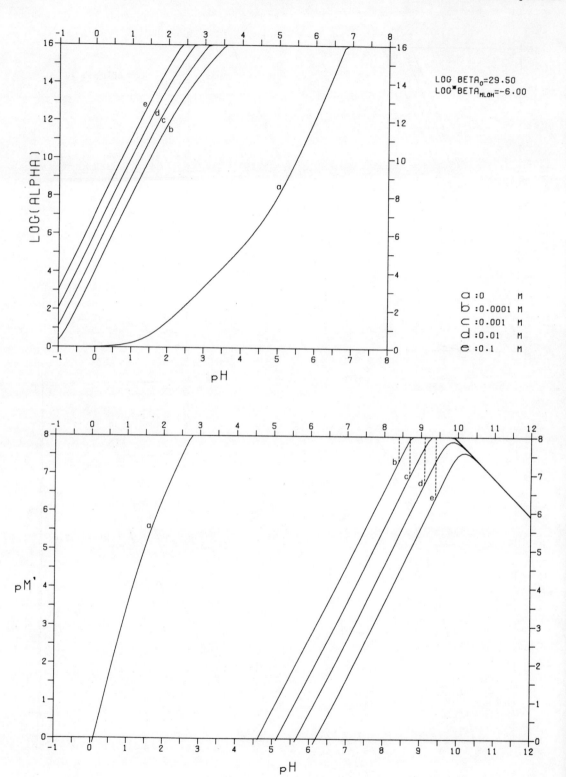

Hafnium(IV)-Fluoride

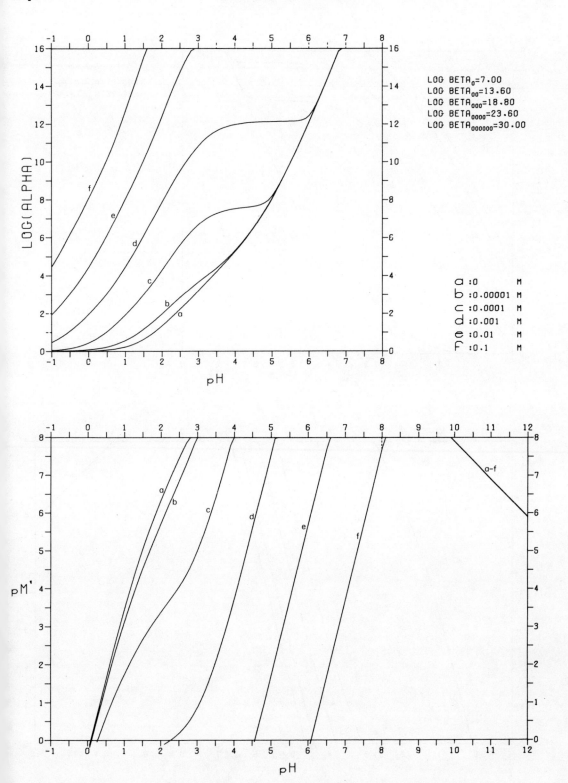

Hafnium(IV)-Oxalate [Ch. 20]

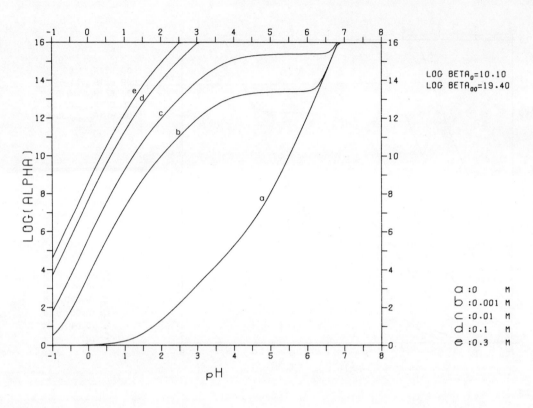

LOG BETA$_0$=10.10
LOG BETA$_{00}$=19.40

a : 0 M
b : 0.001 M
c : 0.01 M
d : 0.1 M
e : 0.3 M

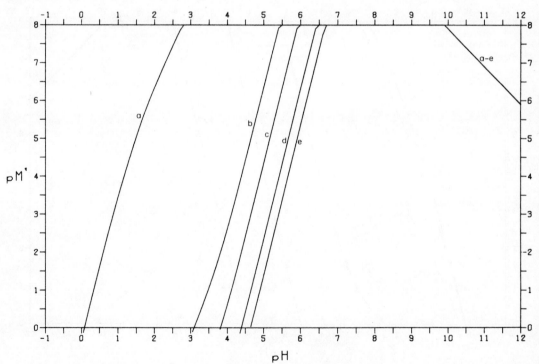

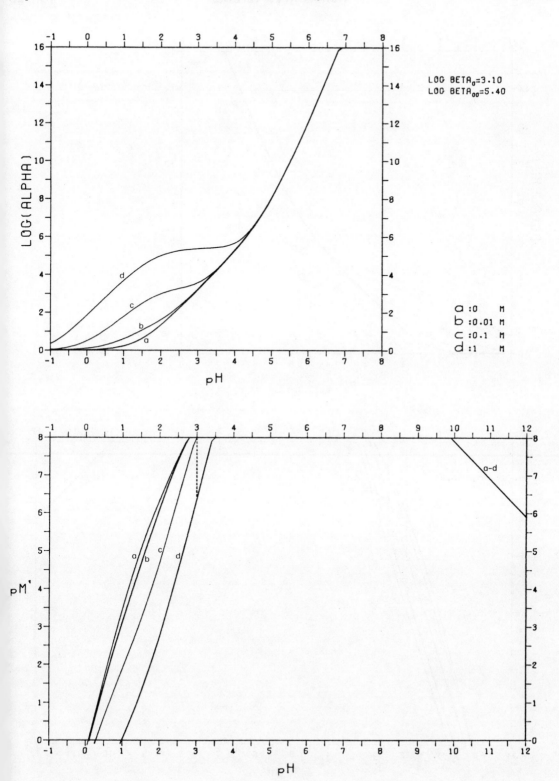

Hafnium(IV)-Tartrate

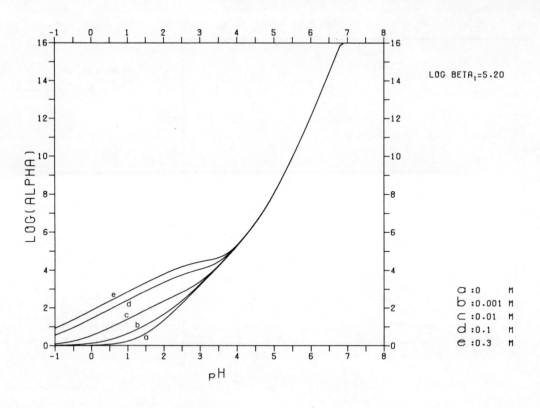

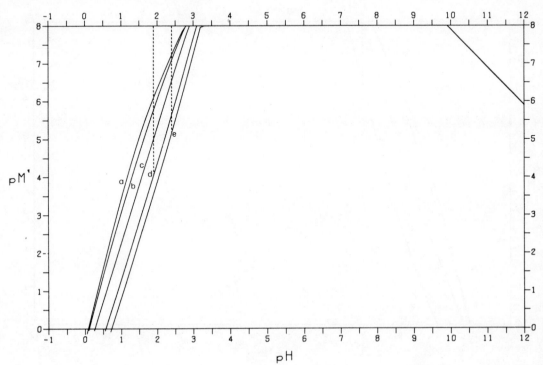

Hafnium(IV)–Tiron

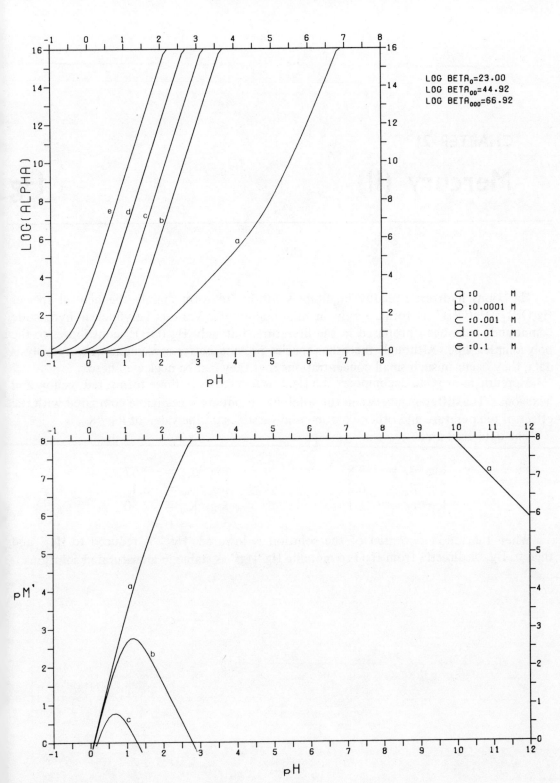

LOG BETA$_0$ = 23.00
LOG BETA$_{00}$ = 44.92
LOG BETA$_{000}$ = 66.92

a : 0   M
b : 0.0001 M
c : 0.001 M
d : 0.01   M
e : 0.1   M

CHAPTER 21

Mercury (II)   Hg

Hg^{2+} ions hydrolyse readily in dilute solutions to form mainly the neutral species $Hg(OH)_2$; $Hg(OH)^+$ is formed only in minor amounts. Various polynuclear hydroxide compounds have been proposed in the literature, but only $Hg_2(OH)^{3+}$ contributes to the polycomplex line. Although the other species provide better agreement with hydrolysis data, they occur in such small concentrations that they can be neglected here.

Mercuric hydroxide decomposes the HgO, which exists in three forms; red, yellow and hexagonal. The difference between the solubility products is negligible compared with the effect of temperature and other experimental variables on the value of log $*K_{s0}$.

The data used are:

$$\log *\beta_1 = -3.8 \qquad \log *\beta_{12} = -2.67$$
$$\log *\beta_2 = -6.2 \qquad \log *\beta_{34} = -6.4$$
$$\log *\beta_3 = -21.1 \qquad \log *K_{s0} = 2.50$$

When the redox potential of the solution is lowered, Hg^{2+} is reduced to Hg_2^{2+} and then to Hg, or directly from HgO to metallic Hg. Hg^{2+} is stable in air-saturated solutions.

Mercury(II)

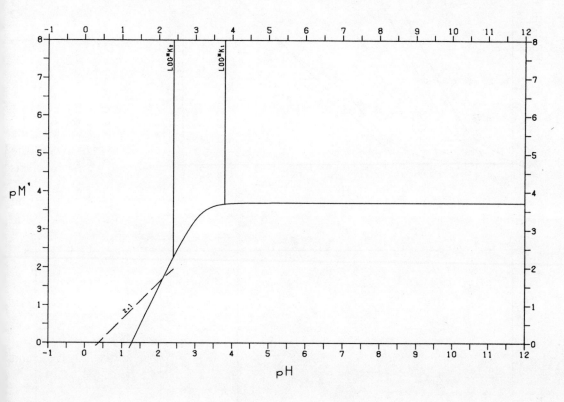

Mercury(II)-Acetate [Ch. 21]

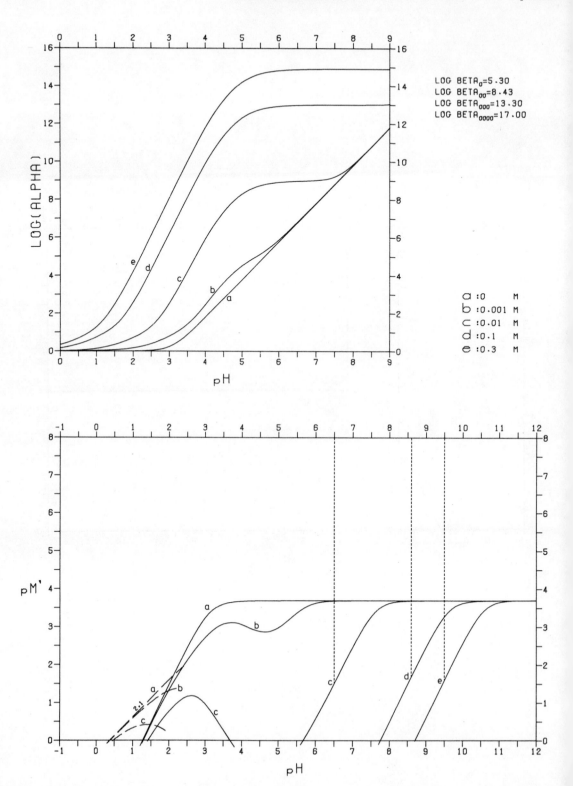

Mercury(II)-Ammonia

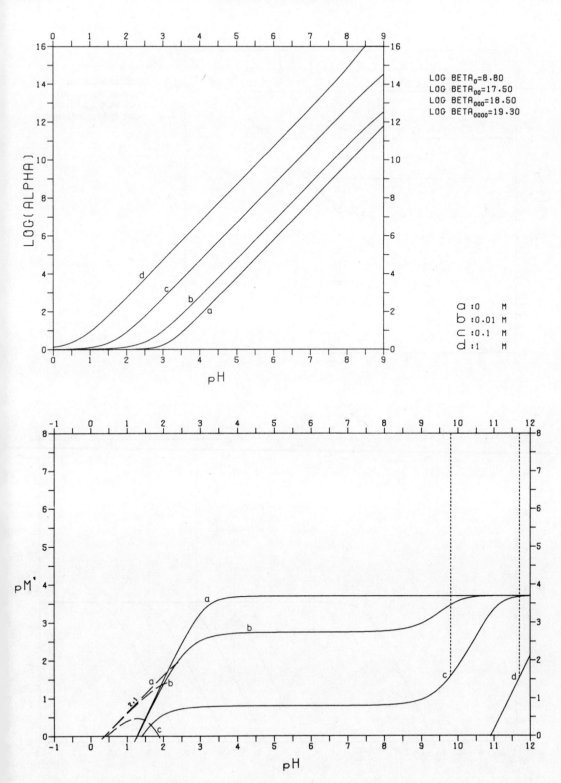

LOG BETA$_0$=8.80
LOG BETA$_{00}$=17.50
LOG BETA$_{000}$=18.50
LOG BETA$_{0000}$=19.30

a : 0   M
b : 0.01 M
c : 0.1 M
d : 1   M

Mercury(II)-Bromide [Ch. 21

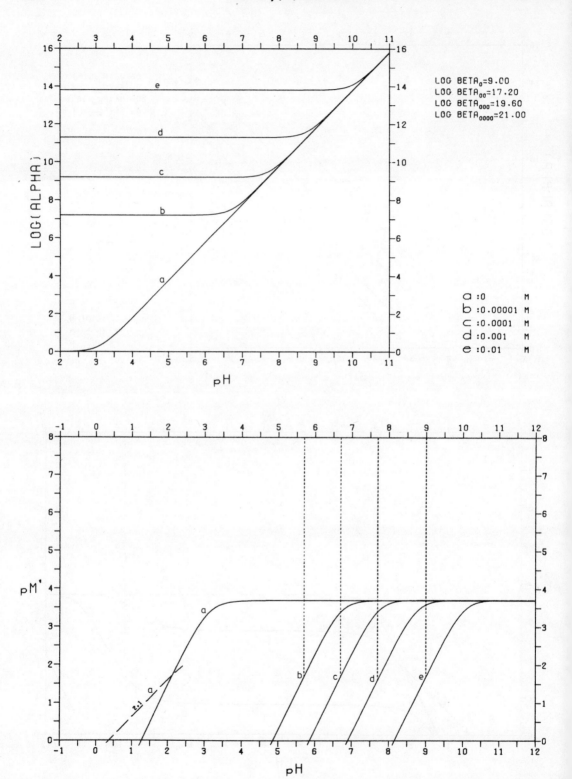

Mercury(II)-Chloride

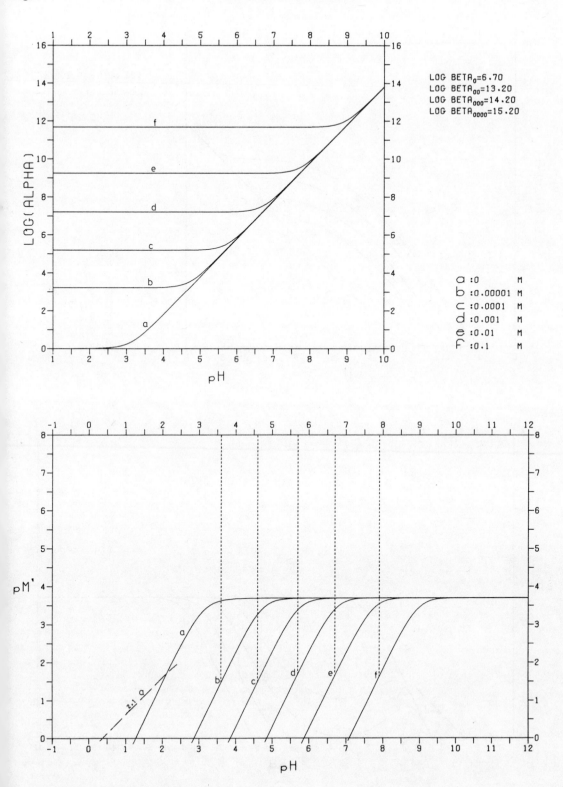

LOG BETA$_0$=6.70
LOG BETA$_{00}$=13.20
LOG BETA$_{000}$=14.20
LOG BETA$_{0000}$=15.20

a : 0   M
b : 0.00001 M
c : 0.0001   M
d : 0.001   M
e : 0.01   M
f : 0.1   M

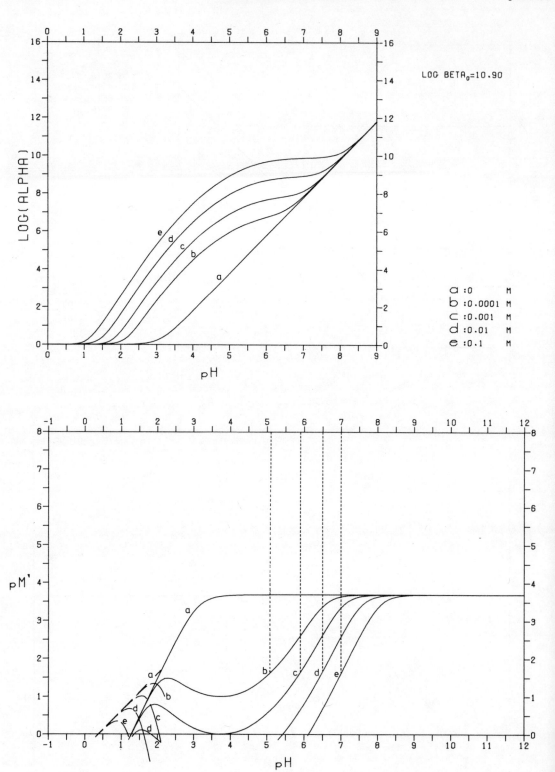

Mercury(II)-Cyanide

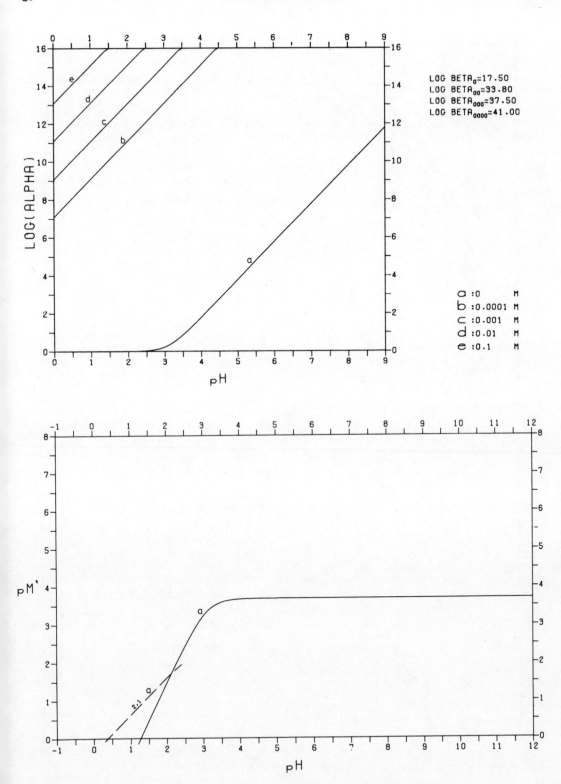

Mercury(II)–DCTA [Ch. 21

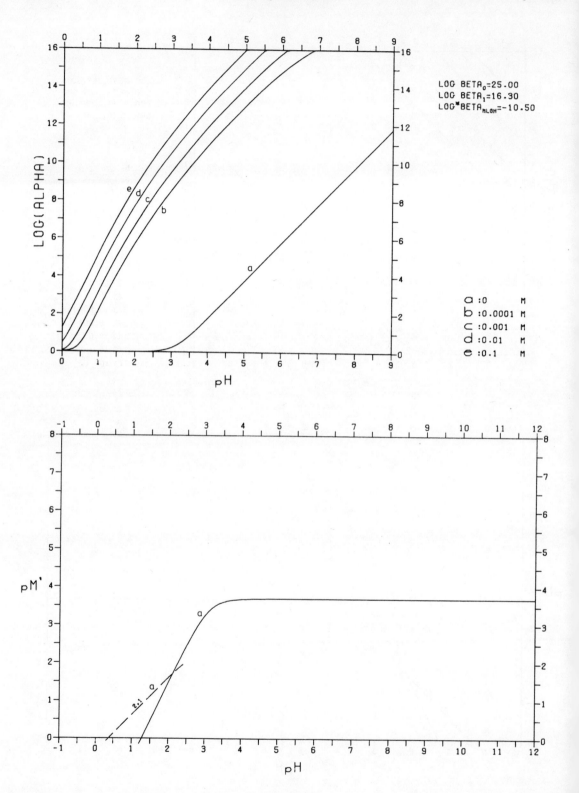

LOG BETA$_0$=25.00
LOG BETA$_1$=16.30
LOG*BETA$_{MLOH}$=-10.50

a : 0   M
b : 0.0001 M
c : 0.001 M
d : 0.01   M
e : 0.1   M

Mercury(II)-DTPA

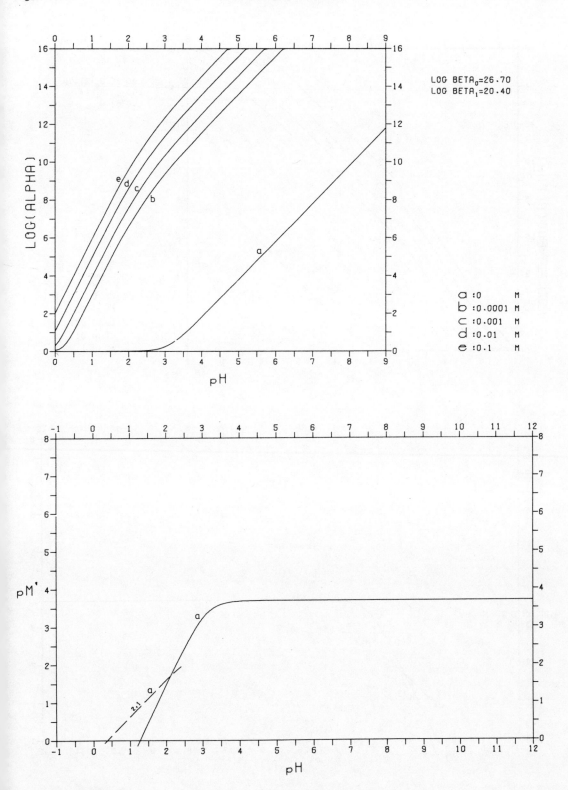

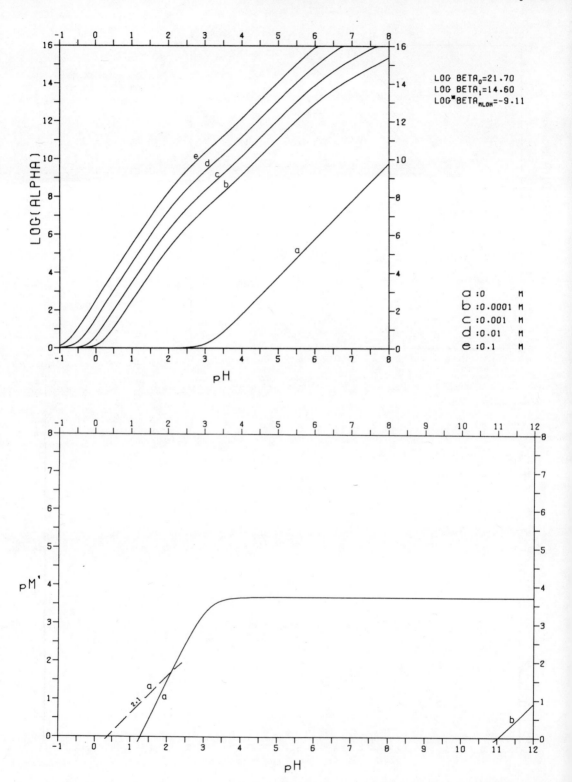

Mercury(II)–EGTA

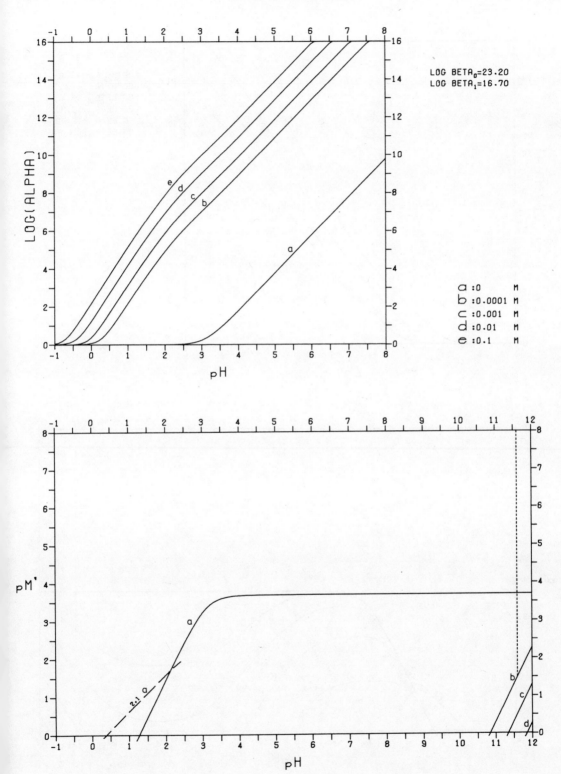

Mercury(II)-Ethylenediamine [Ch. 21]

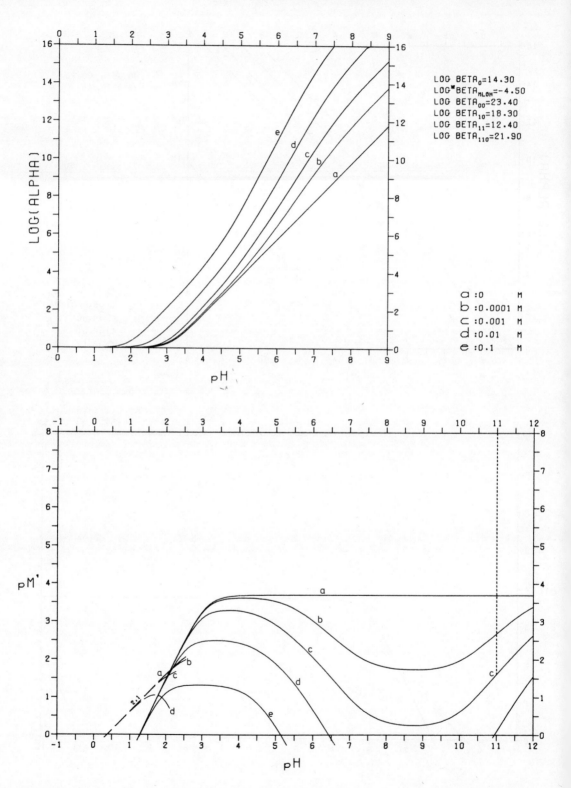

Mercury(II)-Glycine

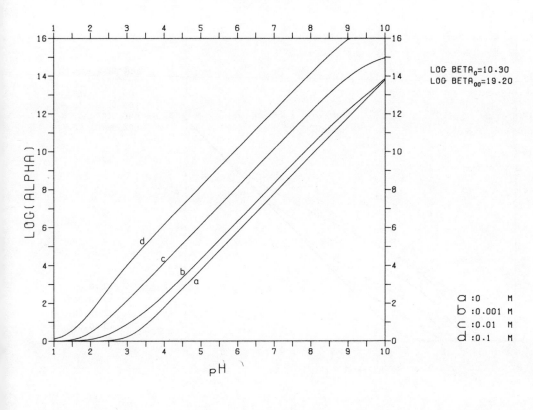

LOG BETA₀=10.30
LOG BETA₀₀=19.20

a : 0   M
b : 0.001 M
c : 0.01 M
d : 0.1   M

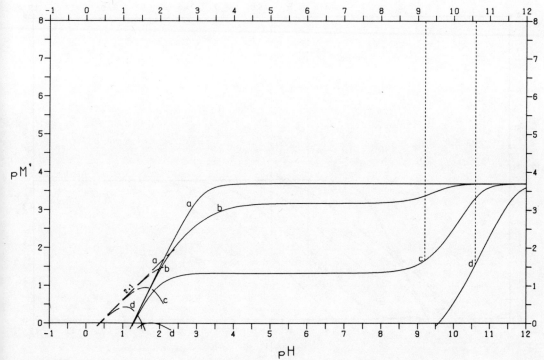

Mercury(II)-Iminodiacetate [Ch. 21]

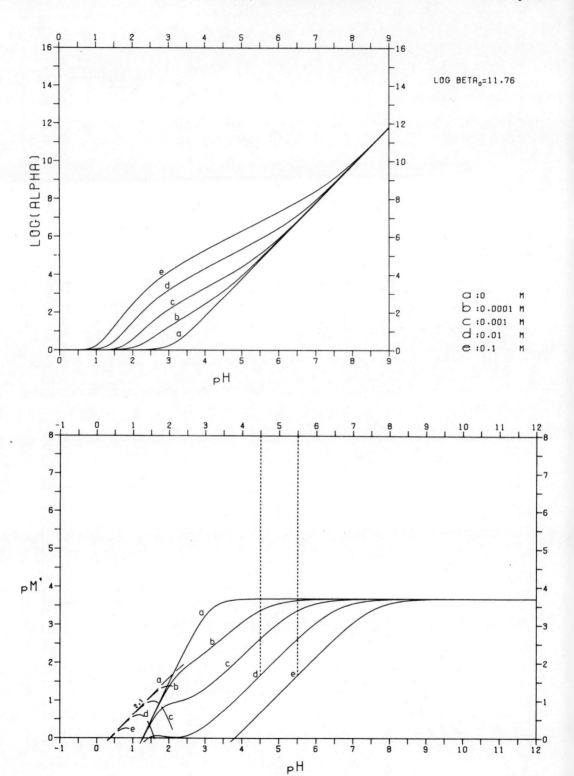

Mercury(II)-Pyridine-2,6-dicarboxylate

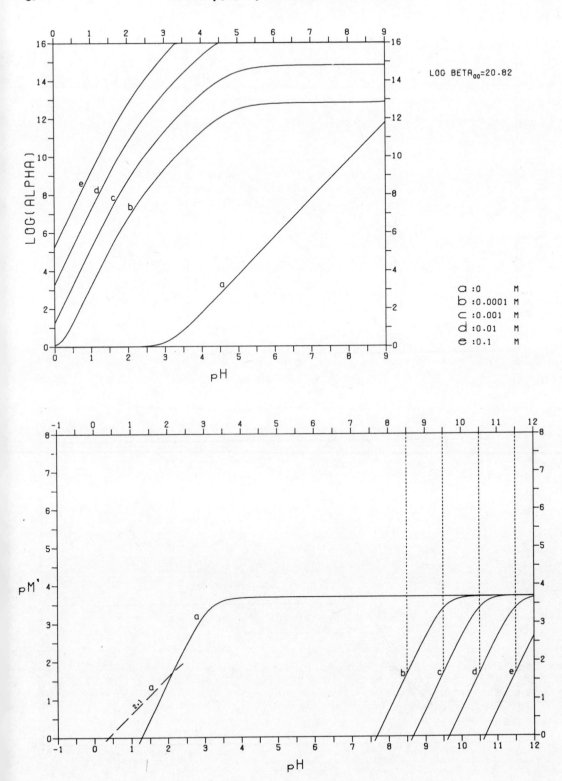

LOG BETA$_{00}$=20.82

a : 0 M
b : 0.0001 M
c : 0.001 M
d : 0.01 M
e : 0.1 M

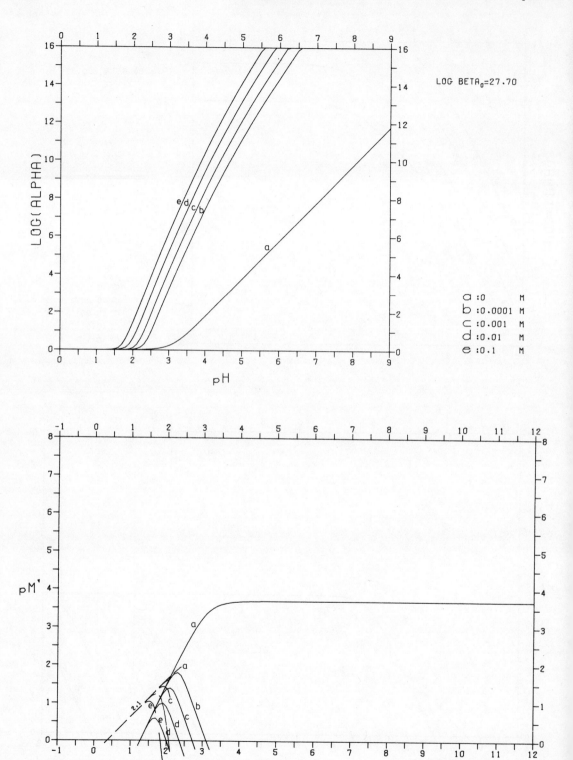

Mercury(II)-Trien

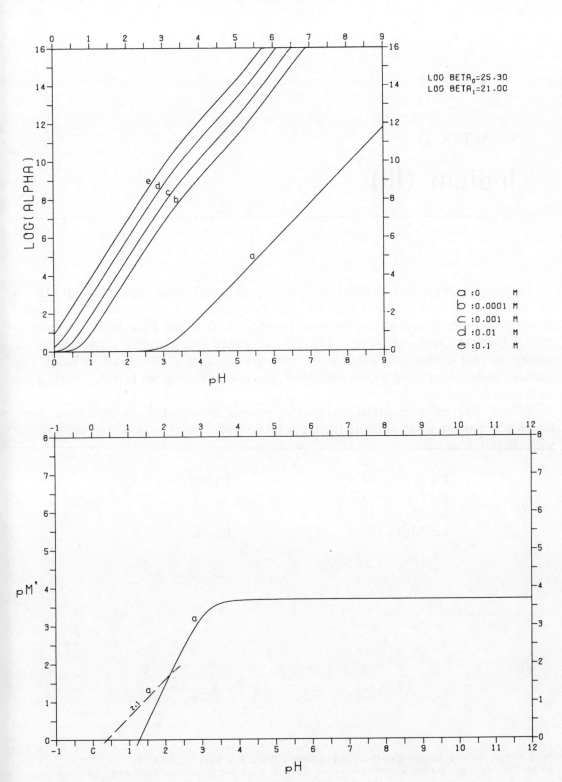

LOG BETA$_0$=25.30
LOG BETA$_1$=21.00

a : 0   M
b : 0.0001 M
c : 0.001   M
d : 0.01   M
e : 0.1   M

CHAPTER 22

Indium (III)   In

Unlike aluminium and gallium, In^{3+} hydrolyses rapidly and reversibly. $In(OH)^{2+}$ is the mononuclear species for which the stability constant is known with the greatest accuracy. Polycomplexation occurs significantly, but it is not clear which complex is most important. Baes and Mesmer [1] have suggested that Occam's razor[†] should be applied in such a situation, and that the scheme consisting of the minimum number of species should be regarded as the best. Here, this reasoning suggests that $In_3(OH)_4^{5+}$ is the most probable species.

Indium hydroxide is precipitated in a reasonably well-defined crystalline form; $\log {}^*K_{s0} = 5.6$ is a good estimate for the solubility product. The plots were constructed with the following data.

$$\log {}^*\beta_1 = -4.42 \qquad \log {}^*\beta_{43} = -9.3$$
$$\log {}^*\beta_2 = -8.34$$
$$\log {}^*\beta_3 = -12.9$$
$$\log {}^*\beta_4 = -22.4 \qquad \log {}^*K_{s0} = 5.6$$

---

[1] C. F. Baes and R. E. Mesmer, *The Hydrolysis of Cations*, Wiley, New York, 1976. (p. 50).
†William of Occam gets the credit for the rule named after him, but it had in fact been stated earlier by several scholastic

Indium(III)

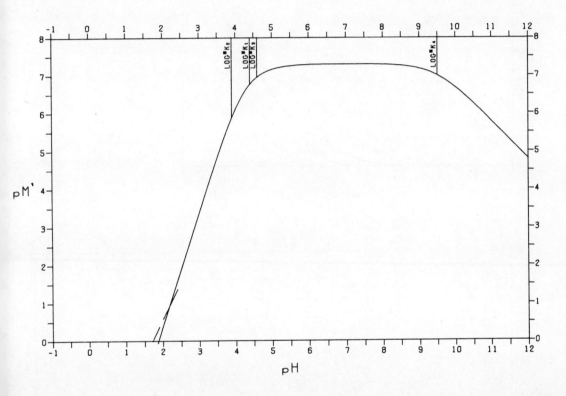

Indium(III)–Acetate

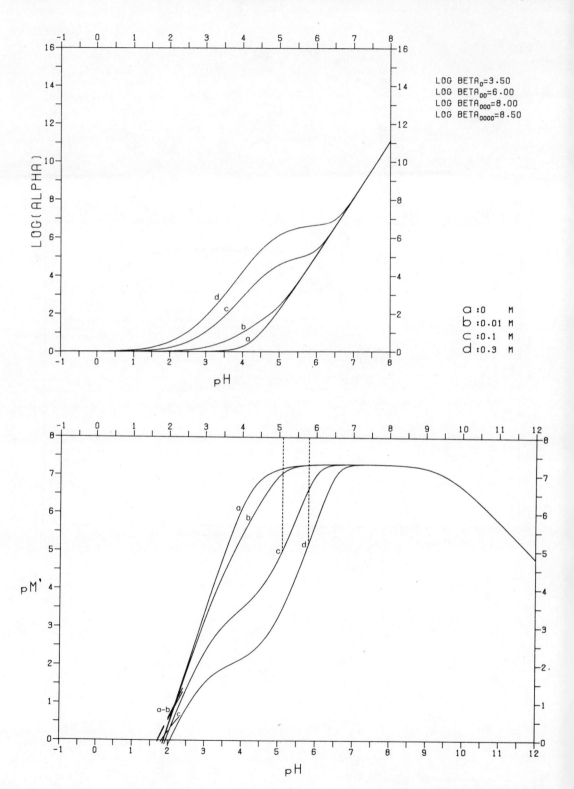

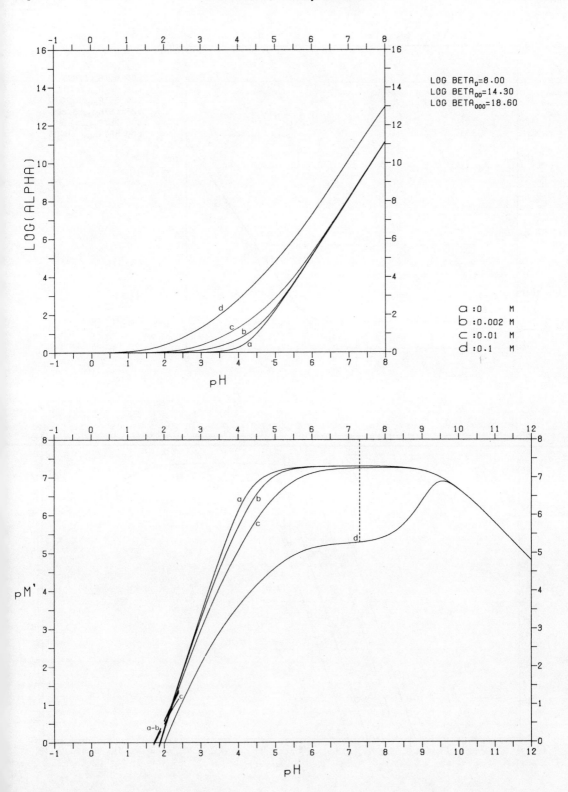

Indium(III)-Bromide

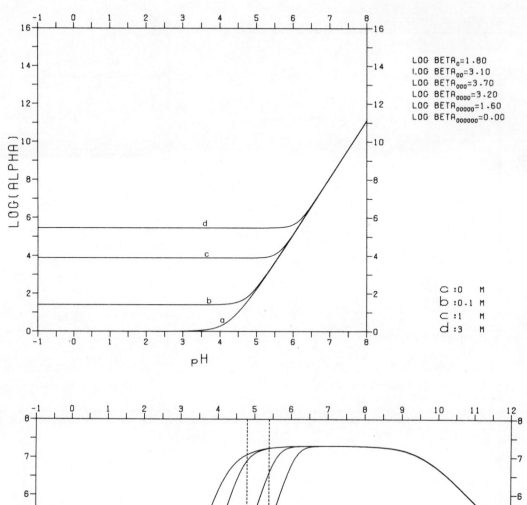

LOG BETA$_0$=1.80
LOG BETA$_{00}$=3.10
LOG BETA$_{000}$=3.70
LOG BETA$_{0000}$=3.20
LOG BETA$_{00000}$=1.60
LOG BETA$_{000000}$=0.00

a : 0 M
b : 0.1 M
c : 1 M
d : 3 M

Indium(III)-Chloride

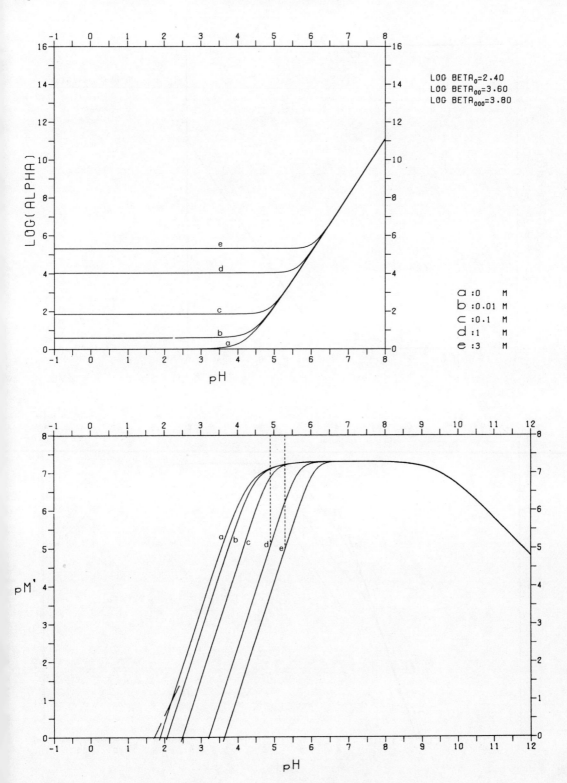

406 Indium(III)–Citrate [Ch. 22

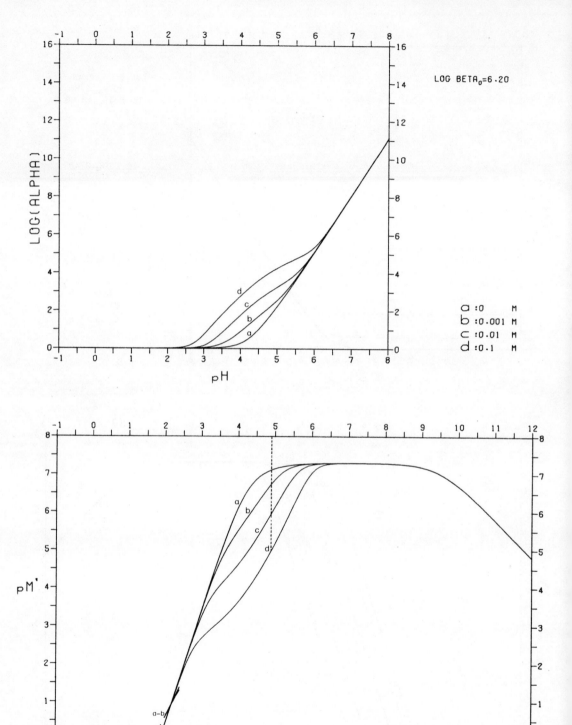

Indium(III)-DCTA

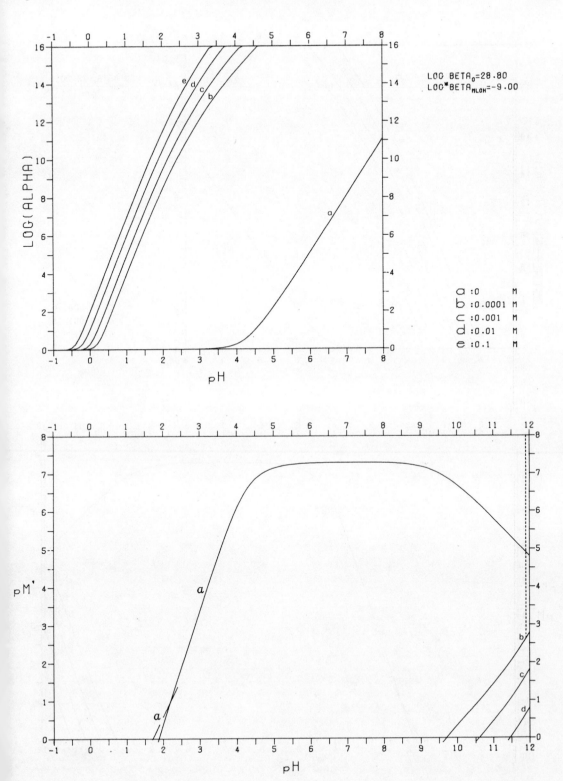

$LOG\ BETA_0 = 28.80$
$LOG^*BETA_{MLOH} = -9.00$

a : 0 M
b : 0.0001 M
c : 0.001 M
d : 0.01 M
e : 0.1 M

Indium(III)-DTPA [Ch. 22

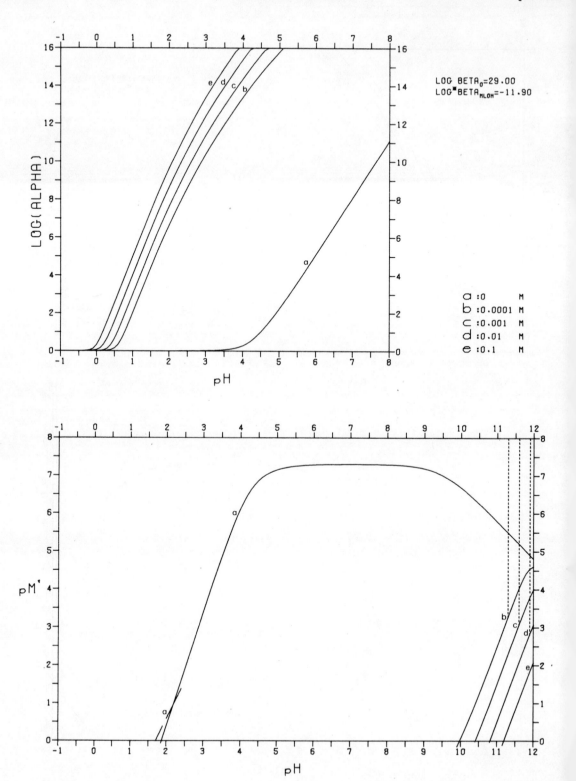

Indium(III)-EDTA

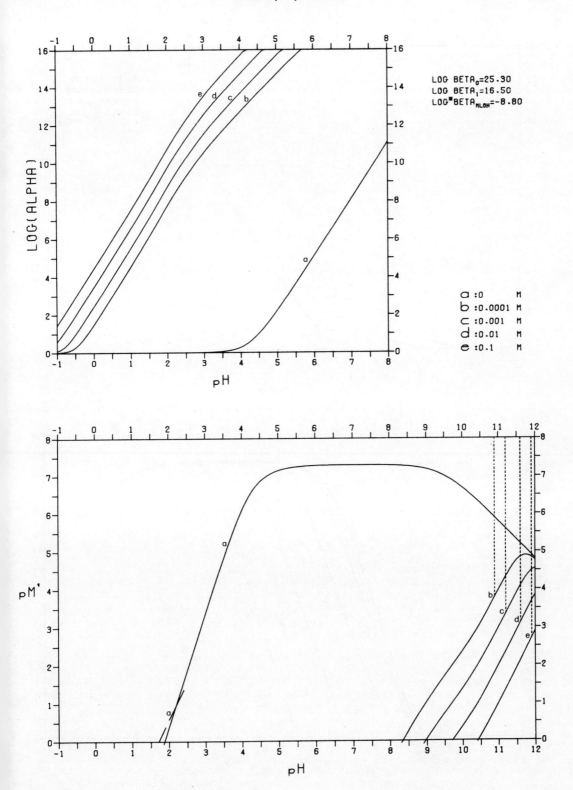

Indium(III)-Fluoride

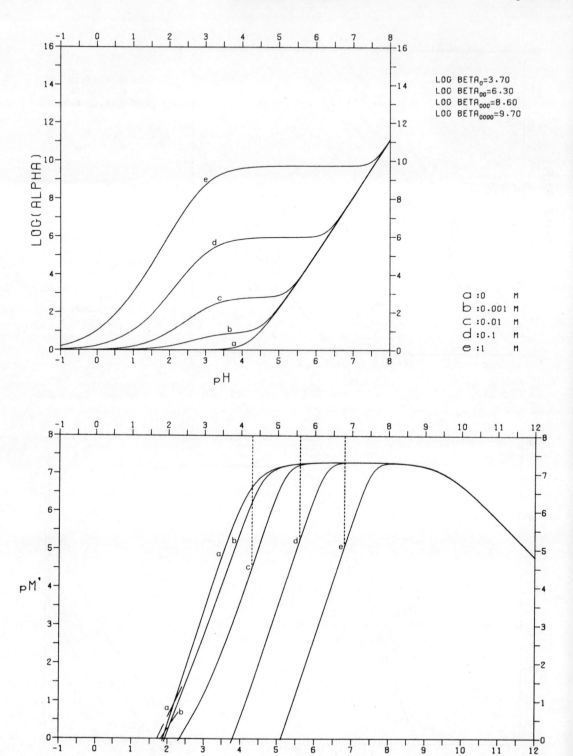

Indium(III)-Formate

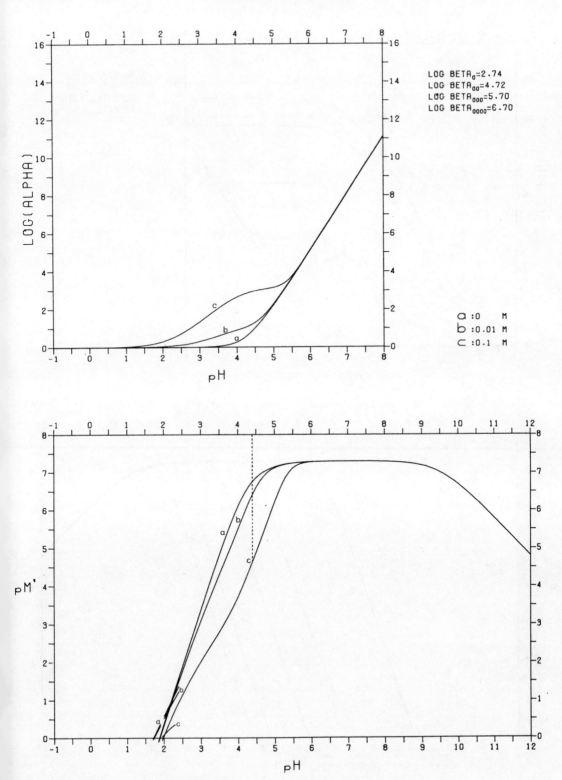

Indium(III)–Oxalate

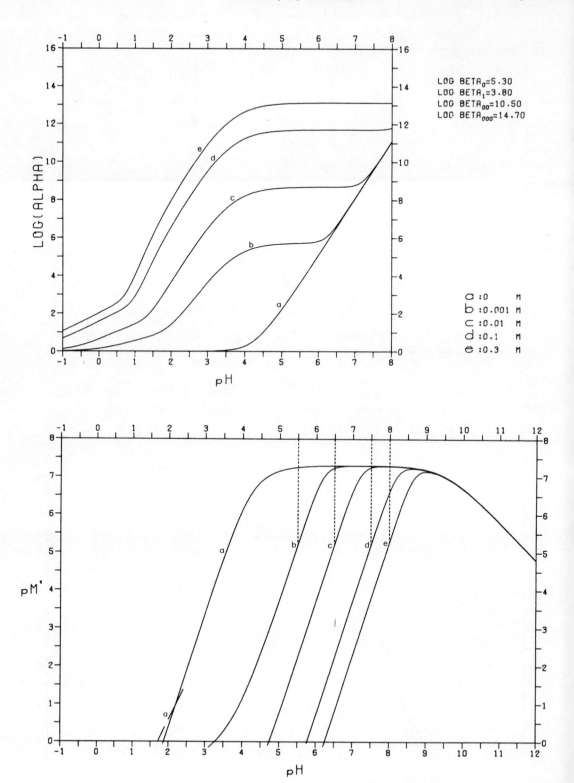

Indium(III)-1,10-Phenanthroline

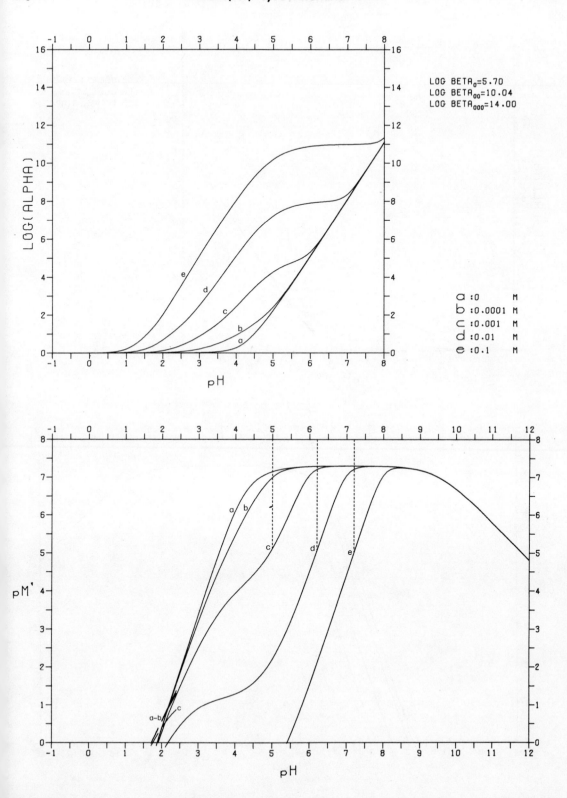

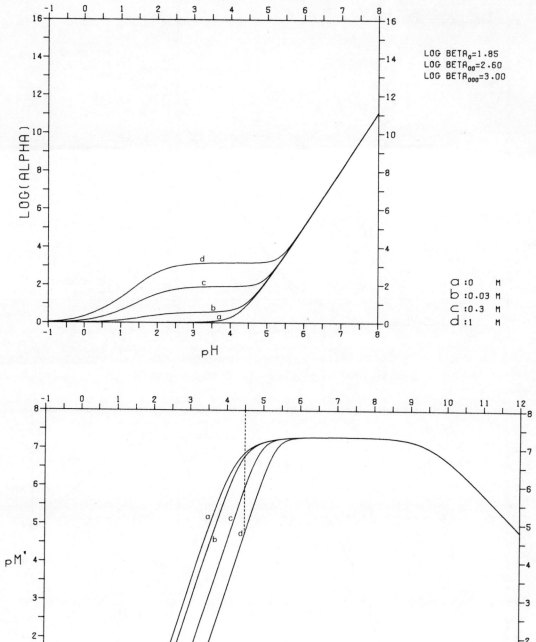

Indium(III)-Tartrate

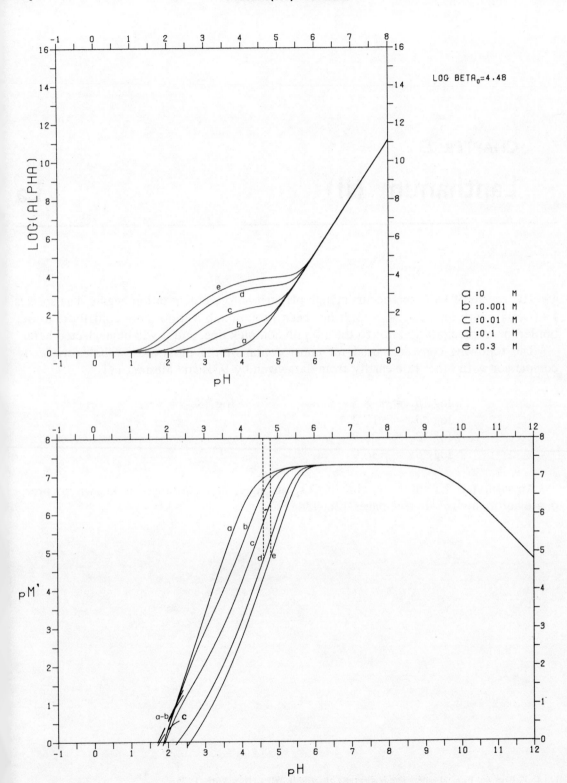

CHAPTER 23

Lanthanum (II) La

Hydrolysis of La^{3+} ions occurs at high pH, just before precipitation begins. The species $La_5(OH)_9^{6+}$, with log $*\beta_{9,5} = -71.2$, has been proposed, but it does not contribute to the borderline of precipitation nor to the 1% polycomplex line, and it will be neglected here.

The following constants used for construction of the plots are estimates made by comparison with other rare earths, from data given by Baes and Mesmer [1].

$$\log *\beta_1 = -8.5 \qquad \log *K_{s0} = 19.7$$
$$\log *\beta_2 = -17.2$$
$$\log *\beta_3 = -25.9$$
$$\log *\beta_4 = -36.9$$

Air-saturated La solutions absorb CO_2 strongly, and carbonate is known to precipitate before hydroxide (see pages 426–434).

[1] C. F. Baes and R. E. Mesmer, *The Hydrolysis of Cations,* Wiley, New York, 1976.

Lanthanum(III)

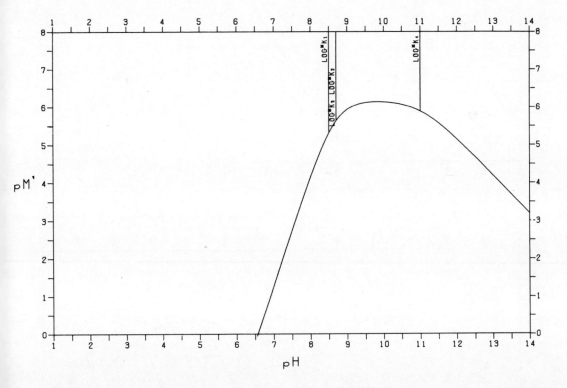

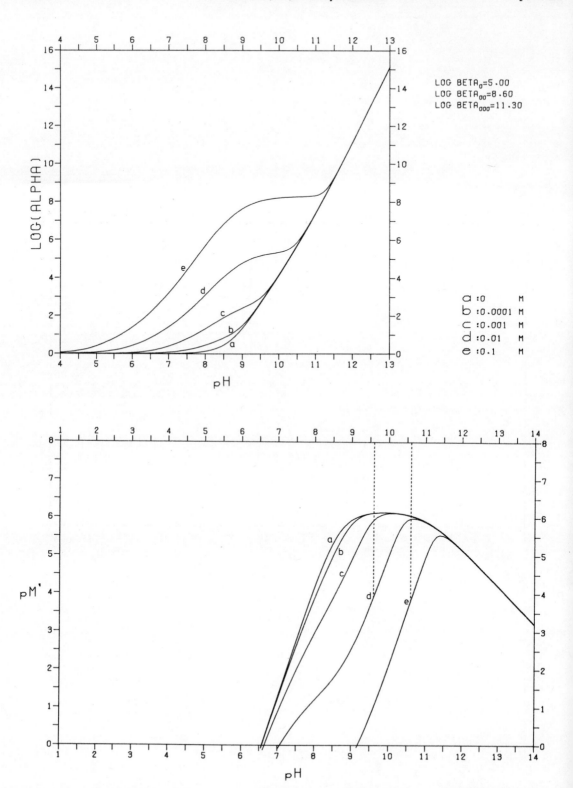

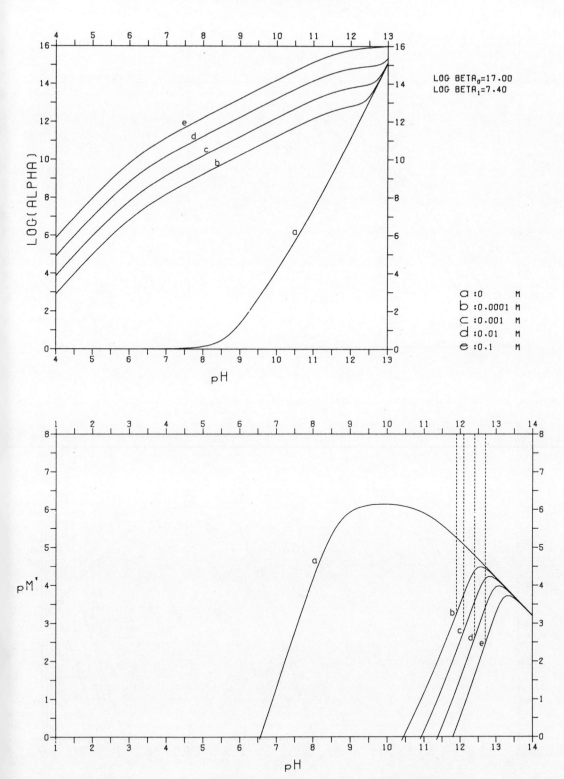

Lanthanum(III)-DTPA [Ch. 23]

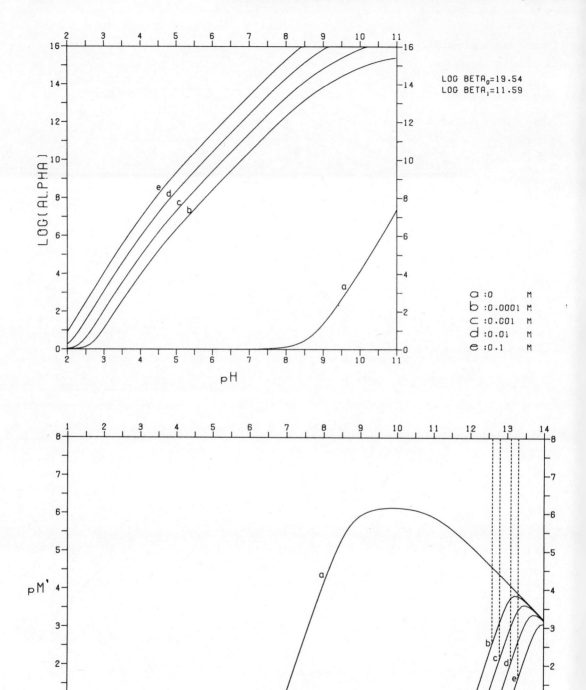

Lanthanum(III)-EDTA

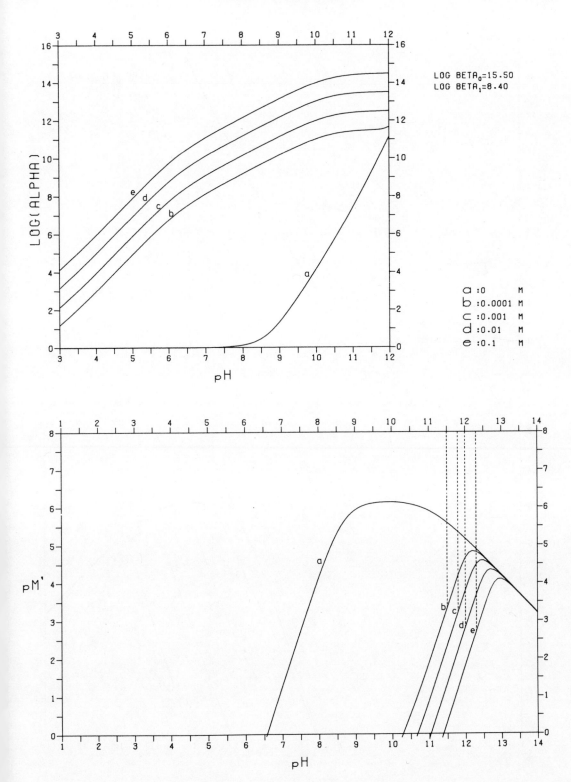

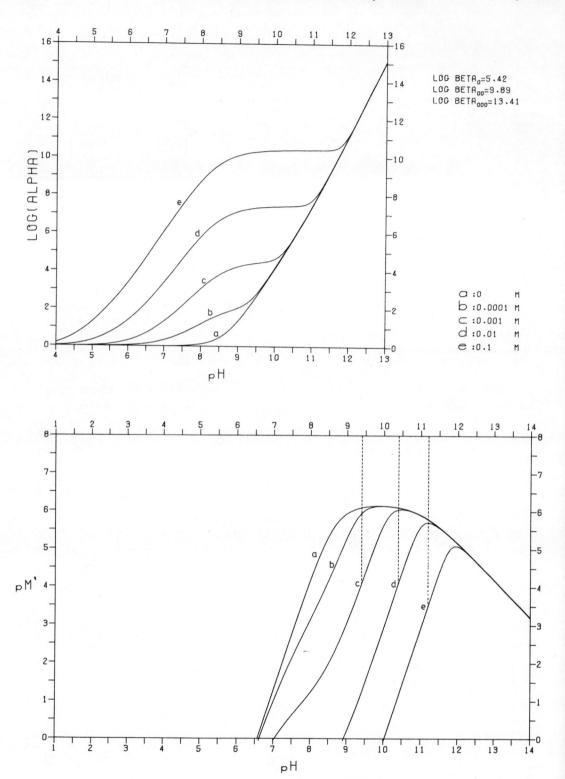

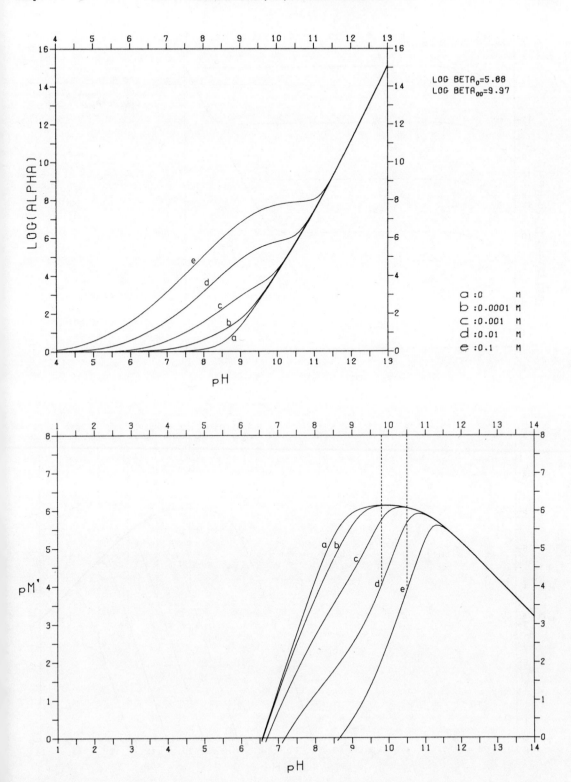

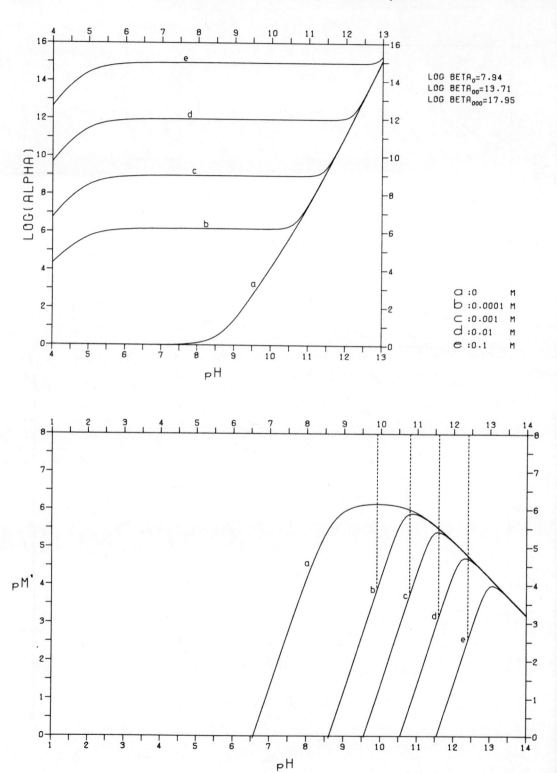

Lanthanum(III)-Tartrate

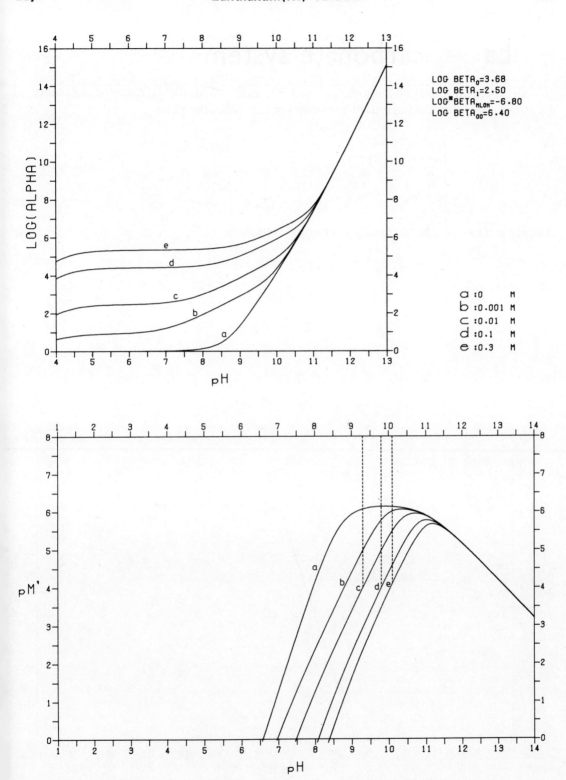

La — carbonate system

An apparent constant, log K_{carb} = 15.9, may be derived from the solubility of $La_2(CO_3)_3$, and this has been used to construct the following plots.

$$\log *\beta_1 = -8.5 \qquad \log K_{carb} = 15.9$$
$$\log *\beta_2 = -17.2$$
$$\log *\beta_3 = -25.9$$
$$\log *\beta_4 = -36.9$$

The log α plots have been repeated for convenience.

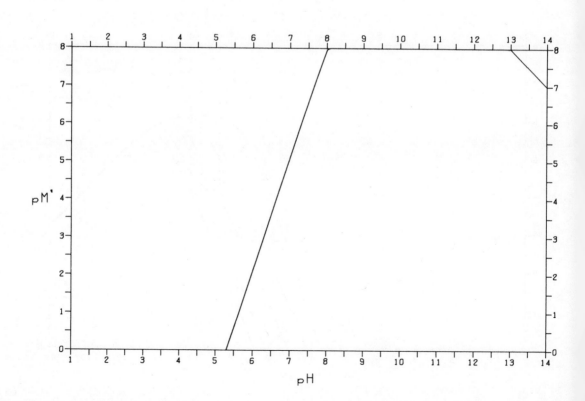

La-Carbonate-Acetylacetone

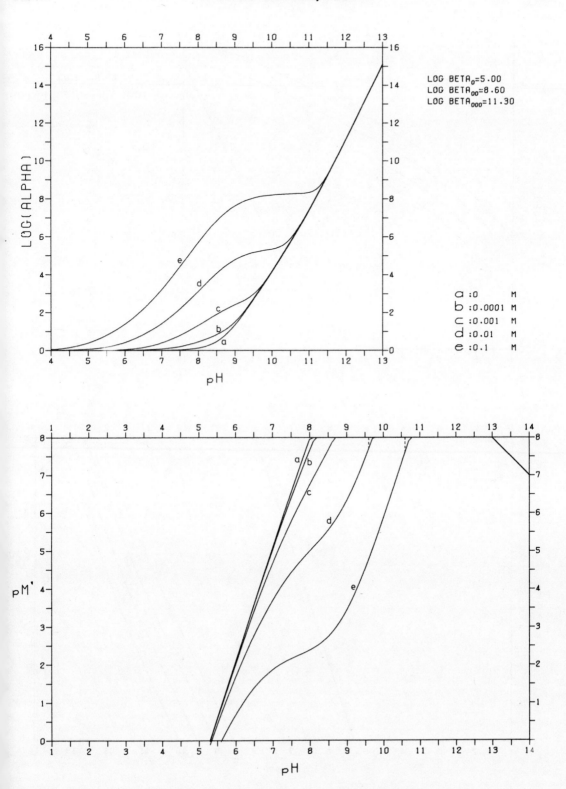

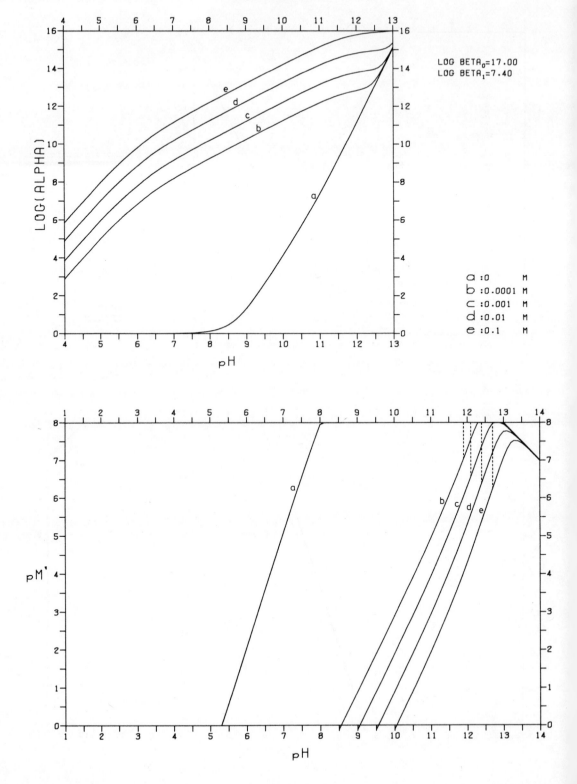

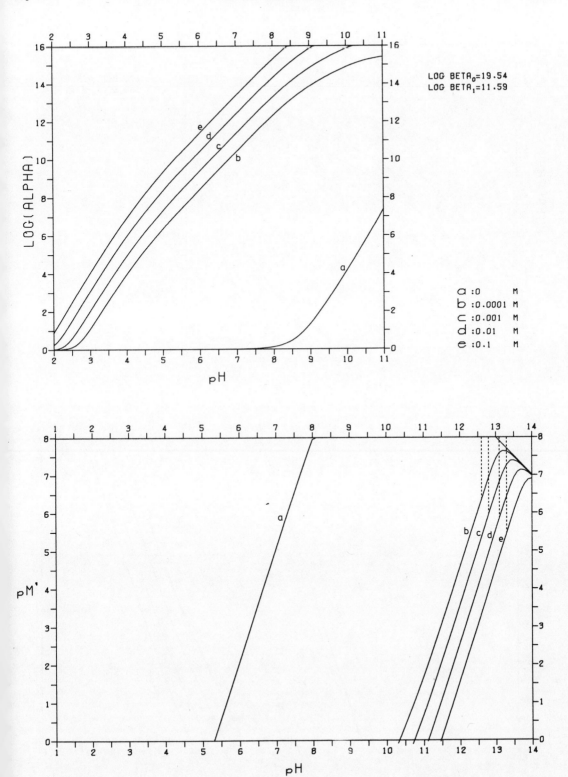

La-Carbonate-EDTA [Ch. 23

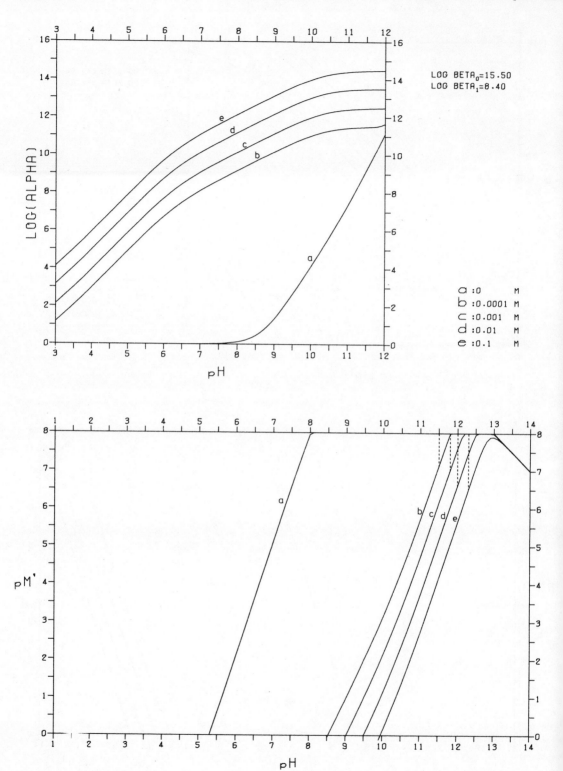

La-Carbonate-(OH)-Quinolinesulphonate

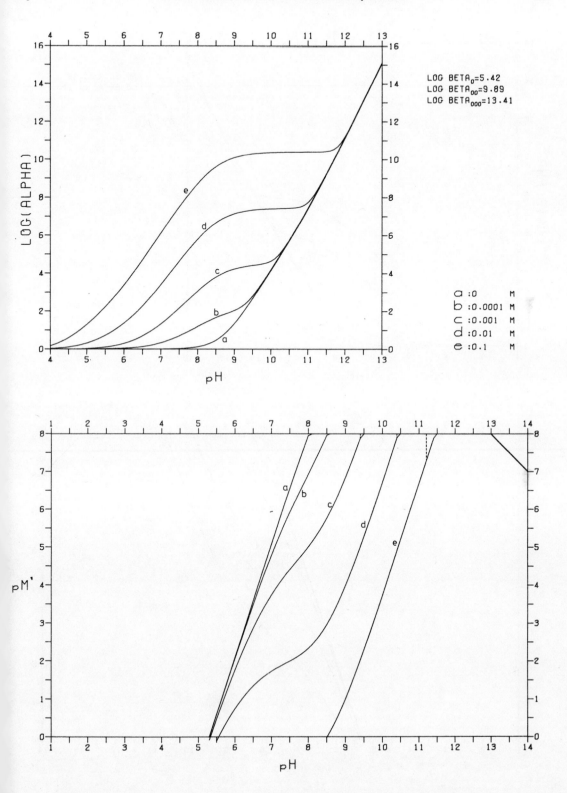

LOG BETA$_0$=5.42
LOG BETA$_{00}$=9.89
LOG BETA$_{000}$=13.41

a : 0   M
b : 0.0001   M
c : 0.001   M
d : 0.01   M
e : 0.1   M

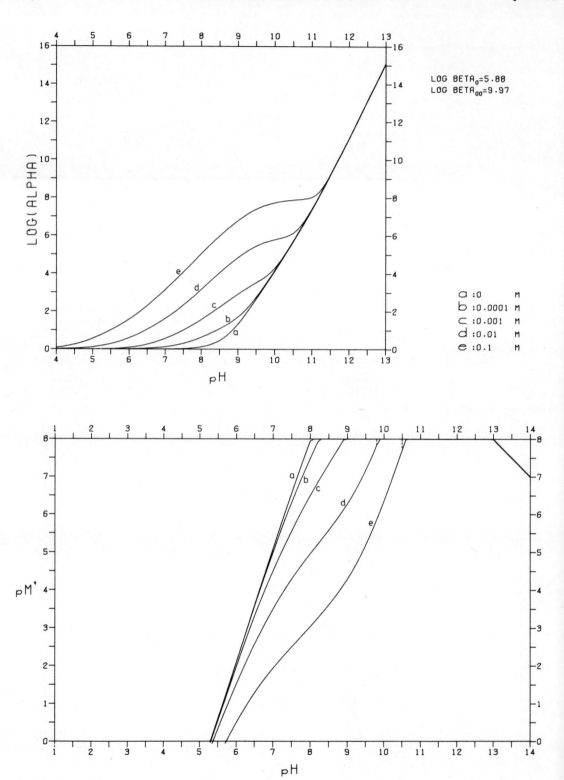

La–Carbonate–Pyridine-2,6-dicarboxylate

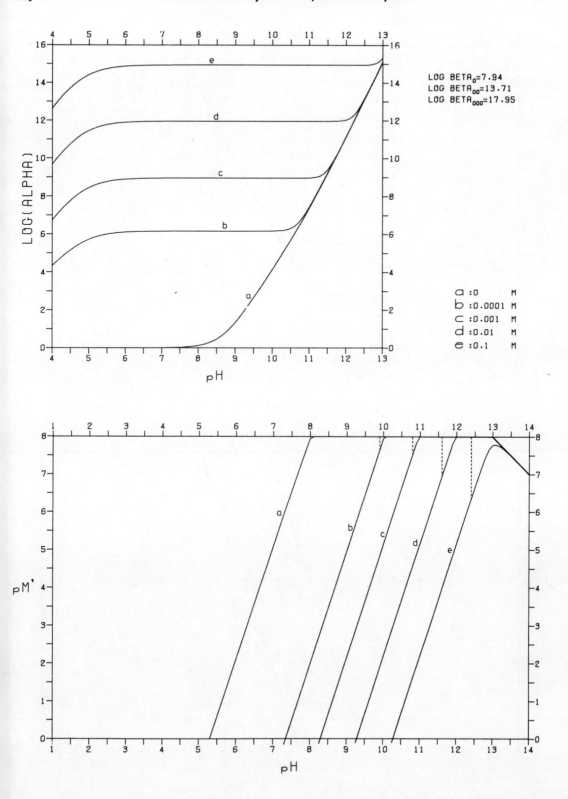

$\log \beta_0 = 7.94$
$\log \beta_{00} = 13.71$
$\log \beta_{000} = 17.95$

a : 0 M
b : 0.0001 M
c : 0.001 M
d : 0.01 M
e : 0.1 M

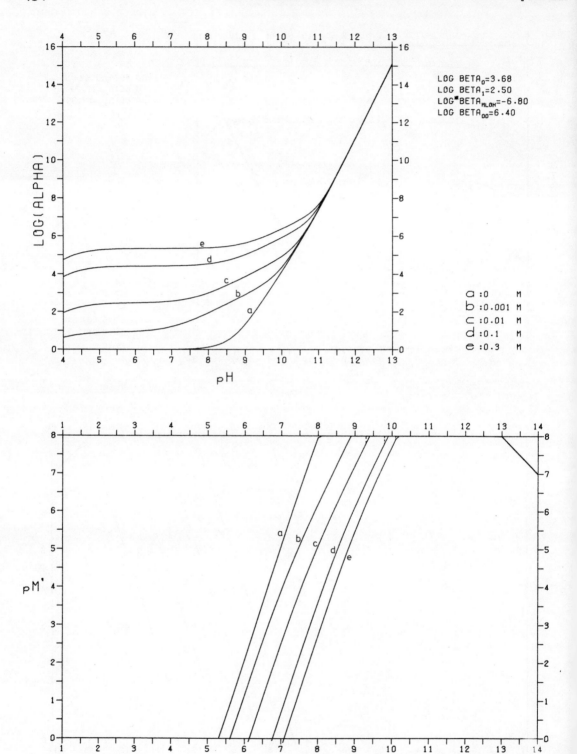

CHAPTER 24

Magnesium (II) Mg

At ordinary concentrations, hydrolysis of Mg^{2+} is detectable just before precipitation of $Mg(OH)_2$ at pH~9.5. $Mg(OH)^+$ is the only product at low concentrations, but its hydrolysis constant is not known accurately. There is some evidence of polycomplexation in concentrated solutions. Although $Mg_4(OH)_4^{4+}$ probably exists, it does not contribute to the 1% polycomplex line.

The $Mg(OH)_2$ precipitated from aqueous solution is reasonably well defined. The solubility product depends to some extent on the particle size, but this may be neglected. The following data were used for construction of the plots.

$$\log {}^*\beta_1 = -12 \qquad \log {}^*\beta_{44} = -39.8$$
$$\log {}^*\beta_2 = -27 \qquad \log {}^*K_{s0} = 18$$

Since the apparent constant for precipitation of $MgCO_3$ has a value close to that of $^*K_{s0}$, the precipitate from air-saturated magnesium solutions may be expected to contain considerable amounts of carbonate, but this does not change the plots.

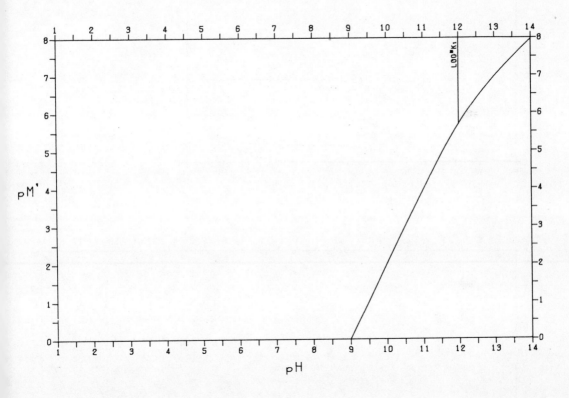

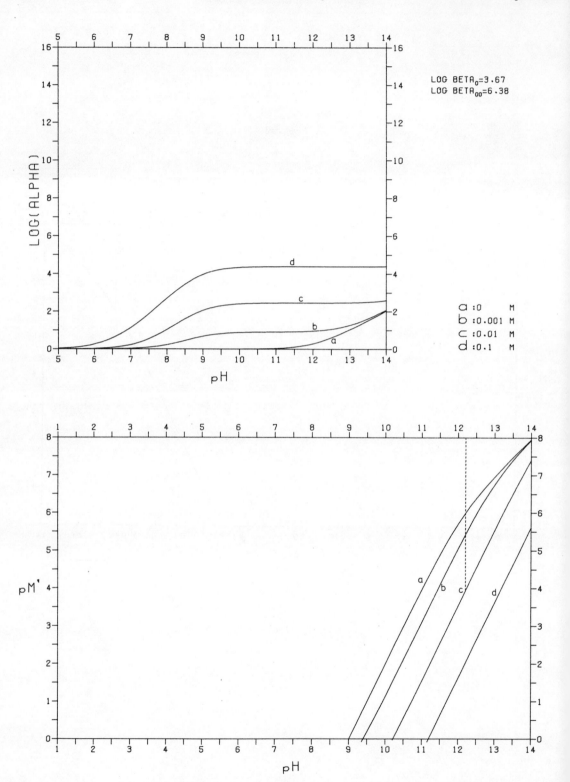

Magnesium(II)-Citrate

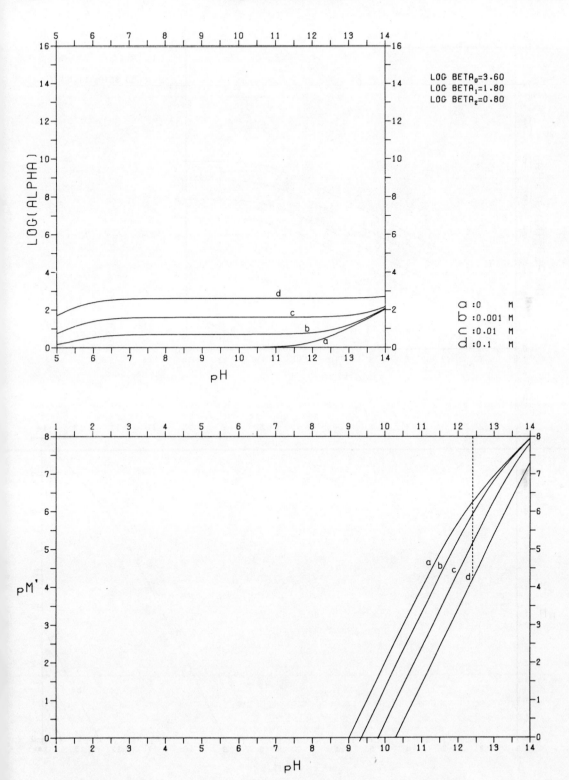

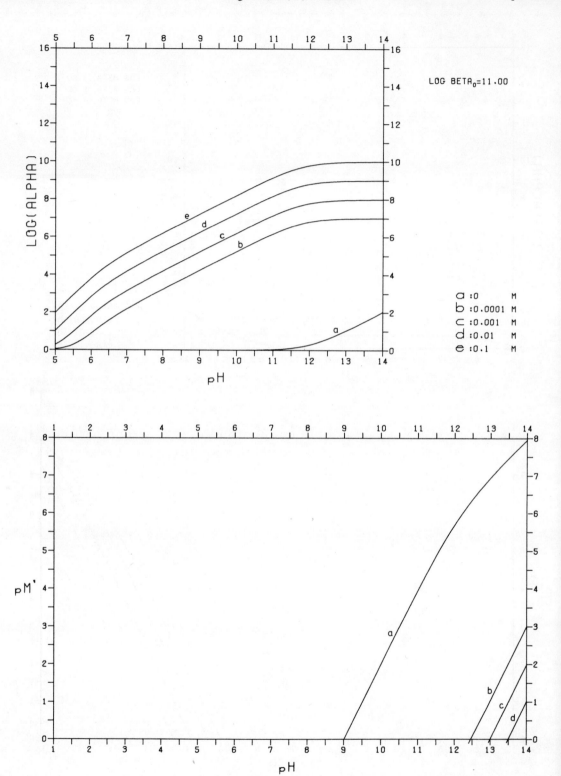

Magnesium(II)-DTPA

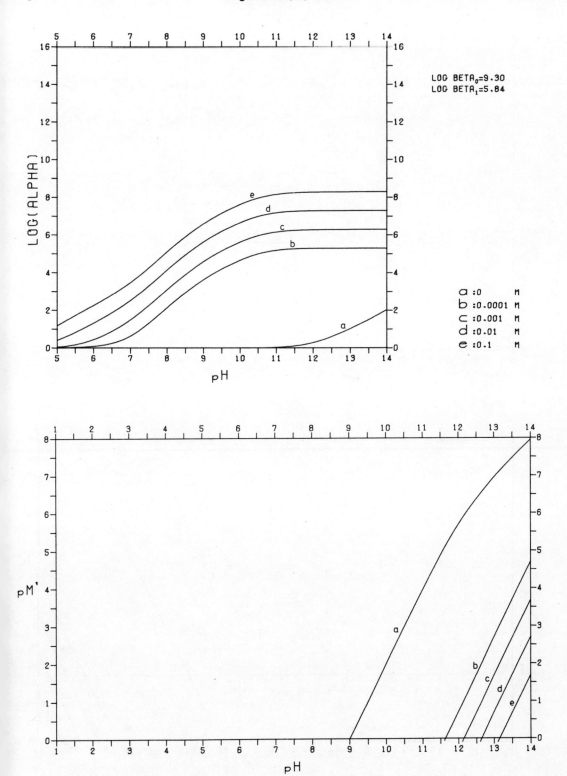

LOG BETA$_0$=9.30
LOG BETA$_1$=5.84

a :0   M
b :0.0001 M
c :0.001   M
d :0.01   M
e :0.1   M

Magnesium(II)–EDTA

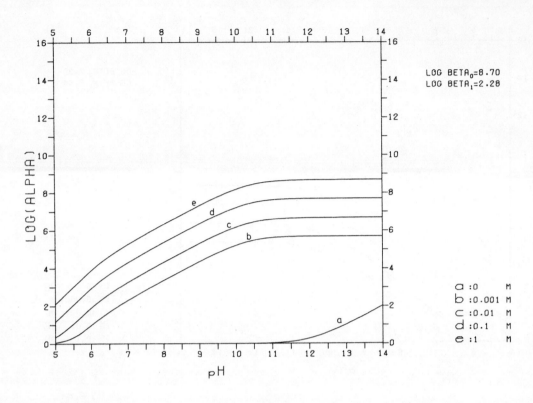

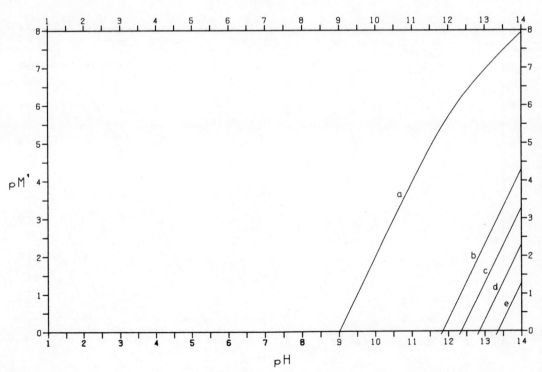

Magnesium(II)–EGTA

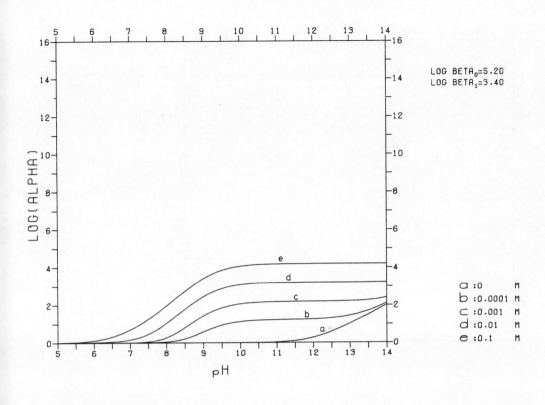

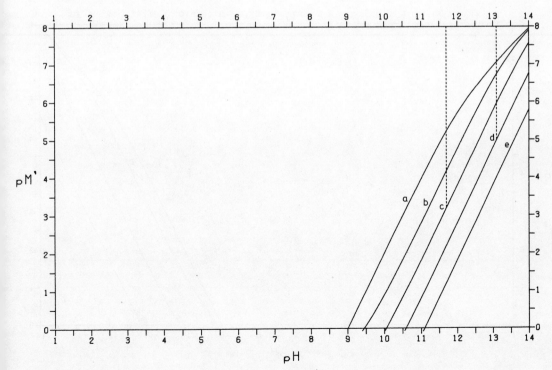

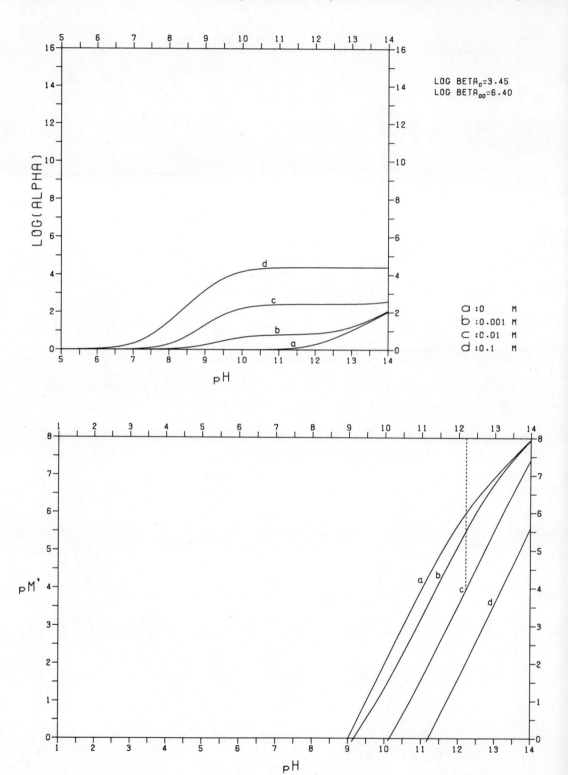

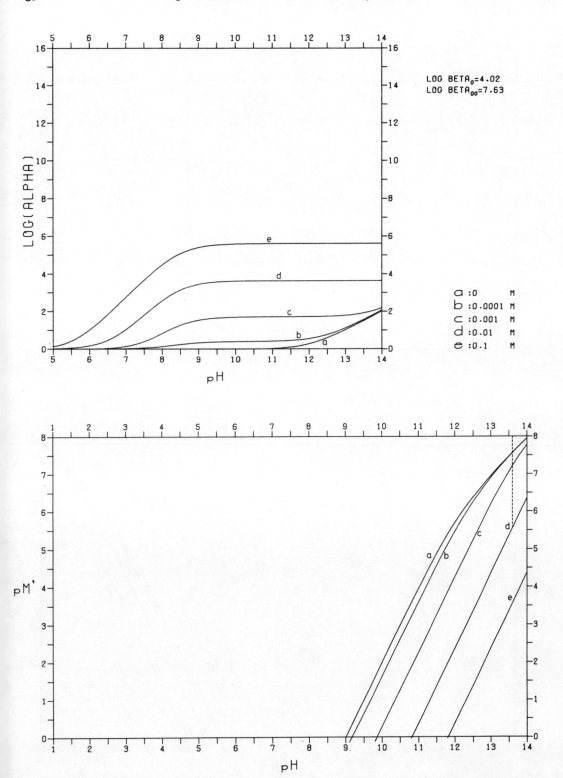

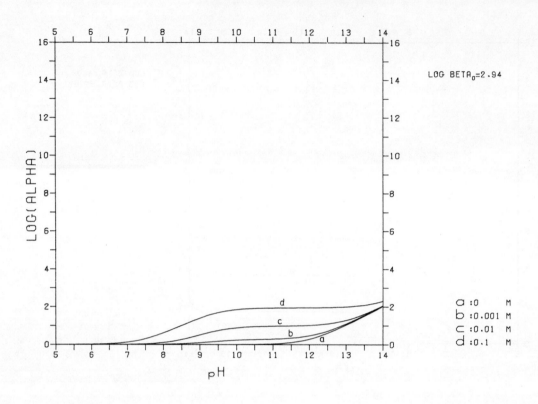
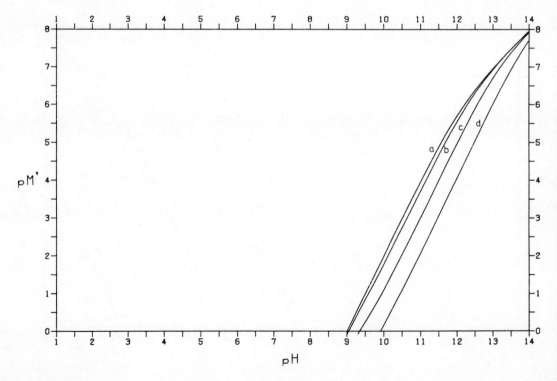

Magnesium(II)-Oxalate

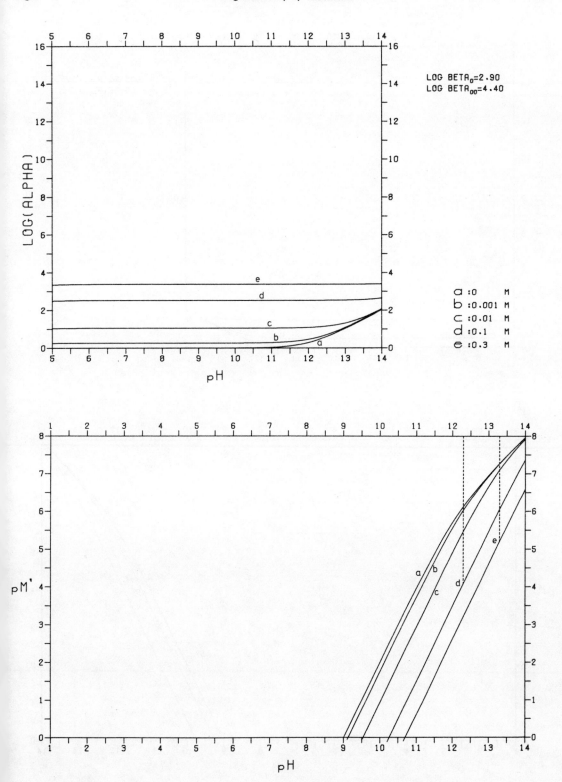

Magnesium(II)–Pyridine-2,6-dicarboxylate [Ch. 24]

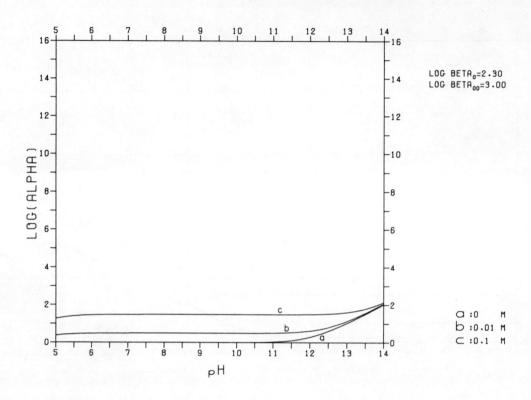

LOG BETA$_0$=2.30
LOG BETA$_{00}$=3.00

a : 0 M
b : 0.01 M
c : 0.1 M

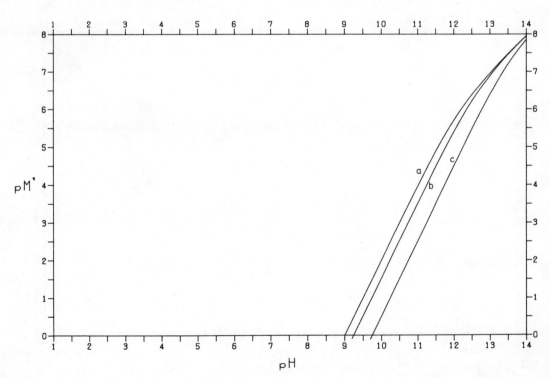

Magnesium(II)-Sulphate

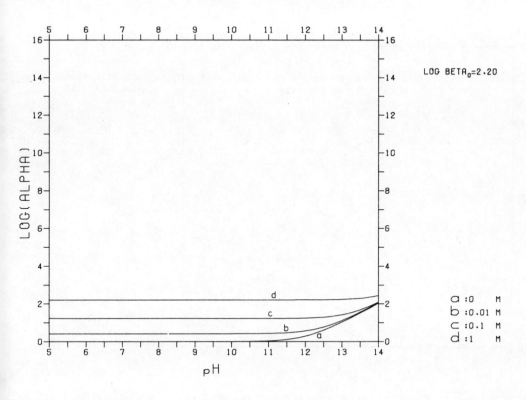

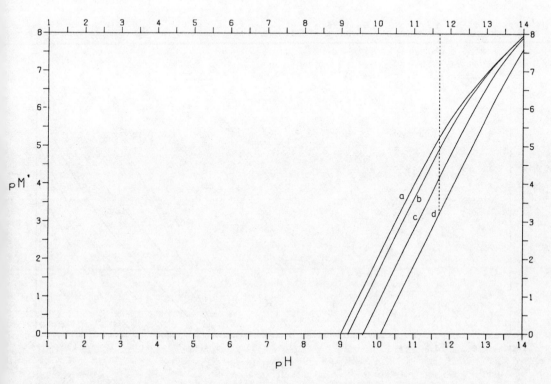

Magnesium(II)–Tiron [Ch. 24

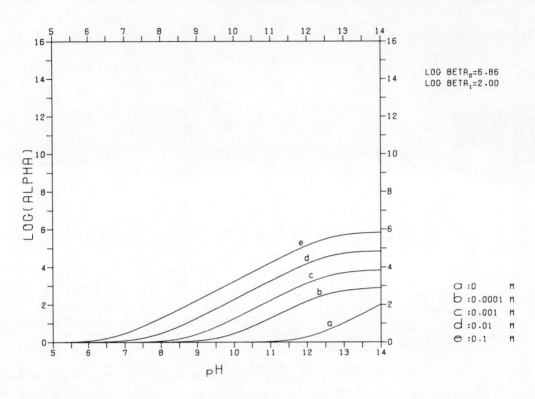

LOG BETA$_0$=6.86
LOG BETA$_1$=2.00

a : 0 M
b : 0.0001 M
c : 0.001 M
d : 0.01 M
e : 0.1 M

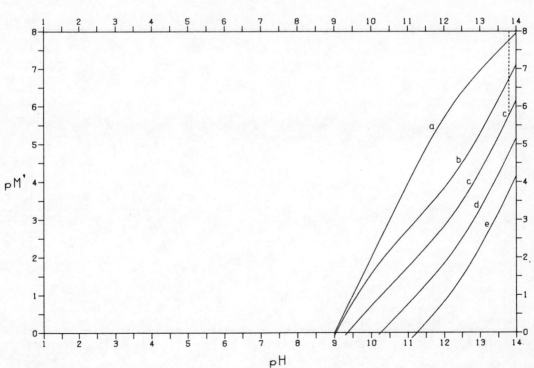

CHAPTER 25

Manganese (II)   Mn

The hydrolysis constant $*\beta_1$ of Mn^{2+} is well established. The value of $*\beta_4$ has been determined from the solubility in alkaline medium. There are no reliable estimates, however, of the stability constants of $Mn(OH)_2(aq)$ and $Mn(OH)_3^-$. The values given below are deduced from $*\beta_1$ and $*\beta_4$ by assuming a regular progression in the stepwise formation constants. Mn^{2+} forms polycomplexes for which the compositions Mn_2OH^{3+} and $Mn_2(OH)_3^+$ have been proposed, with log $*\beta_{12} = -10.5$ and log $*\beta_{32} = -24.5$. However, if this is correct, $Mn_2(OH)_3^+$ should predominate near pH 11 in very dilute solutions (10^{-7}–10^{-10} M). This suggests a serious error in the assignments, so $Mn_2(OH)_3^+$ will not be considered here. Examination of the original measurement [1] has shown that the observations can just as well be explained by presence of the species $Mn_3(OH)_3^{3+}$ and this species is adopted here as being the most consistent with the evidence.

Manganous hydroxide precipitated from solution is ill-defined, but its solubility product seems to be independent of its state and is known accurately. The following data were used:

$$\log *\beta_1 = -10.8 \qquad \log *\beta_{33} = -24.4$$
$$\log *\beta_2 = -22.4$$
$$\log *\beta_3 = -34.8$$
$$\log *\beta_4 = -47.9 \qquad \log *K_{s0} = 15.4$$

In air-saturated solutions, manganous carbonate precipitates at lower pH than manganous hydroxide (for the relevant plots, see p. 474).

Mn^{2+} is the most stable oxidation state of the element in aqueous medium, but at some values of the redox potential of the solution, Mn^{2+} ions can be in equilibrium with $Mn(OH)_2$, Mn_3O_4, Mn_2O_3, and MnO_2. Because of supersaturation, Mn_3O_4 and Mn_2O_3 rarely precipitate from solution. Apart from $Mn(OH)_2$, only the very stable MnO_2 is formed frequently.

Manganese(II)

In the absence of complexing ligands the equilibrium $Mn^{2+} \rightleftharpoons Mn(OH)_2(s)$ can be established at potentials below $E = (774 - 59.1 \text{ pH})$ mV. In air-saturated solutions, an apparent solubility product of $\log {}^*K_{s0} = 0$ can be assumed, indicating that only in strongly acidic solution will Mn^{2+} not be precipitated as MnO_2 (pH<0). The formation of MnO_2 in weakly acidic media is usually very slow, but it may be catalysed by other ions (see also page 27).

The limits of potential given above shift when complexing ligands are present, but in theory these shifts are insufficient to prevent the precipitation of MnO_2. Precipitation does not occur in practice because the formation reaction is very slow in the presence of excess of DCTA and EDTA.

[1] S. Fontana and F. Brito, *Inorg. Chim. Acta* 1968, 2, 179.

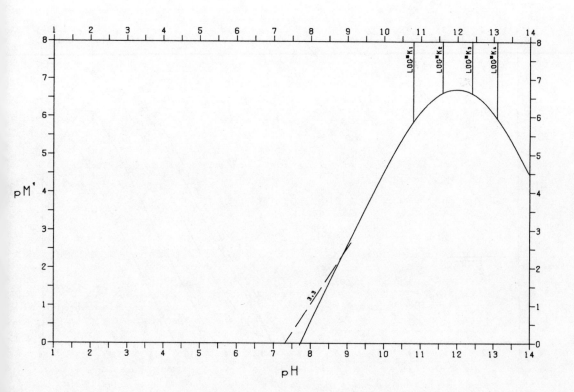

454   **Manganese(II)-Acetylacetone**   [Ch. 25

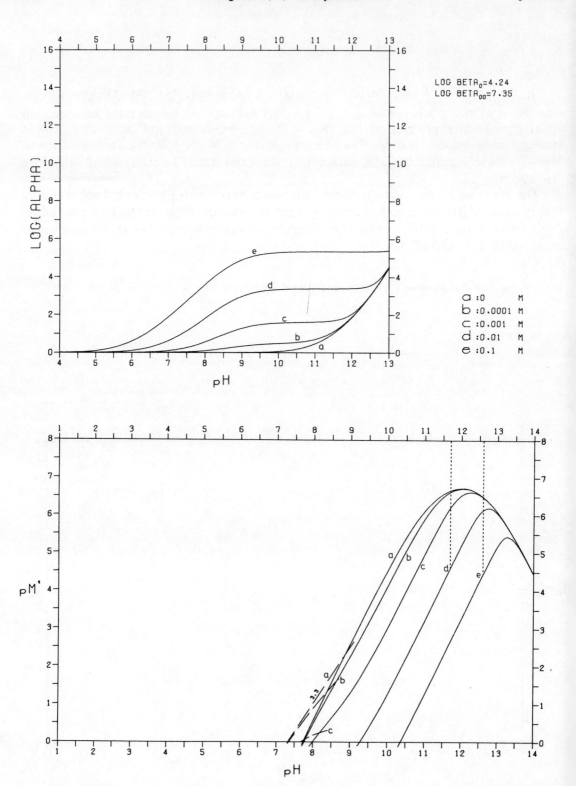

Manganese(II)−Ammonia

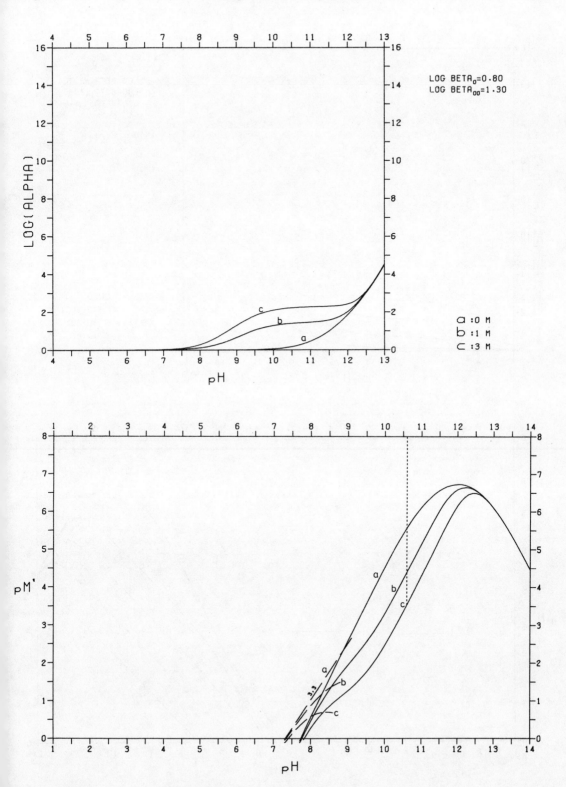

LOG BETA₀=0.80
LOG BETA₀₀=1.30

a : 0 M
b : 1 M
c : 3 M

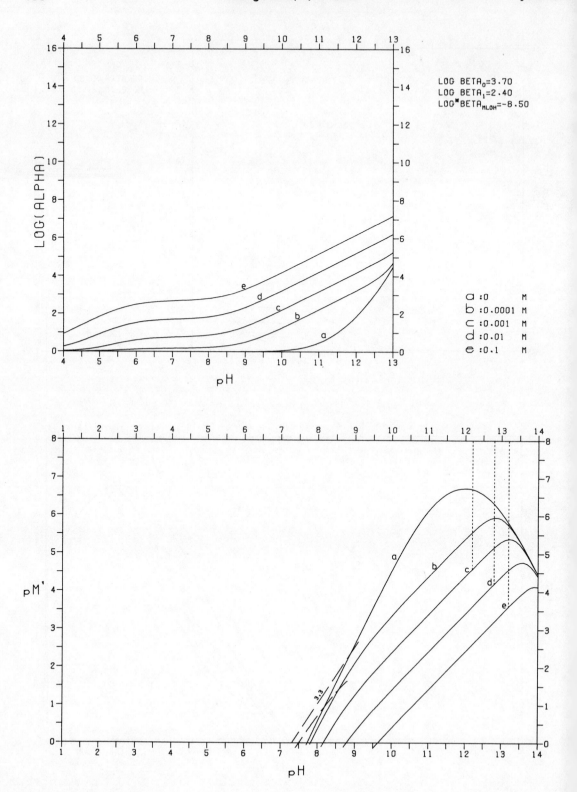

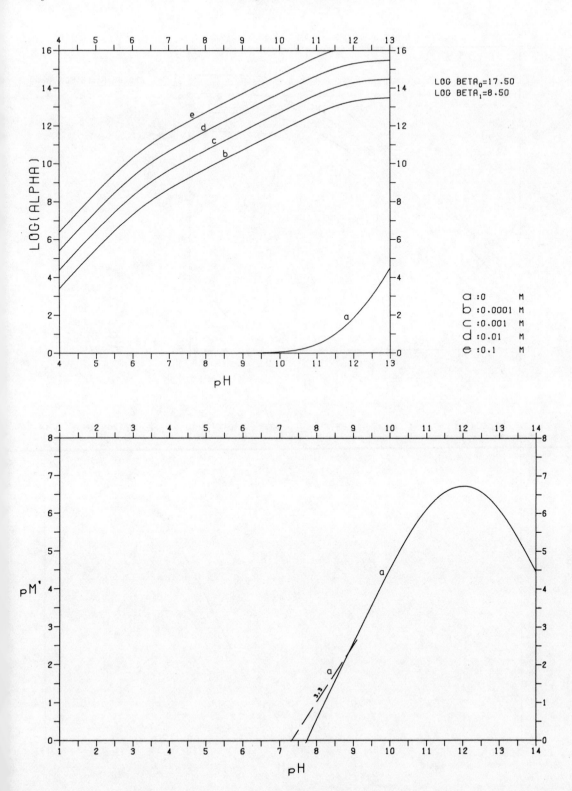

Manganese(II)-DTPA [Ch. 25

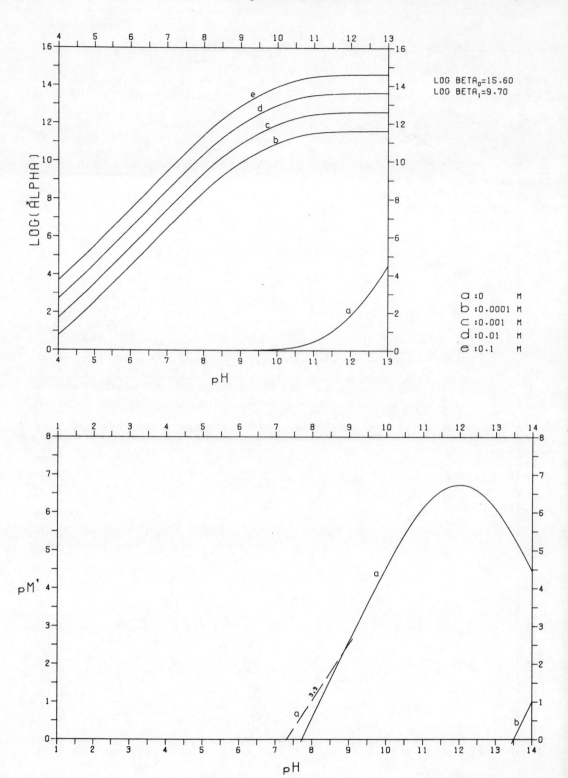

Manganese(II)–EDTA

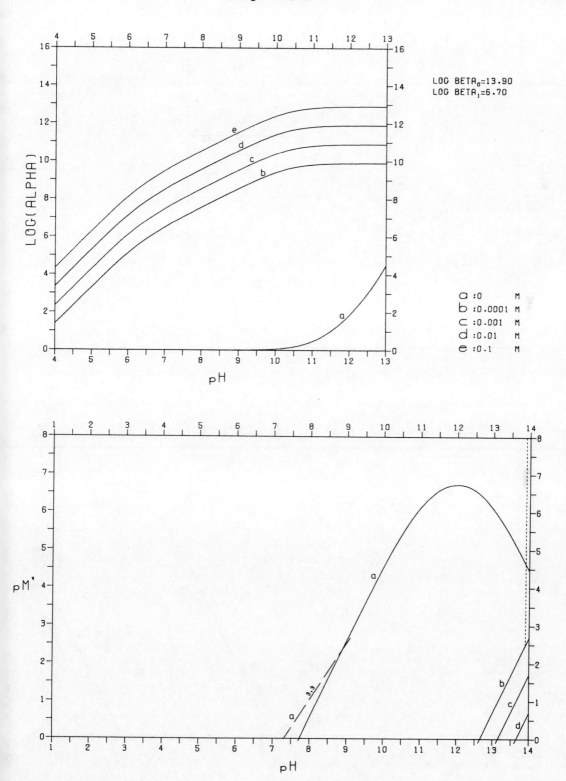

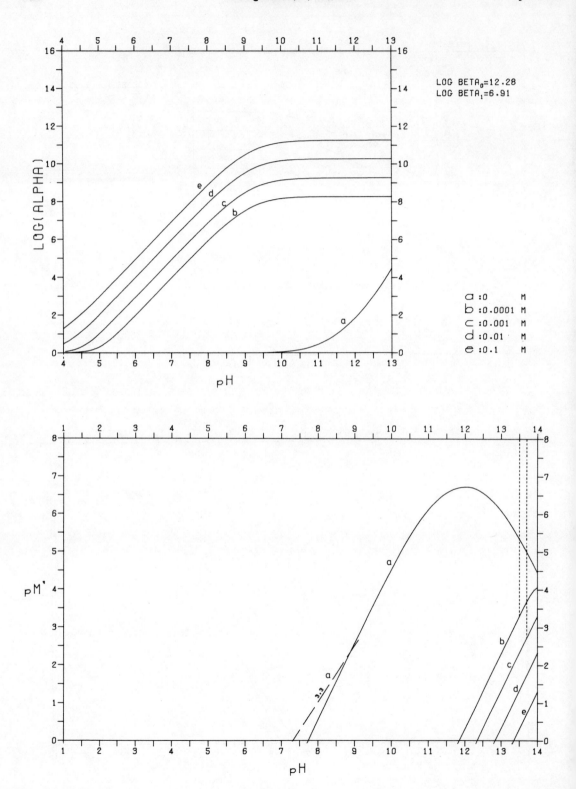

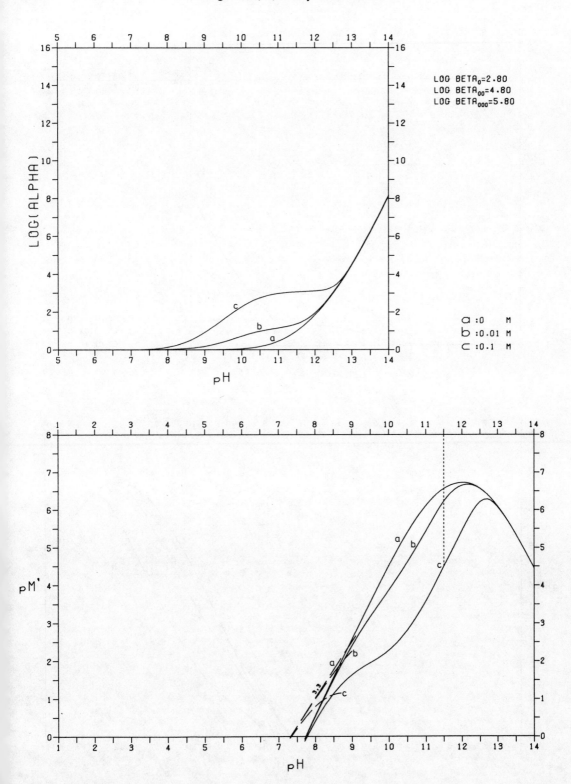

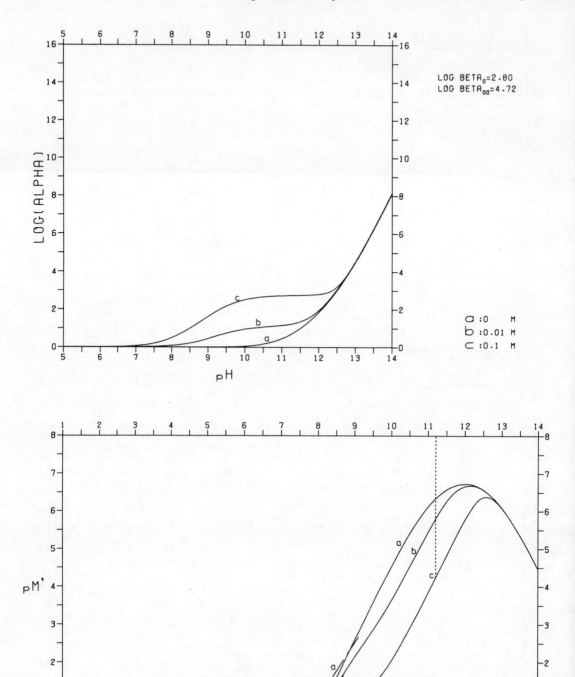

Manganese(II)-(OH)-Quinolinesulphonate

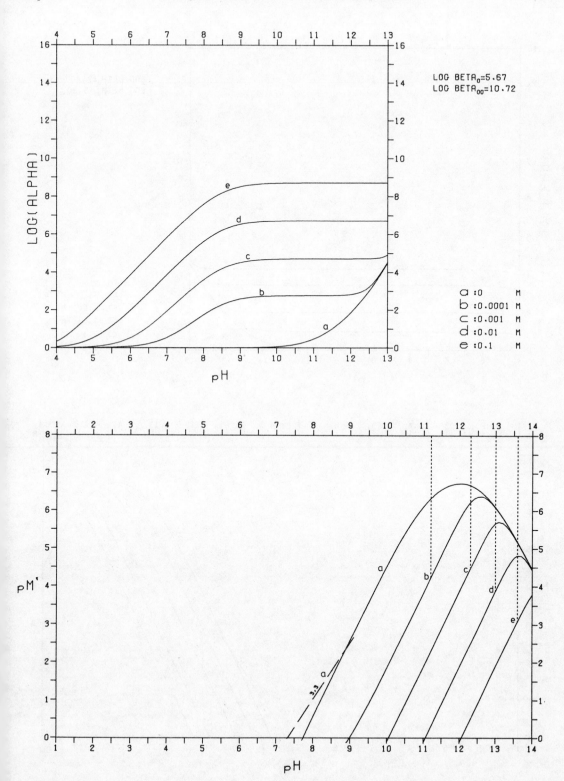

464   **Manganese(II)–Oxalate**   [Ch. 25

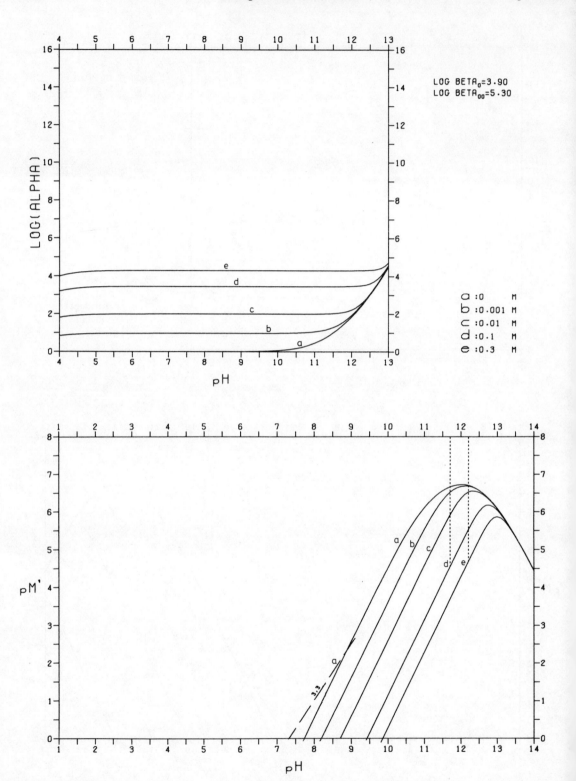

Manganese(II)-Pyridine-2,6-dicarboxylate

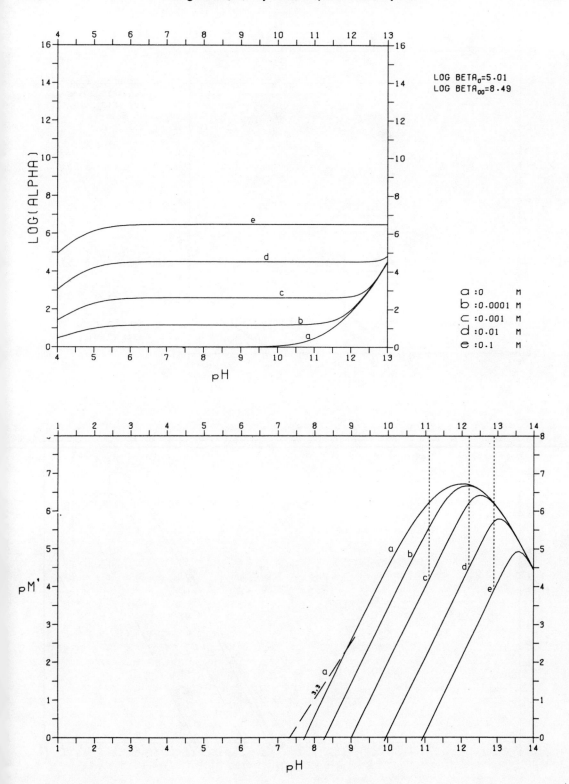

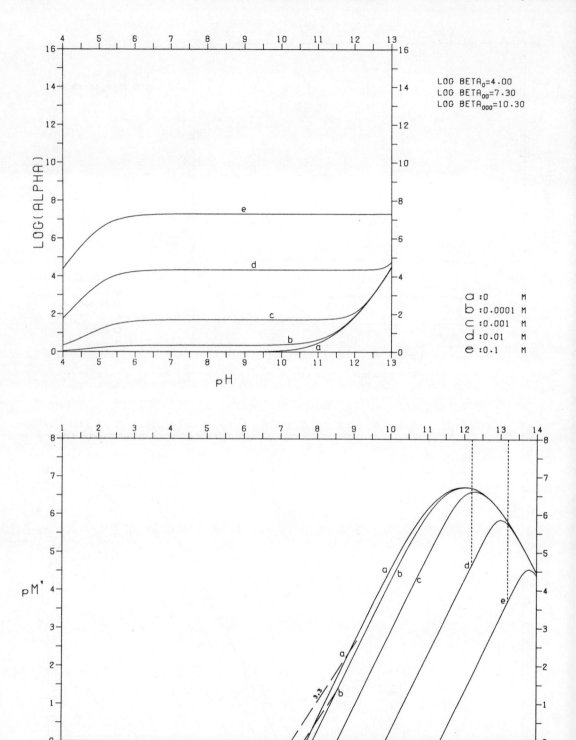

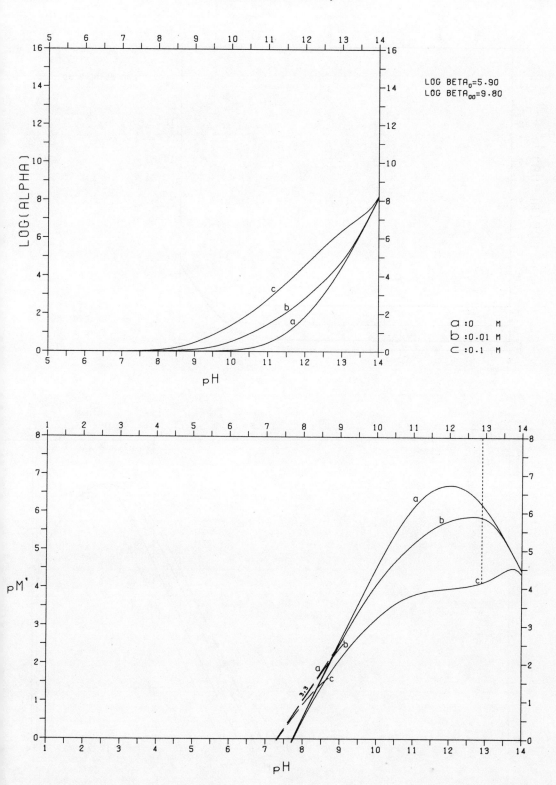

Manganese(II) Sulphate

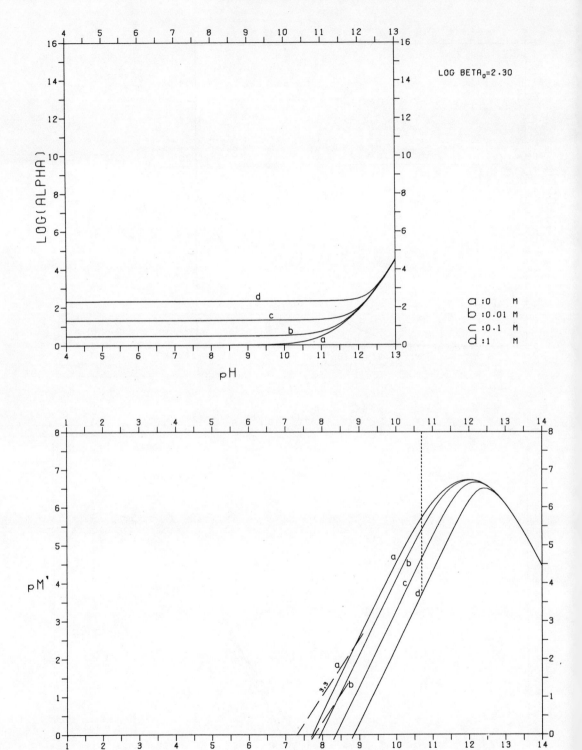

Manganese(II)-5-Sulphosalicylate

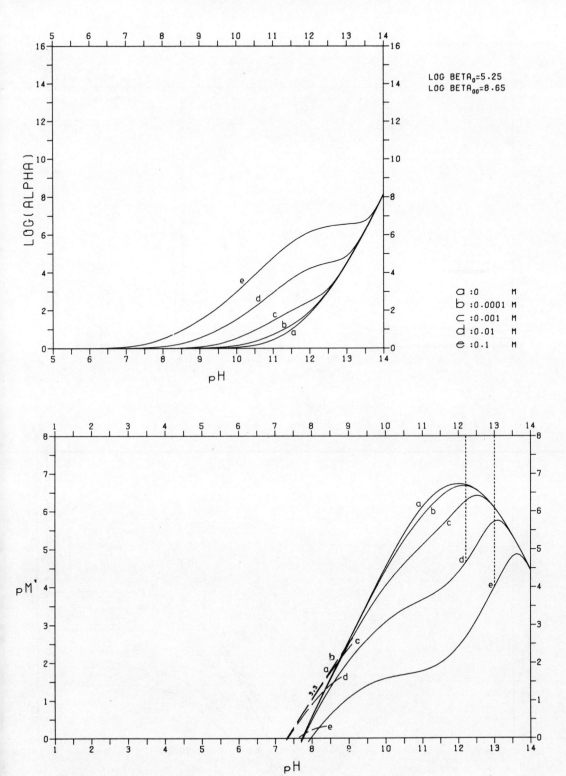

Manganese(II)-Tartrate

[Ch. 25

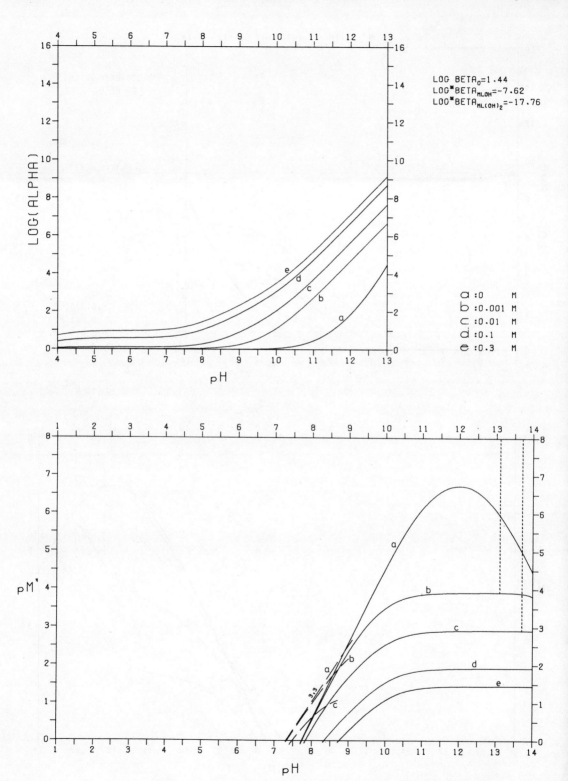

Manganese(II)-Tetren

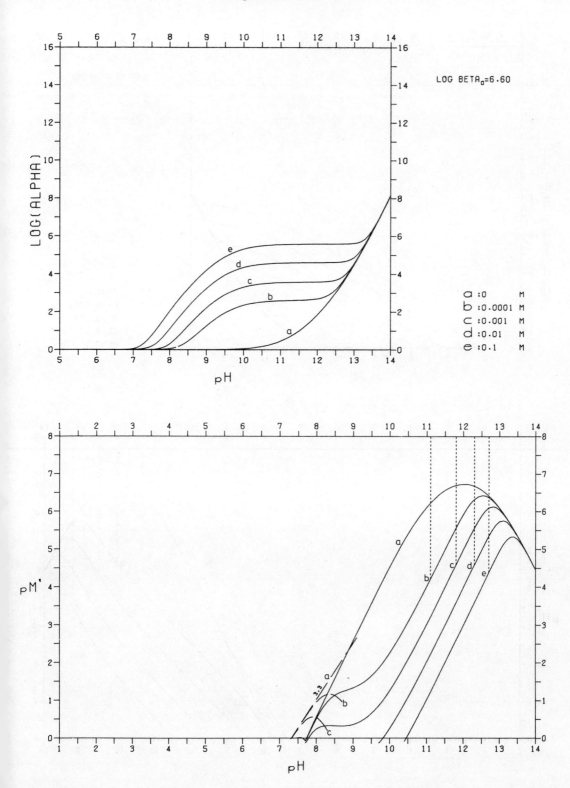

LOG BETA$_0$=6.60

a : 0 M
b : 0.0001 M
c : 0.001 M
d : 0.01 M
e : 0.1 M

472 **Manganese(II)-Tiron** [Ch. 25

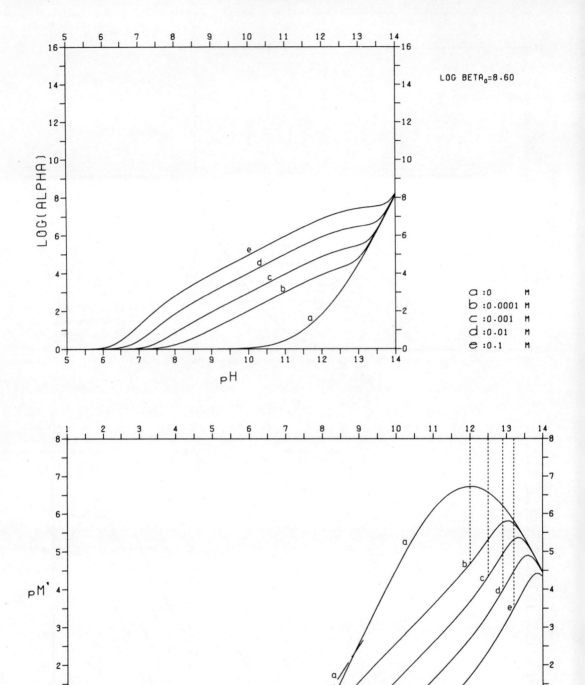

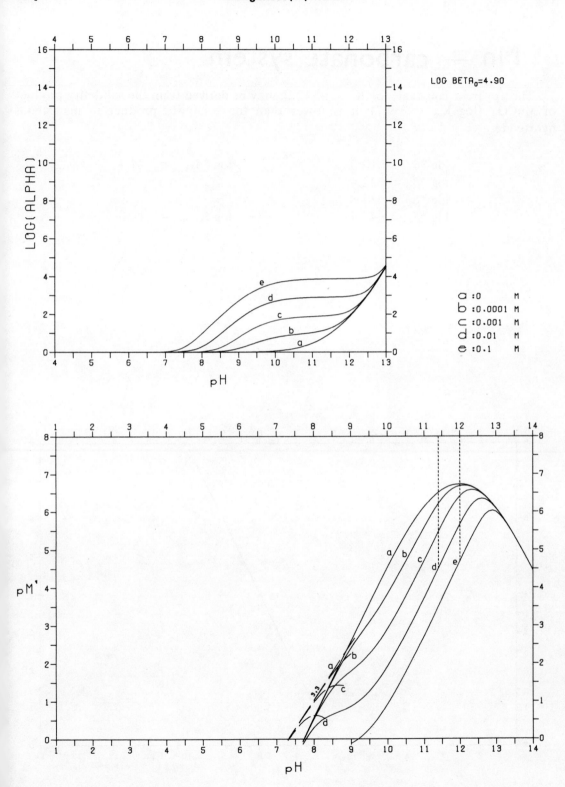

Mn — carbonate system

The apparent constant, log K_{carb} = 12.0, may be derived from the solubility product of $MnCO_3$ (log K_{s0} = −9.7); it is lower than the solubility product of manganous hydroxide.

$$\log {}^*\beta_1 = -10.8 \qquad \log {}^*\beta_{33} = -24.4$$
$$\log {}^*\beta_2 = -22.4$$
$$\log {}^*\beta_3 = -34.8$$
$$\log {}^*\beta_4 = -47.9 \qquad \log K_{carb} = 12.0$$

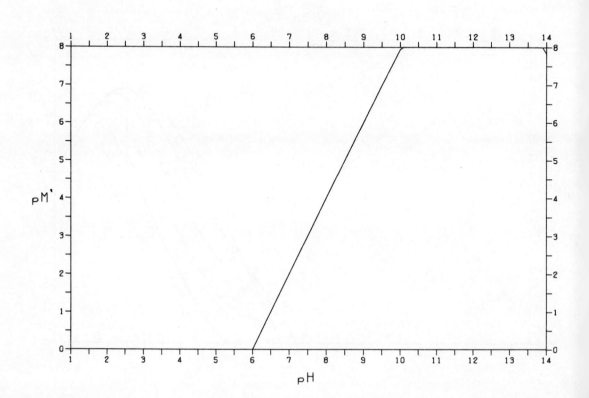

Mn-Carbonate-Acetylacetone

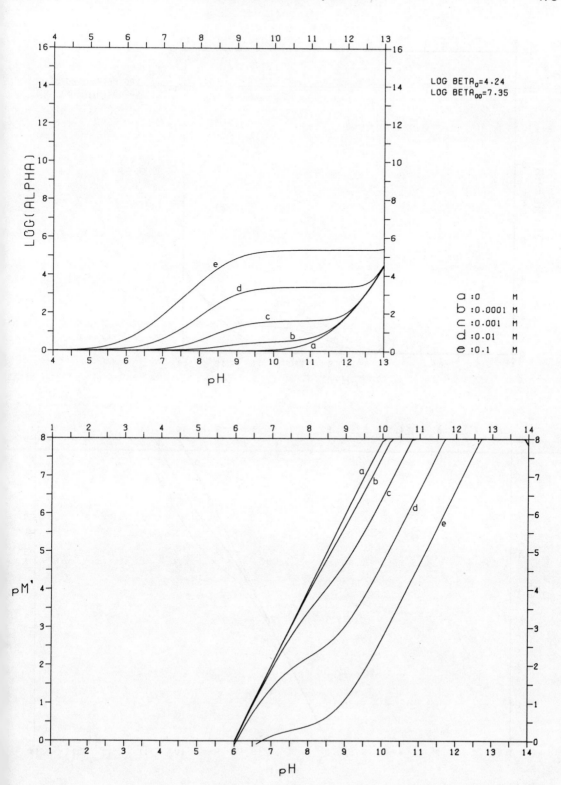

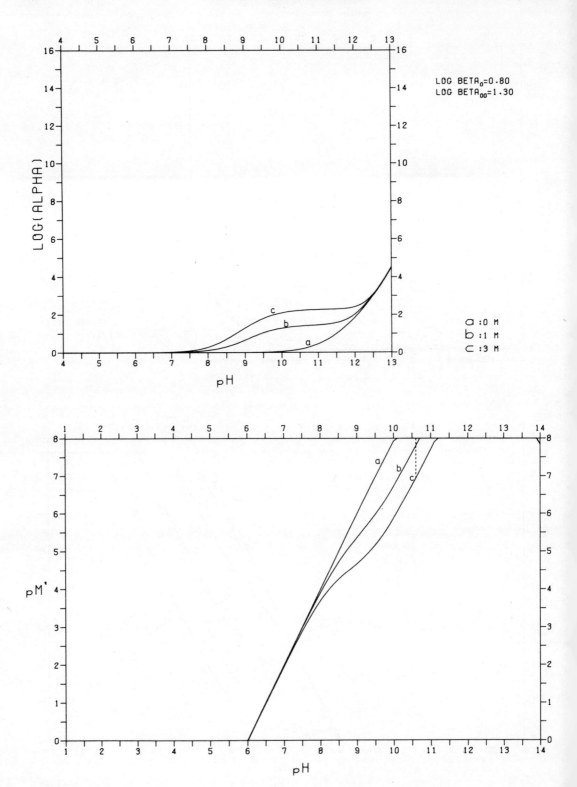

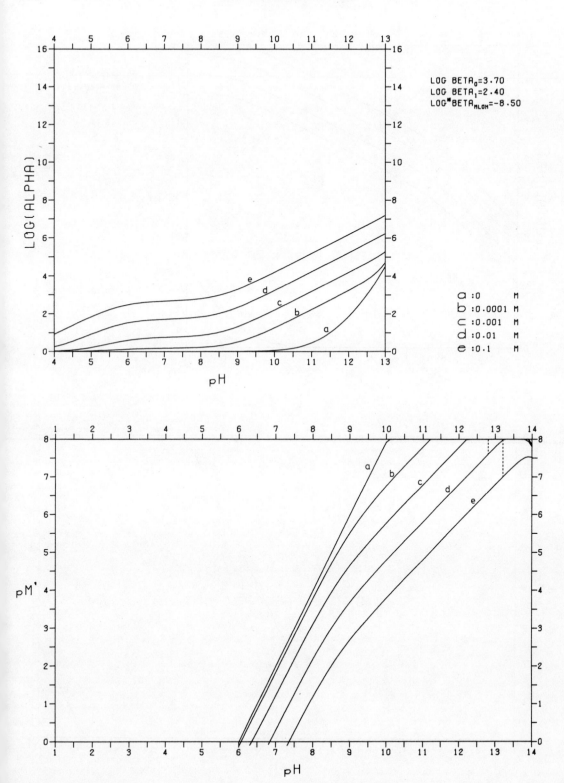

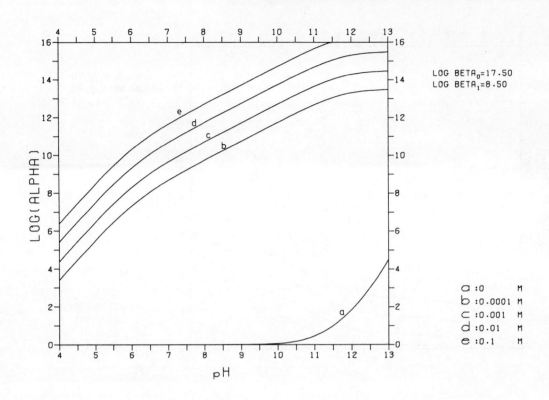

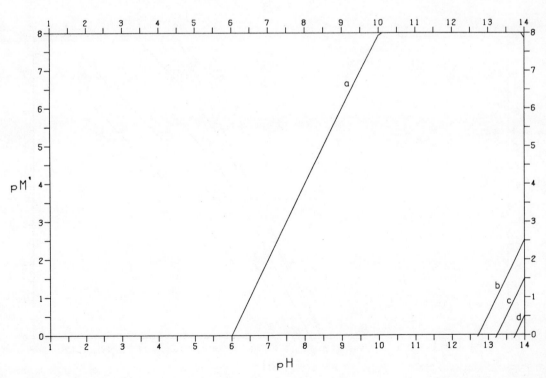

Mn-Carbonate-DTPA

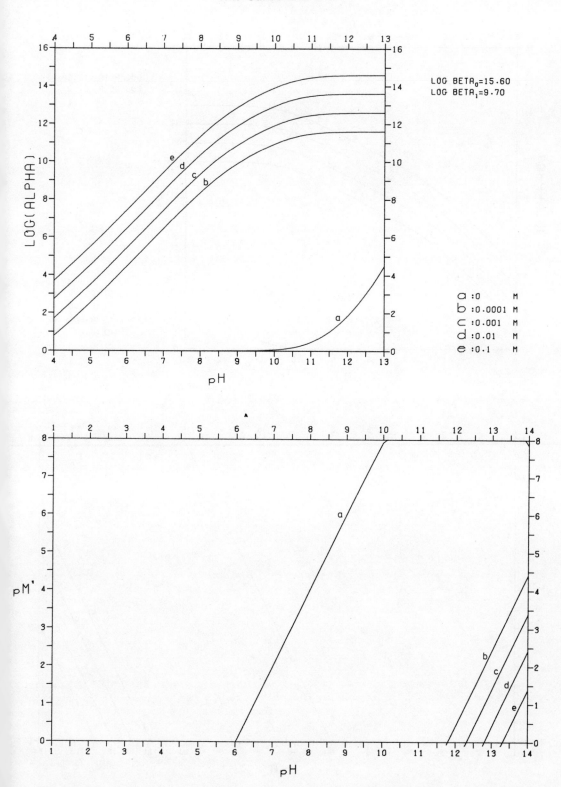

Mn-Carbonate-EDTA [Ch. 25

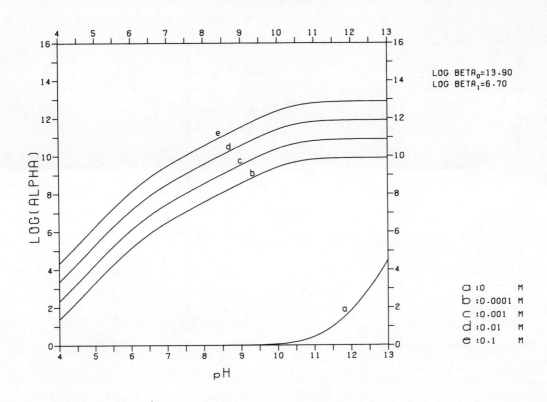

LOG BETA$_0$=13.90
LOG BETA$_1$=6.70

a : 0   M
b : 0.0001 M
c : 0.001 M
d : 0.01   M
e : 0.1   M

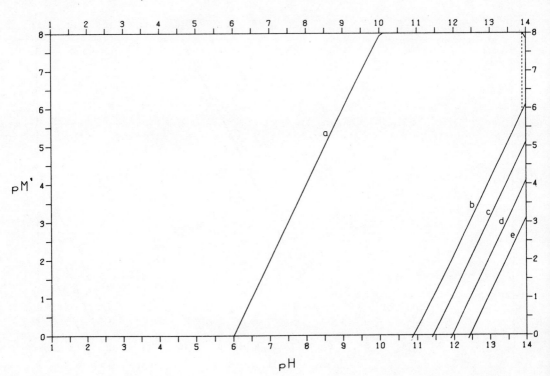

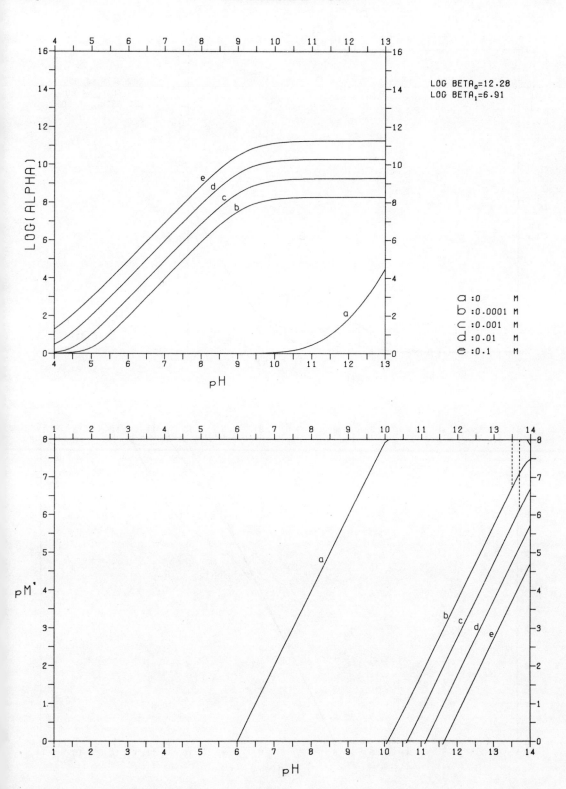

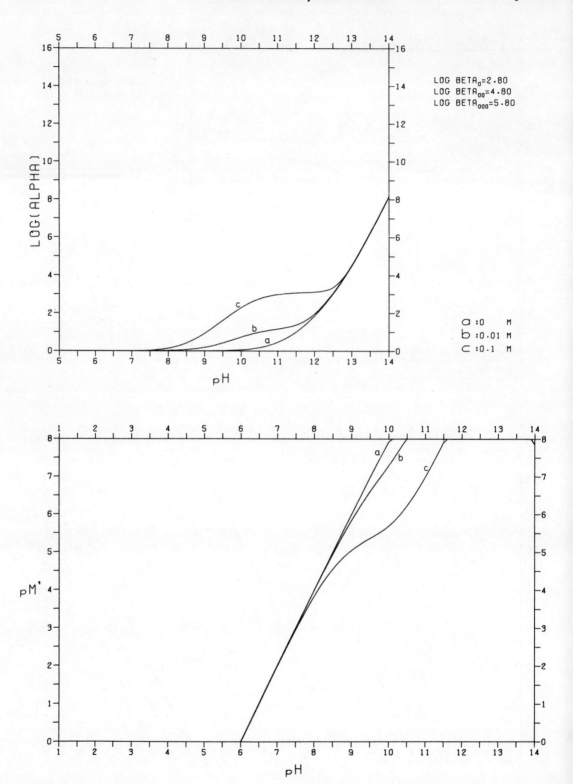

Mn-Carbonate-Glycine

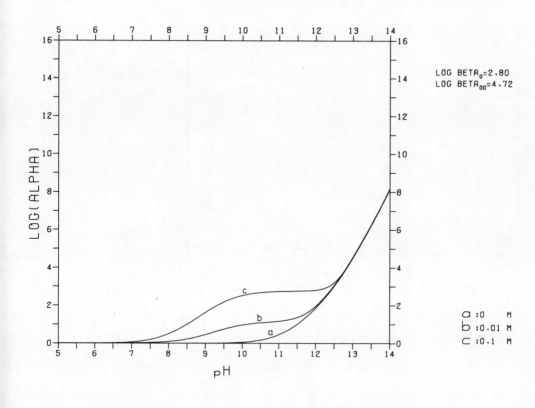

LOG BETA$_0$=2.80
LOG BETA$_{00}$=4.72

a : 0   M
b : 0.01 M
c : 0.1 M

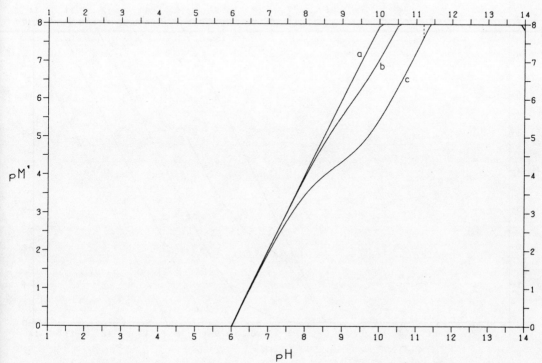

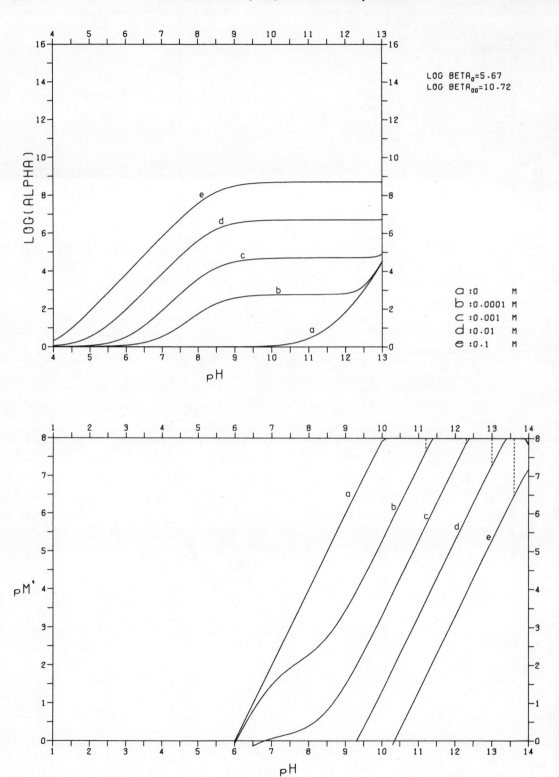

Mn-Carbonate-Oxalate

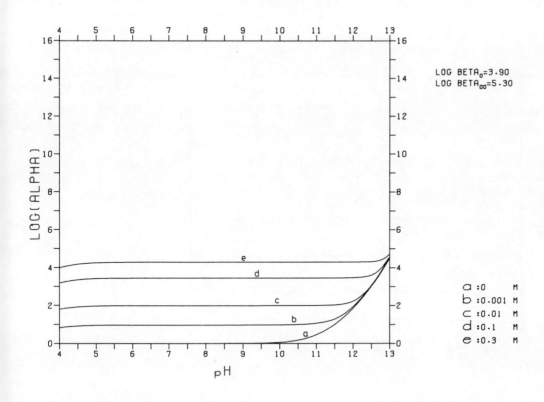

LOG BETA$_0$=3.90
LOG BETA$_{00}$=5.30

a : 0 M
b : 0.001 M
c : 0.01 M
d : 0.1 M
e : 0.3 M

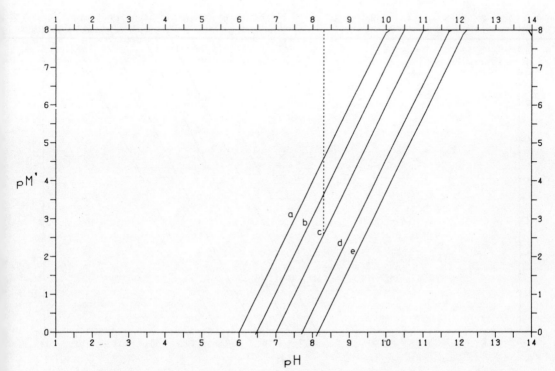

Mn–Carbonate–Pyridine-2,6-dicarboxylate

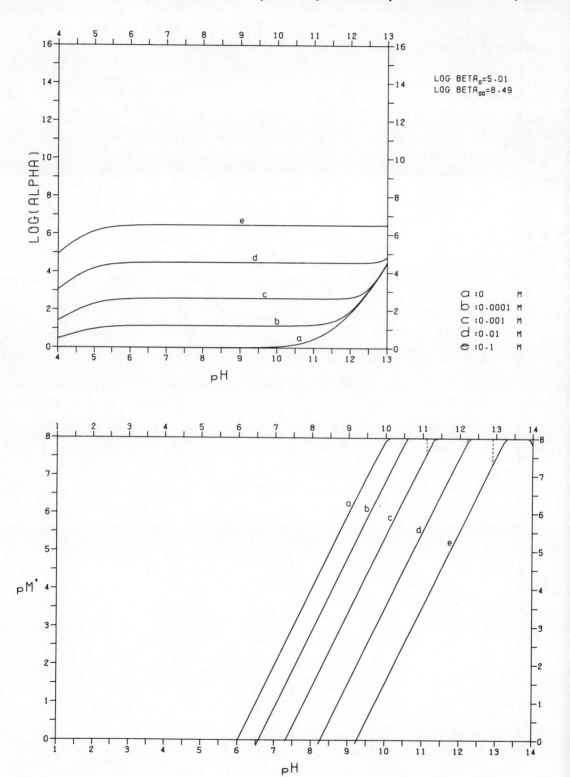

Mn-Carbonate-1,10-Phenanthroline

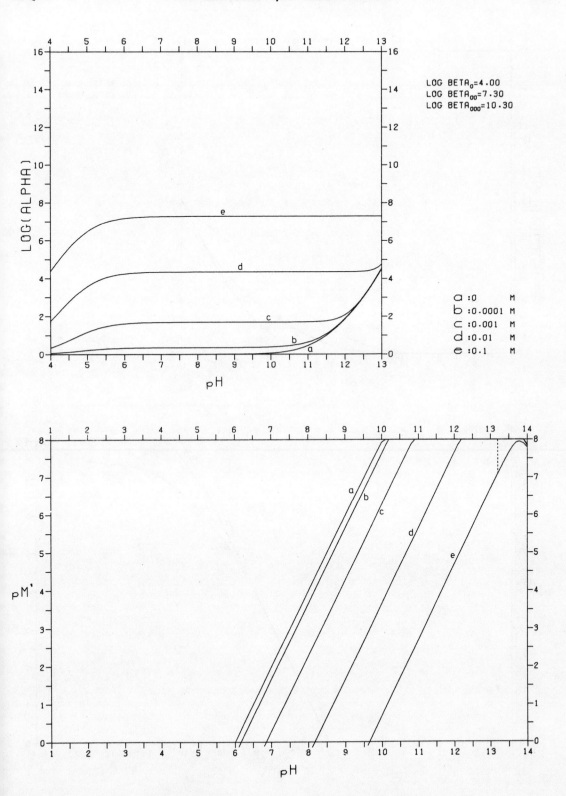

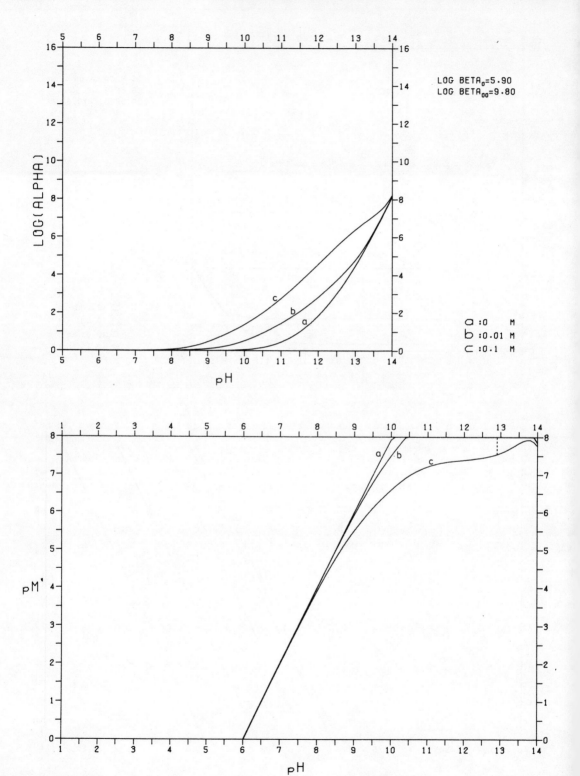

Mn–Carbonate–Sulphate

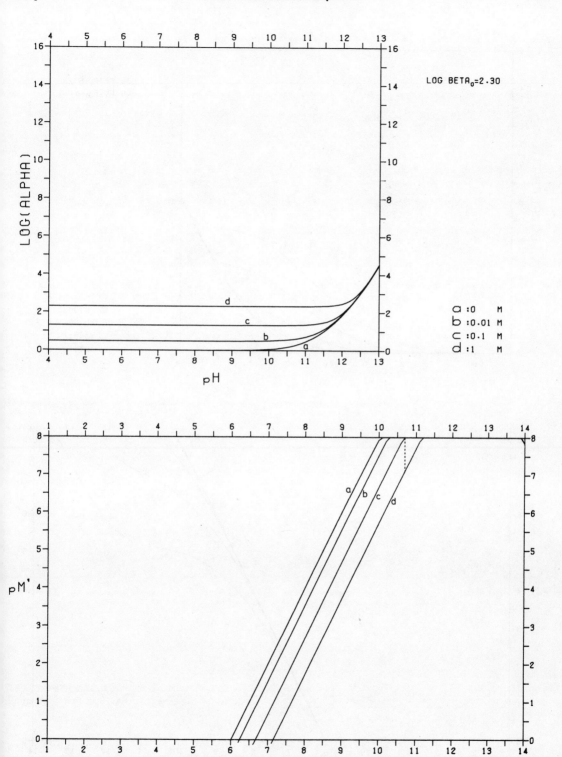

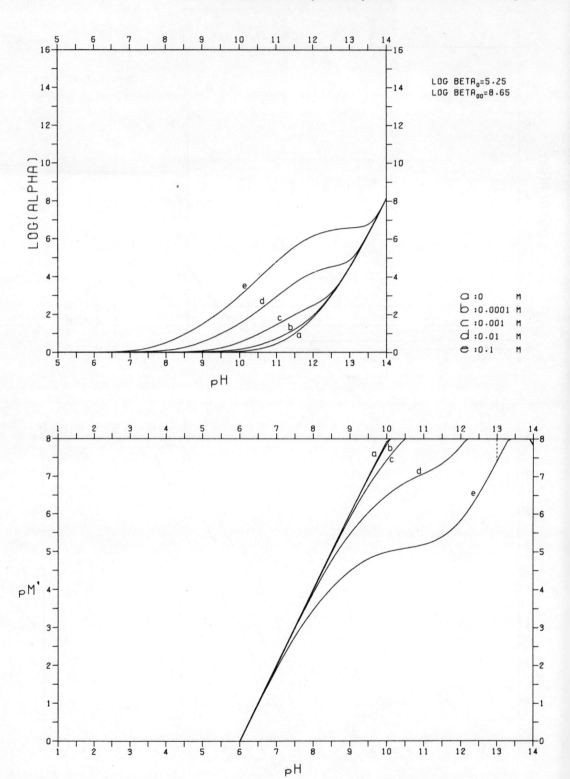

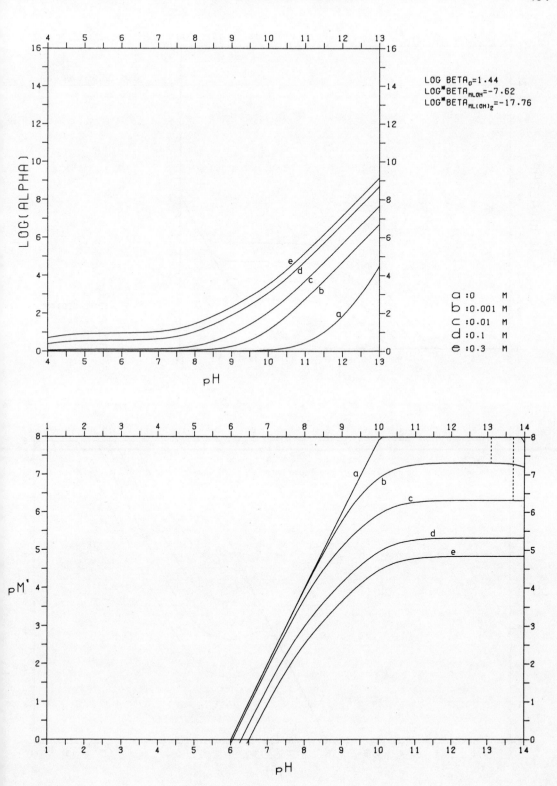

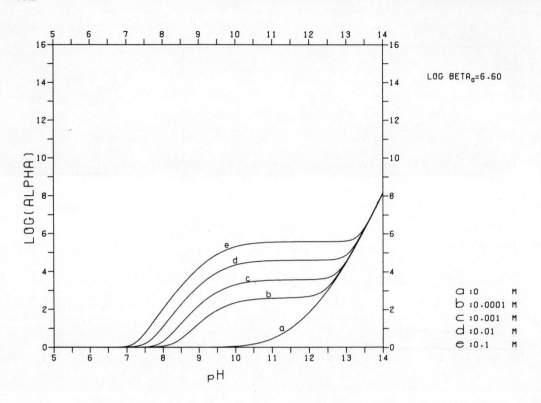

LOG BETA₀ = 6.60

a : 0 M
b : 0.0001 M
c : 0.001 M
d : 0.01 M
e : 0.1 M

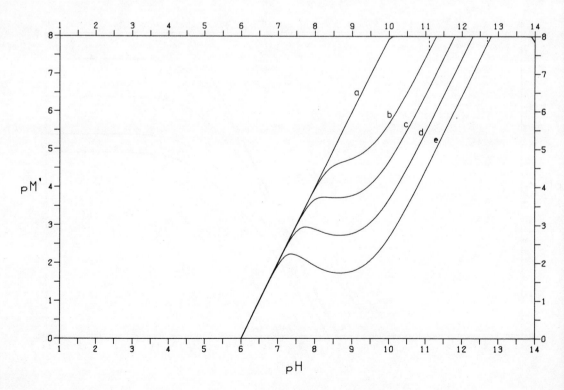

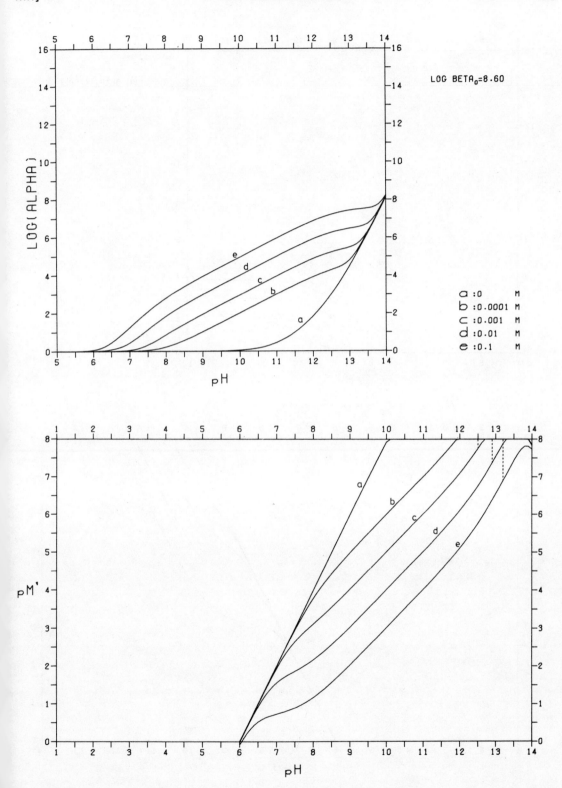

Mn-Carbonate-Trien [Ch. 25

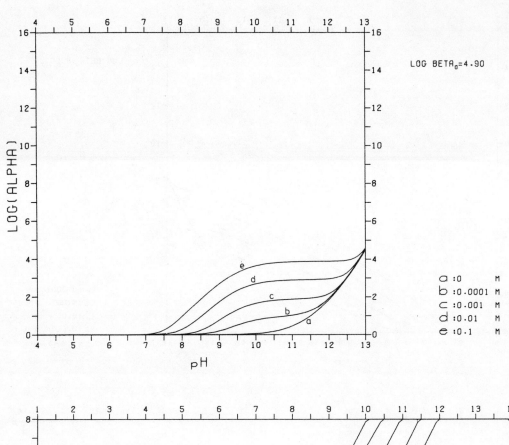

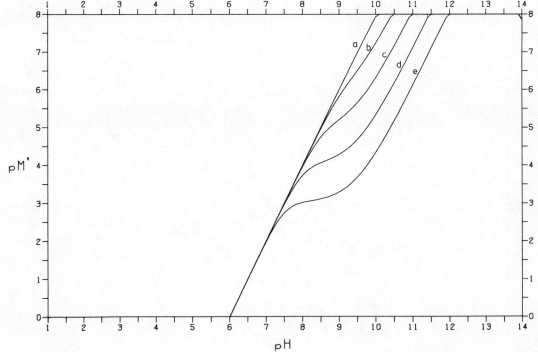

CHAPTER 26

Neodymium (III)   Nd

Hydrolysis of the rather large Nd^{3+} ion does not start until fairly high pH. At ordinary concentrations, hydroxide precipitation accompanies hydrolysis; precipitation begins when only very small amounts are hydrolysed. From the limited amount of reliable information, there is conclusive evidence that $Nd_2(OH)_2^{4+}$ is formed.

Study of hydroxide precipitation is difficult because of supersaturation, colloid formation and the dependence of $^*K_{s0}$ on particle size, and rapid recrystallization. These phenomena make it difficult to define a useful equilibrium situation.

The following data have been adopted; the solubility product applies to a one-hour-old precipitate.

$$\log {}^*\beta_1 = -\ 8.0 \qquad \log {}^*\beta_4 = -36.0$$
$$\log {}^*\beta_2 = -16.3 \qquad \log {}^*\beta_{22} = -13.9$$
$$\log {}^*\beta_3 = -24.5 \qquad \log {}^*K_{s0} = \ \ 18.7$$

There is evidence that a hydroxide–carbonate mixture is precipitated from air-saturated solutions, but this does not measurably influence the pM′–pH plot below pH 11.

Neodymium(III)

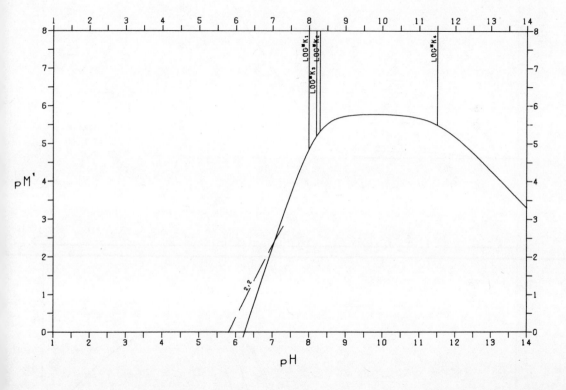

Neodymium(III)-Acetate

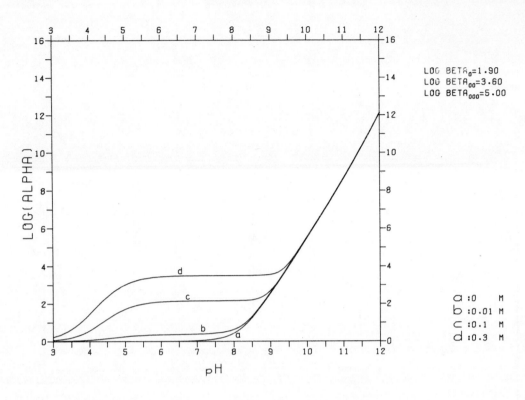

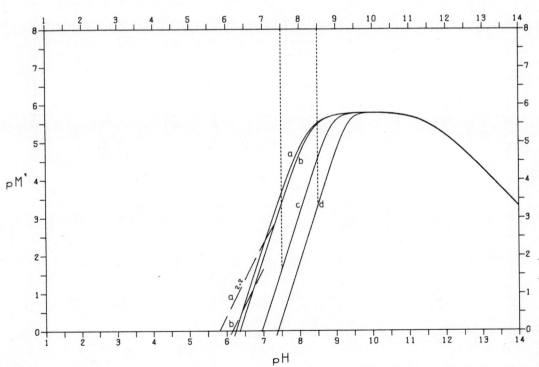

Neodymium(III)-Acetylacetone

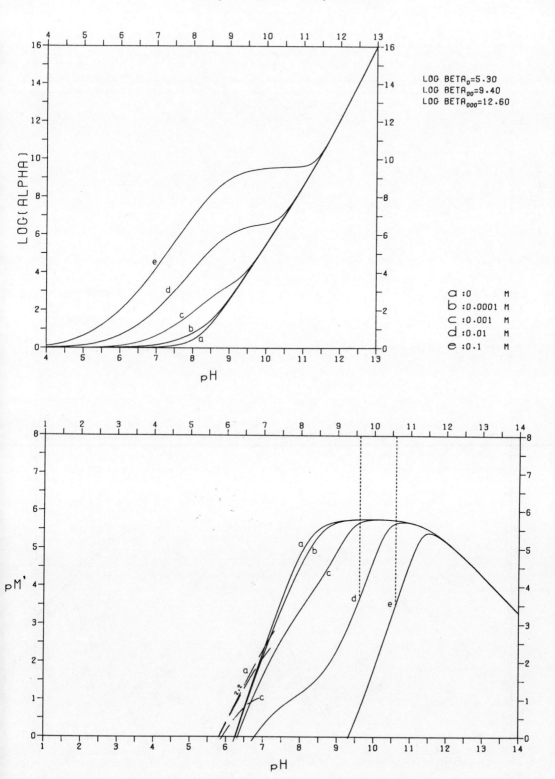

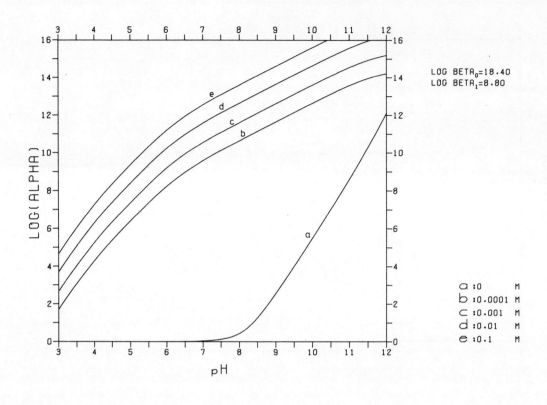
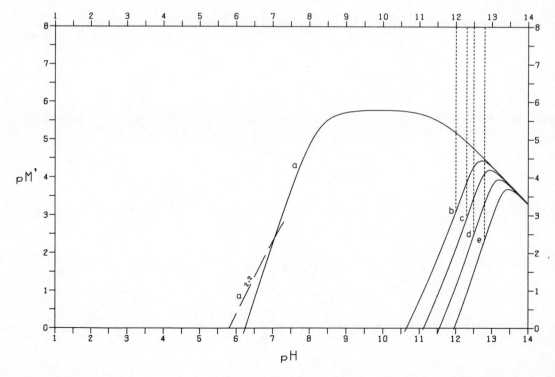

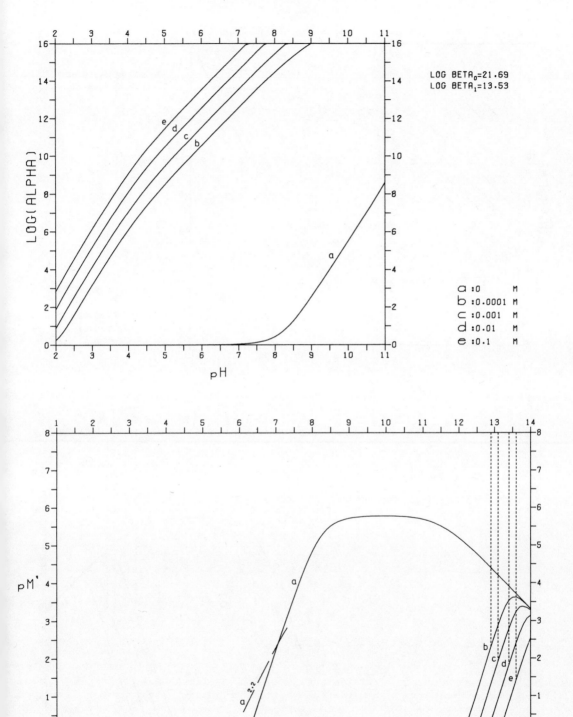

Neodymium(III)–EDTA [Ch. 26]

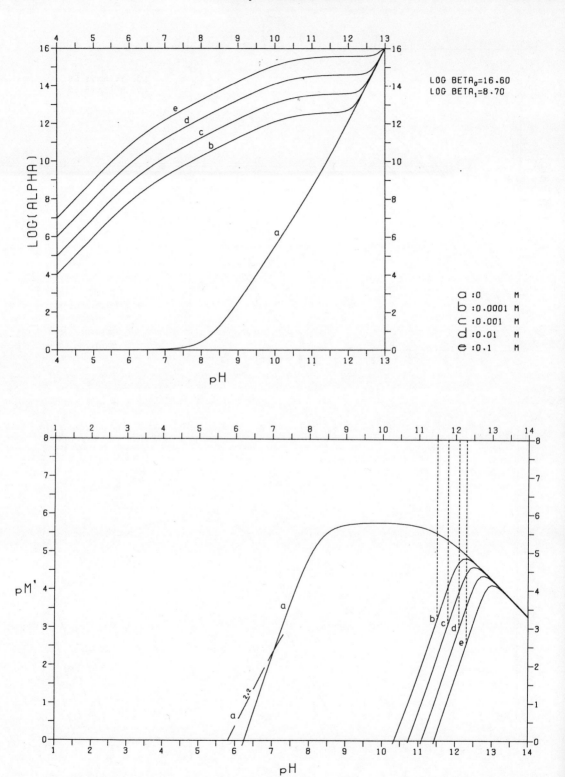

LOG BETA$_0$=16.60
LOG BETA$_1$=8.70

a : 0 M
b : 0.0001 M
c : 0.001 M
d : 0.01 M
e : 0.1 M

Neodymium(III)-(OH)-Quinolinesulphonate

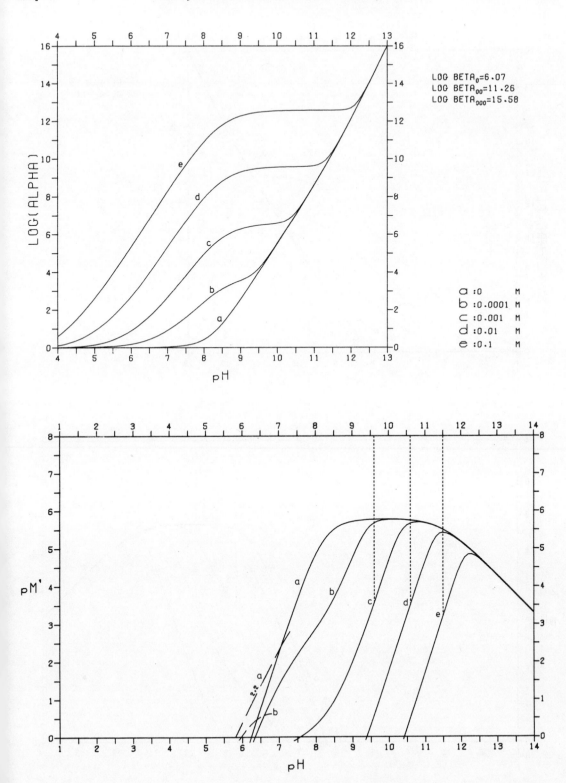

LOG BETA$_0$=6.07
LOG BETA$_{00}$=11.26
LOG BETA$_{000}$=15.58

a : 0   M
b : 0.0001 M
c : 0.001 M
d : 0.01   M
e : 0.1   M

Neodymium(III)-Iminodiacetate [Ch. 26]

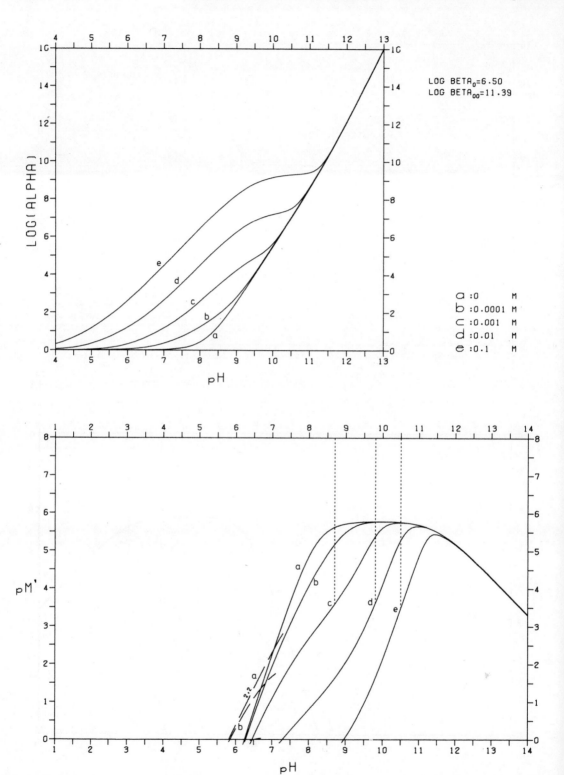

Neodymium(III)-Pyridine-2,6-dicarboxylate

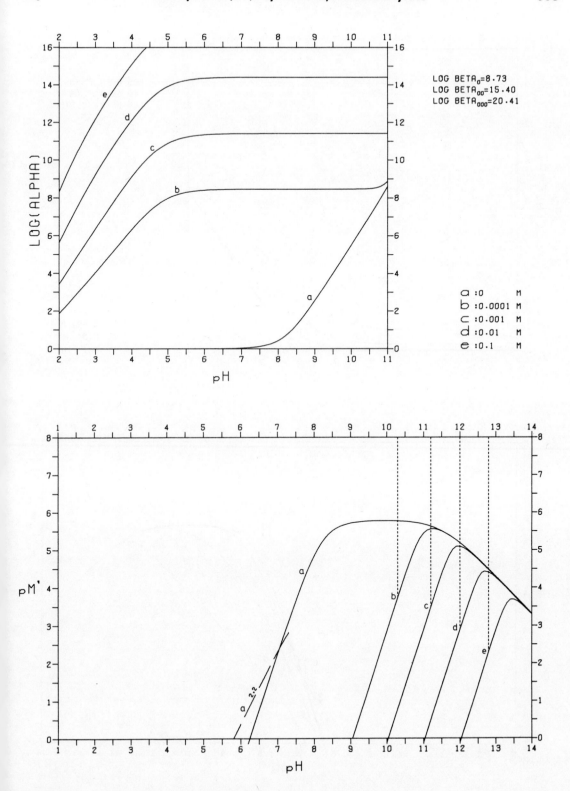

Neodymium(III)-5-Sulphosalicylate

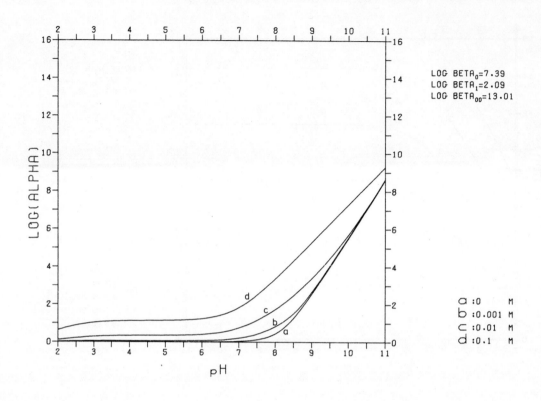

LOG BETA$_0$=7.39
LOG BETA$_1$=2.09
LOG BETA$_{00}$=13.01

a : 0   M
b : 0.001 M
c : 0.01   M
d : 0.1   M

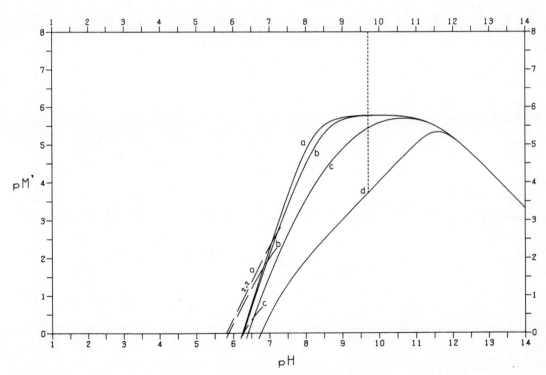

Neodymium(III)-Tartrate

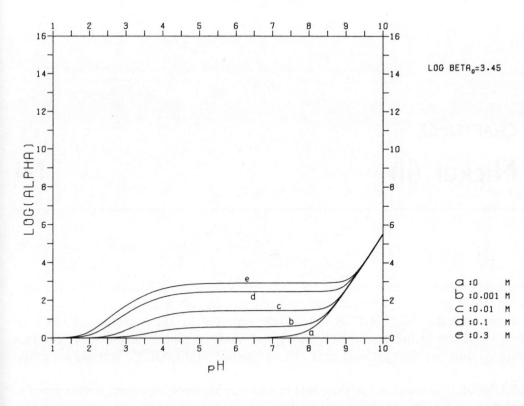

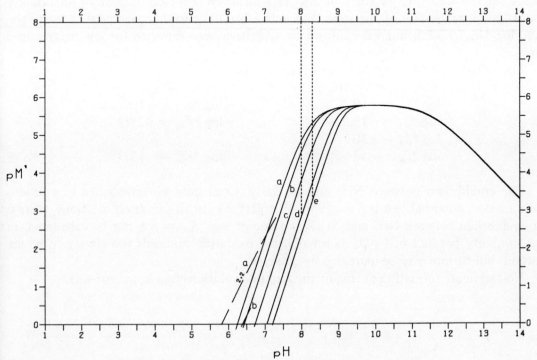

CHAPTER 27

Nickel (II)   Ni

The first hydrolysis constant, log $*\beta_1$, of Ni^{2+} is accurately known. The other constants have been estimated from solubility studies, and are less certain.

The polycomplex $Ni_4(OH)_4^{4+}$ is extensively reported in the literature; it has been found at moderate nickel concentrations at high ionic strengths, and it gives rise to a polycomplex line on the pM'–pH plot. The existence of $Ni_2(OH)^{3+}$ is assumed tentatively.

Solid $Ni(OH)_2$ is more stable than NiO in aqueous medium. However, a wide range of values for the solubility product of $Ni(OH)_2$ has been reported because of uncertainty about the physical state of the solid. For active $Ni(OH)_2$, freshly precipitated from solution, log $*K_{s0}$ = 13.3, but values down to 10.8 have been reported for the inactive aged form.

The values used for construction of the plots are:

$$\log *\beta_1 = -10.2 \qquad \log *\beta_{12} = -10.5$$
$$\log *\beta_2 = -19.2 \qquad \log *\beta_{44} = -27.3$$
$$\log *\beta_3 = -30$$
$$\log *\beta_4 = -44 \qquad \log *K_{s0} = 13.3$$

The equilibrium between Ni^{2+} and $Ni(OH)_2(s)$ can only be established in solutions with a redox potential below E = (897 − 59.1 pH) mV. In air-saturated solutions, there is an equilibrium between Ni^{2+} and Ni_3O_4; a value of log $*K_{s0}$ = 8.5 can be calculated for the solubility product of Ni_3O_4. Normally, the oxidation proceeds too slowly to be important, but its rate may be increased by catalysis.

No carbonate formation occurs in air-saturated solutions (log K_{carb} = 14.8).

Nickel(II)

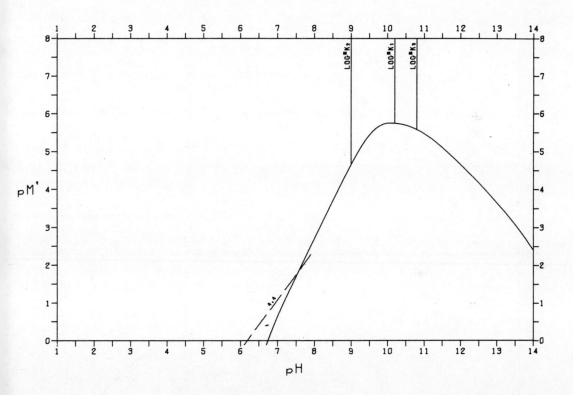

Nickel(II)-Acetylacetone

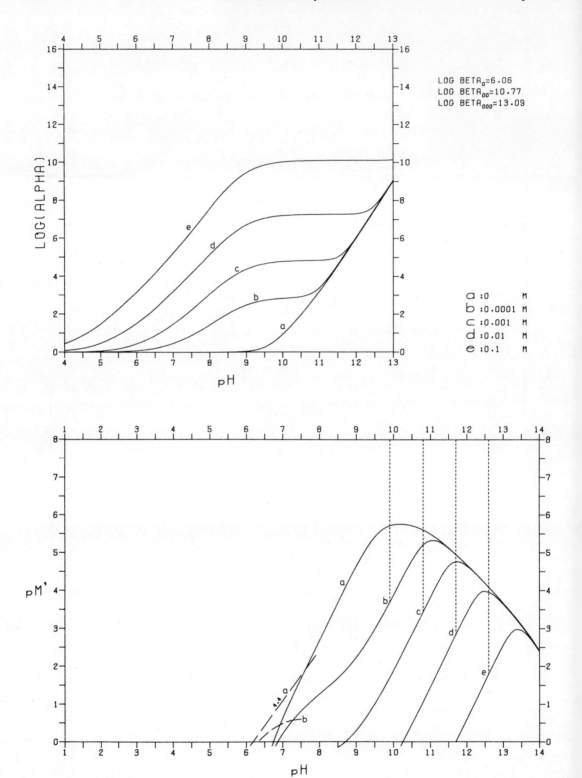

$\log \beta_0 = 6.06$
$\log \beta_{00} = 10.77$
$\log \beta_{000} = 13.09$

a : 0 M
b : 0.0001 M
c : 0.001 M
d : 0.01 M
e : 0.1 M

Nickel(II)-Ammonia

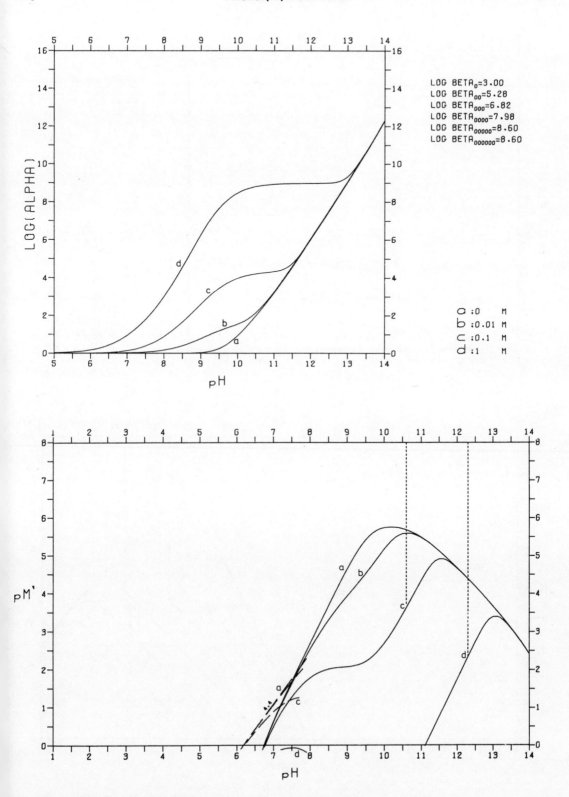

Nickel(II)-Citrate

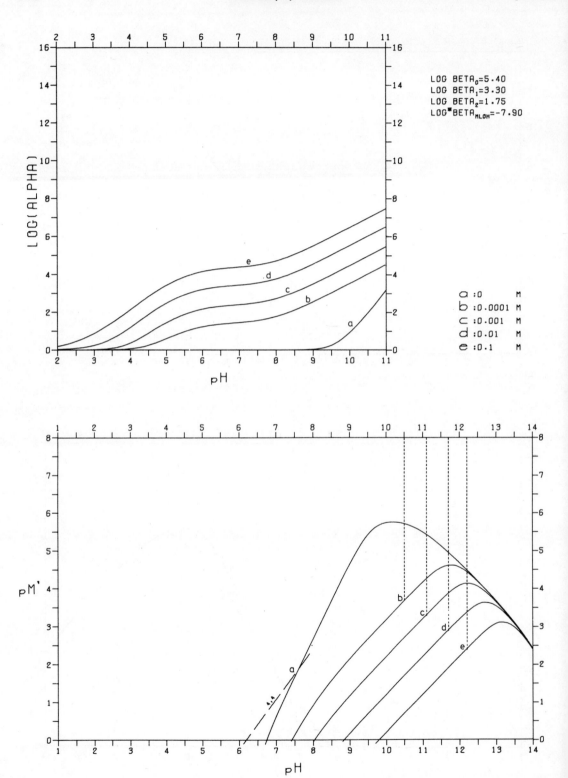

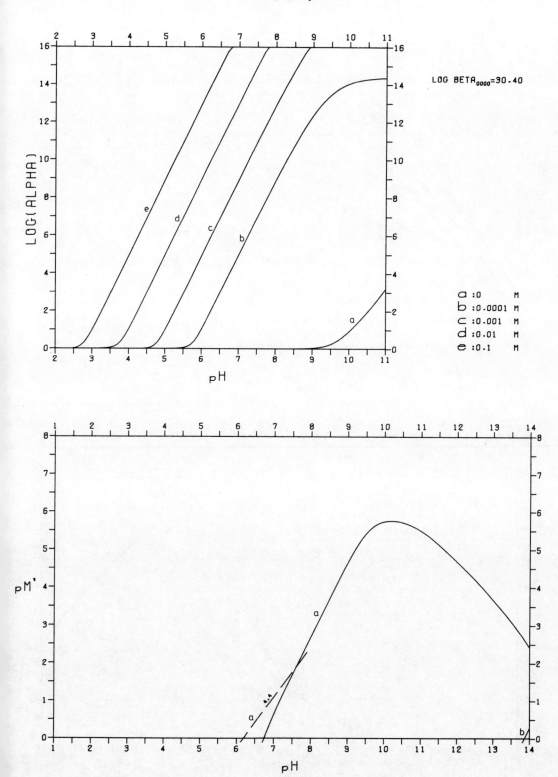

Nickel(II)-DCTA

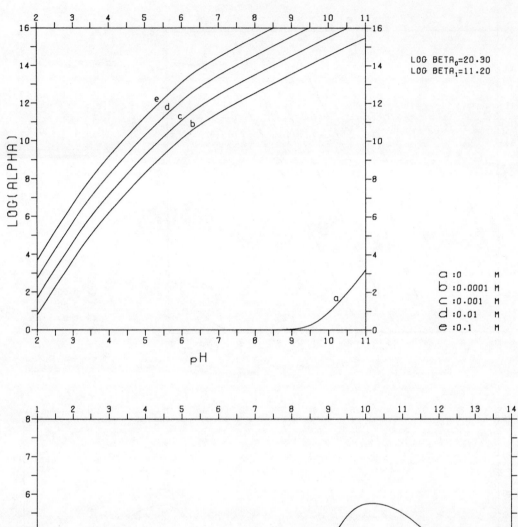

Nickel(II)-DTPA

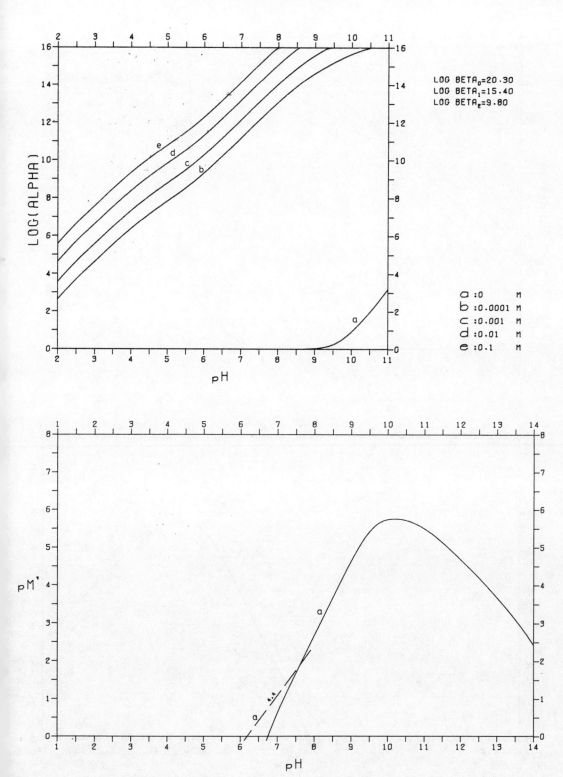

LOG BETA$_0$=20.30
LOG BETA$_1$=15.40
LOG BETA$_2$=9.80

a : 0   M
b : 0.0001 M
c : 0.001   M
d : 0.01   M
e : 0.1   M

Nickel(II)–EDTA

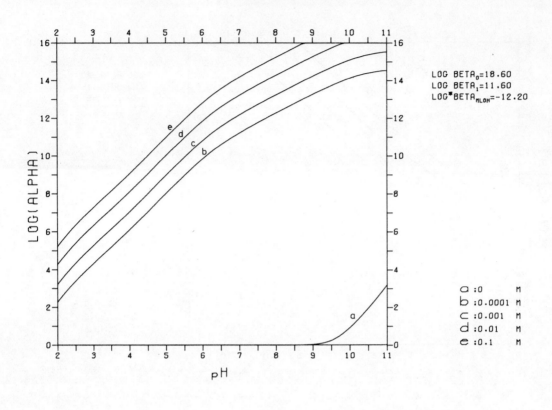

LOG BETA$_0$=18.60
LOG BETA$_1$=11.60
LOG*BETA$_{MLOH}$=-12.20

a : 0 M
b : 0.0001 M
c : 0.001 M
d : 0.01 M
e : 0.1 M

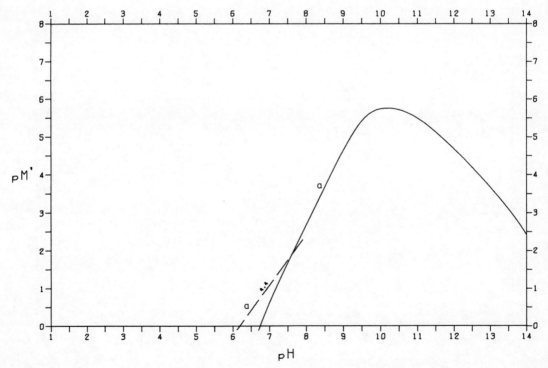

Nickel(II)-EGTA

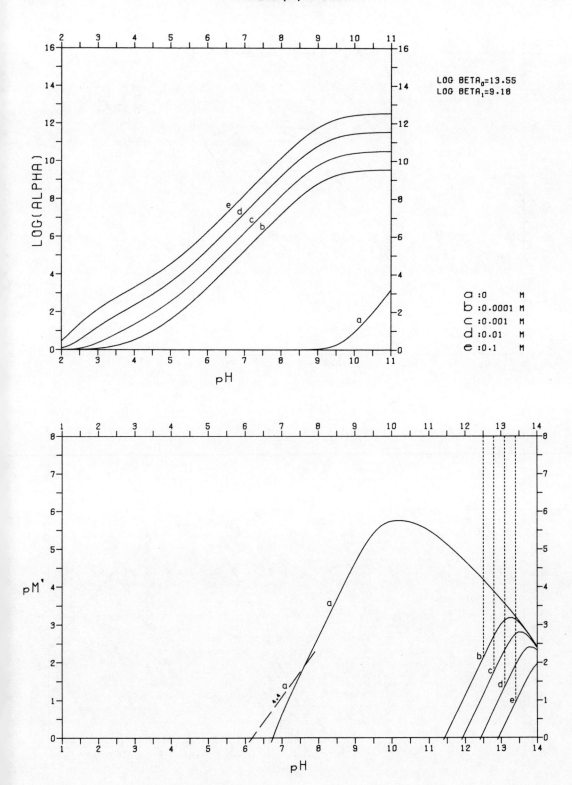

LOG BETA$_0$=13.55
LOG BETA$_1$=9.18

a : 0 M
b : 0.0001 M
c : 0.001 M
d : 0.01 M
e : 0.1 M

Nickel(II)-Ethylenediamine [Ch. 27]

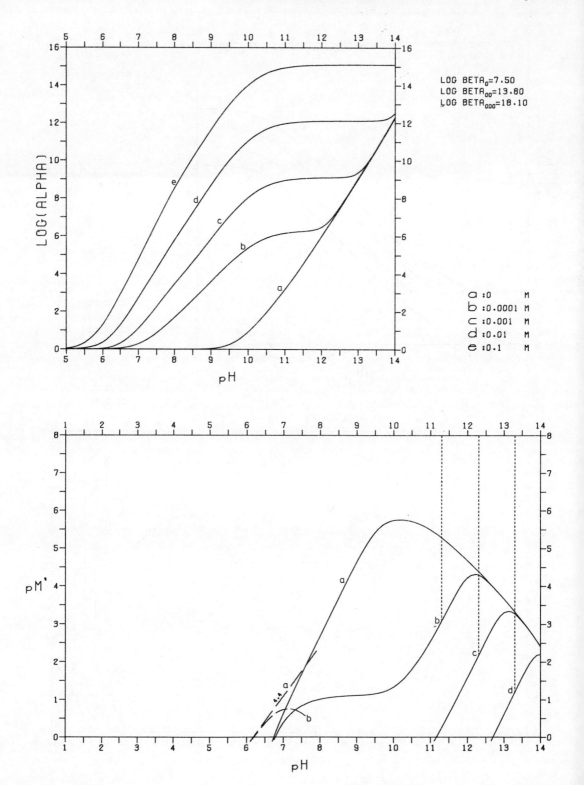

Nickel(II)-Glycine

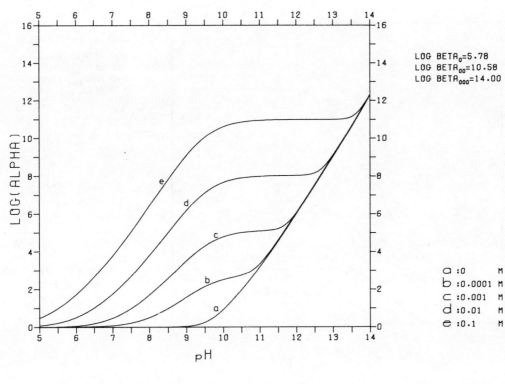

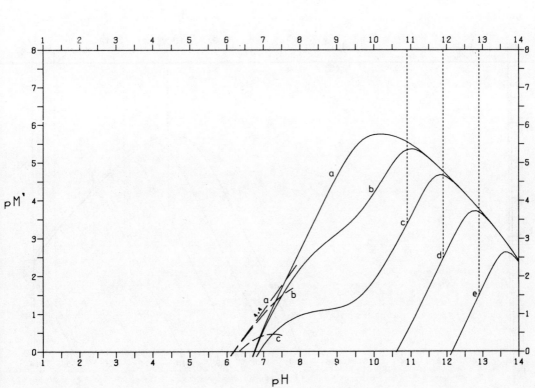

Nickel(II)–Hydroxylamine

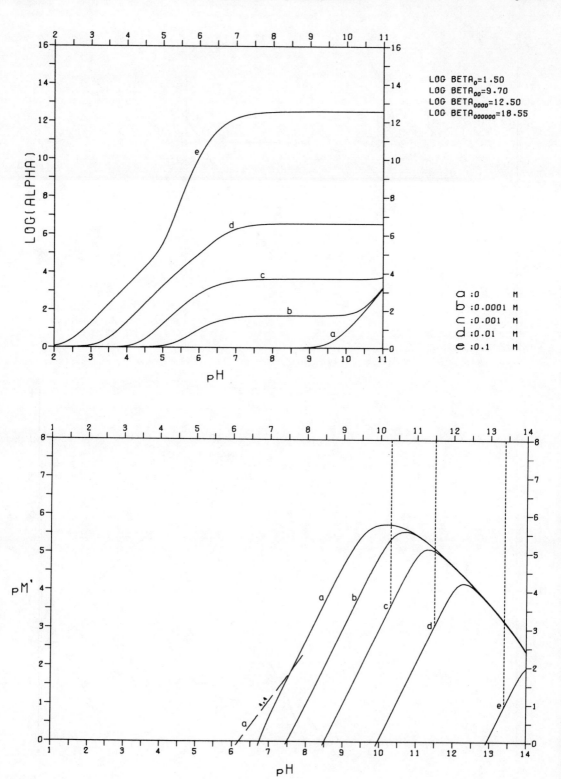

LOG BETA$_0$=1.50
LOG BETA$_{00}$=9.70
LOG BETA$_{0000}$=12.50
LOG BETA$_{000000}$=18.55

a : 0 M
b : 0.0001 M
c : 0.001 M
d : 0.01 M
e : 0.1 M

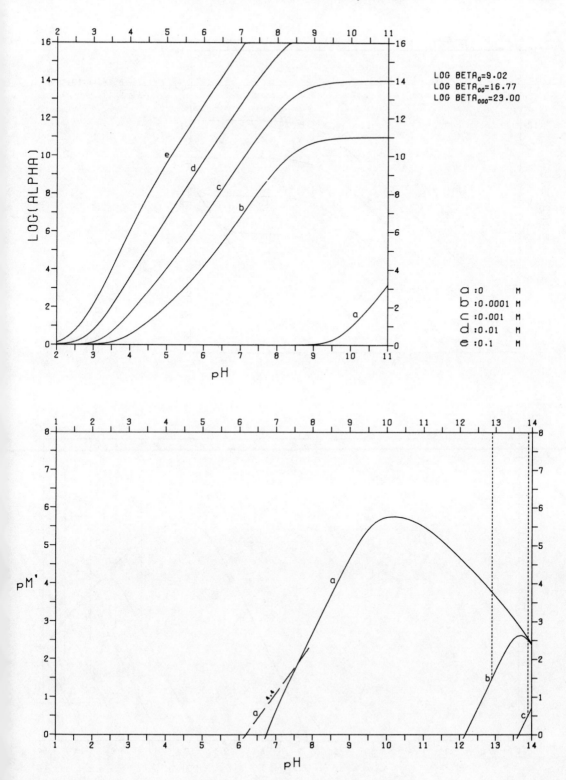

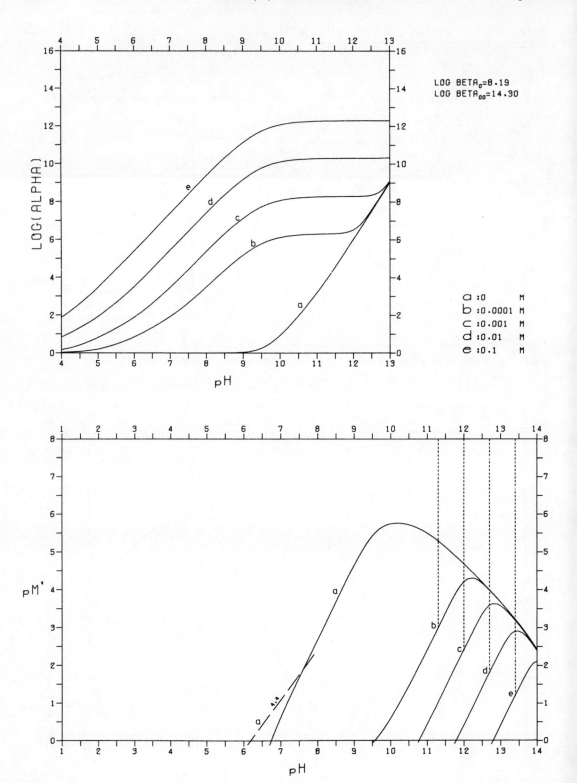

Nickel(II)-Oxalate

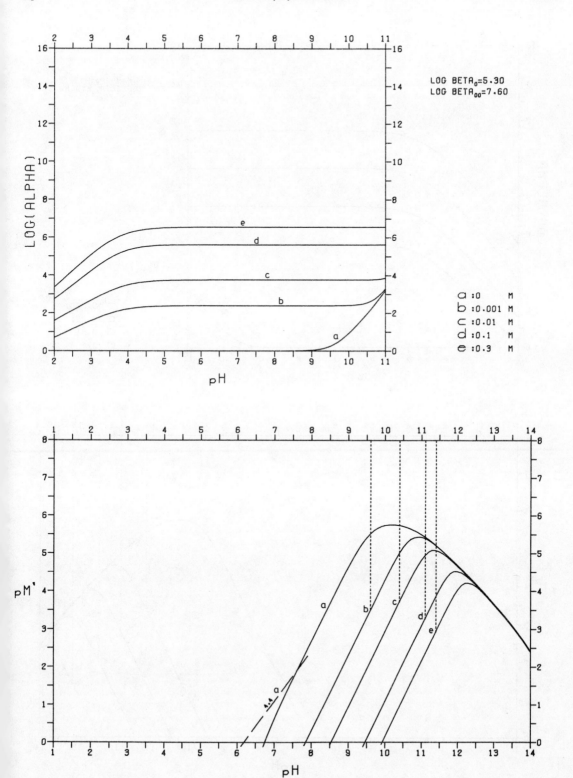

Nickel(II)-Pyridine-2,6-dicarboxylate

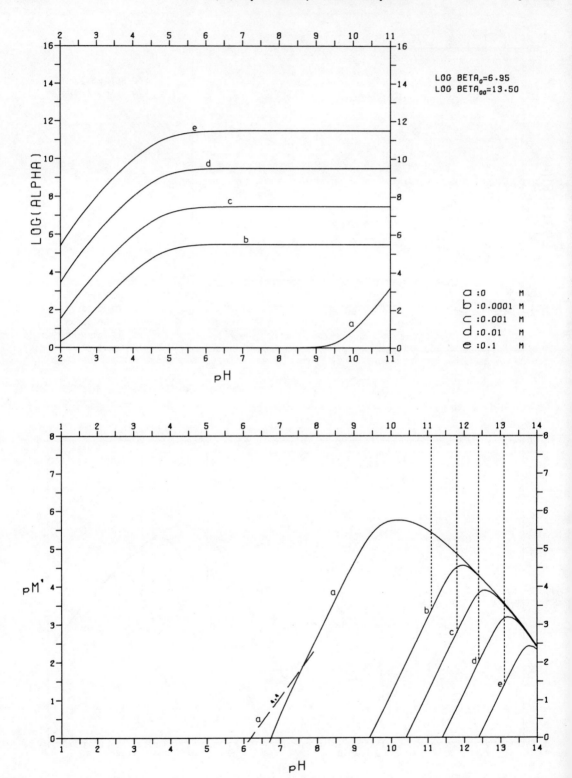

LOG BETA$_0$=6.95
LOG BETA$_{00}$=13.50

a : 0   M
b : 0.0001 M
c : 0.001   M
d : 0.01   M
e : 0.1   M

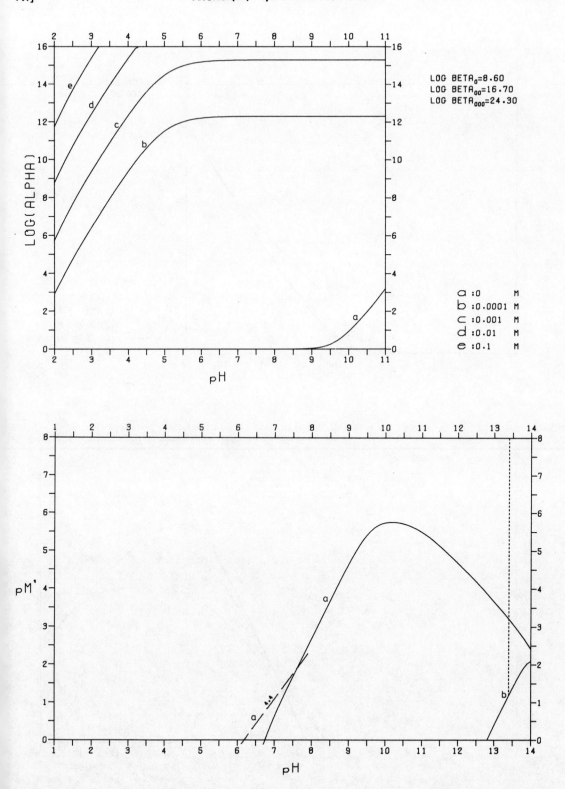

Nickel(II)-Salicylate [Ch. 27]

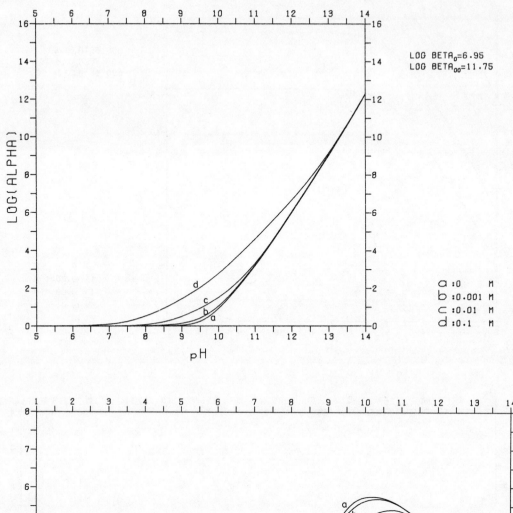

LOG BETA$_0$=6.95
LOG BETA$_{00}$=11.75

a : 0 M
b : 0.001 M
c : 0.01 M
d : 0.1 M

Nickel(II)-Sulphate

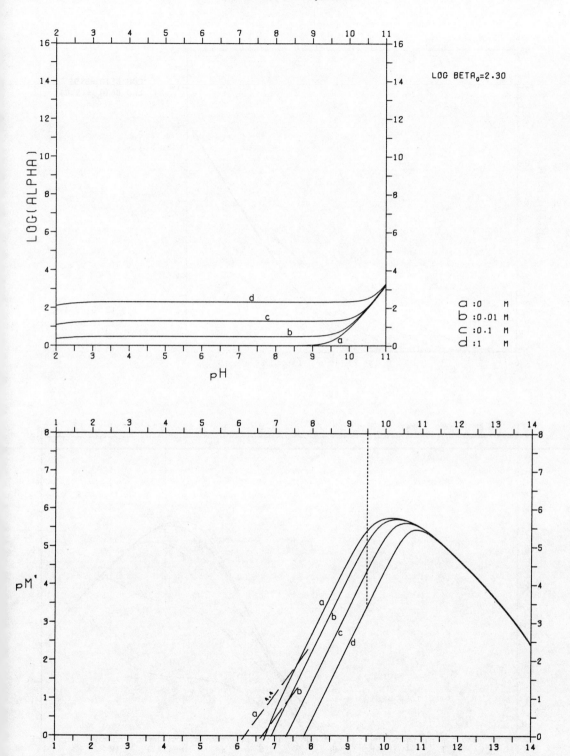

LOG BETA₀=2.30

a : 0 M
b : 0.01 M
c : 0.1 M
d : 1 M

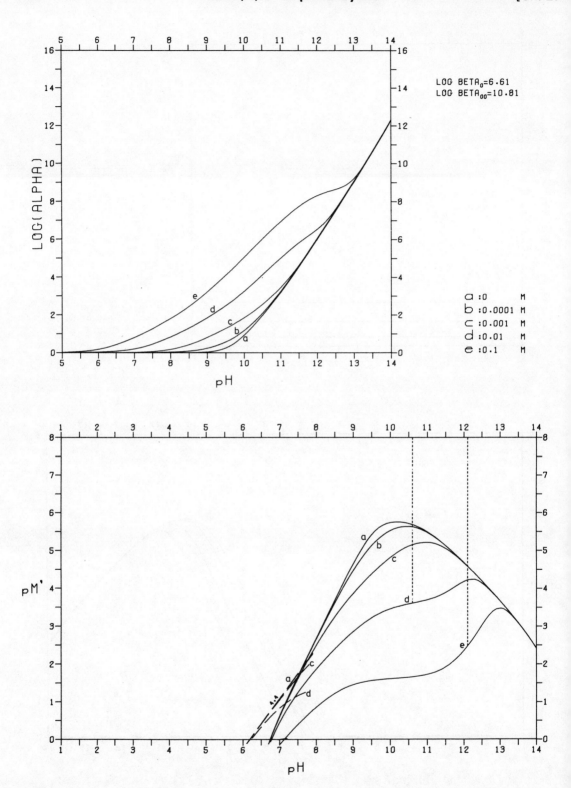

Nickel(II)-Tartrate

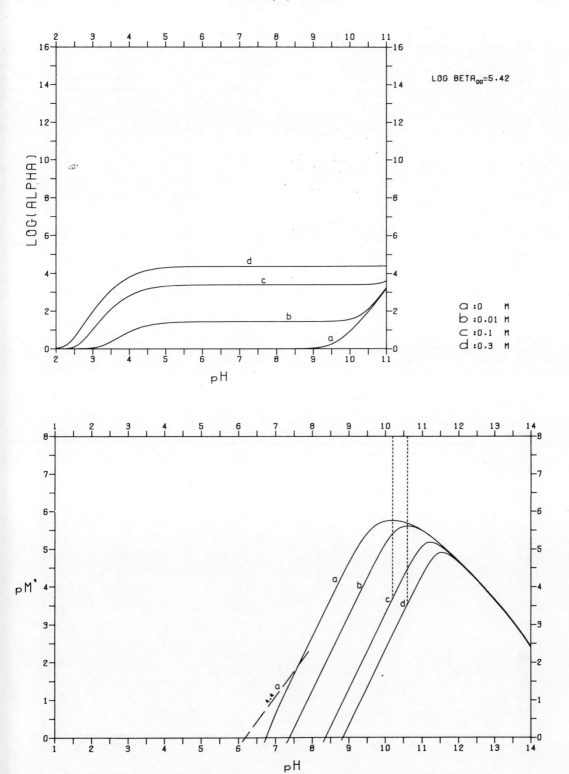

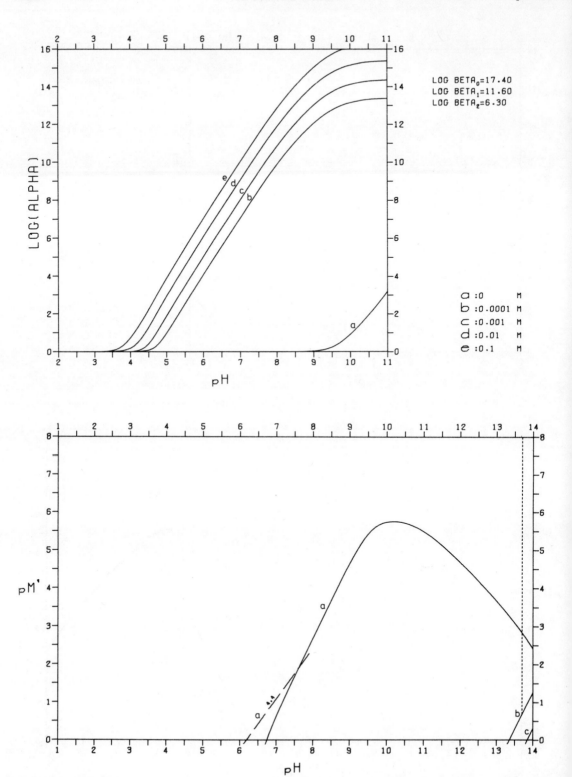

Nickel(II)-Tiron

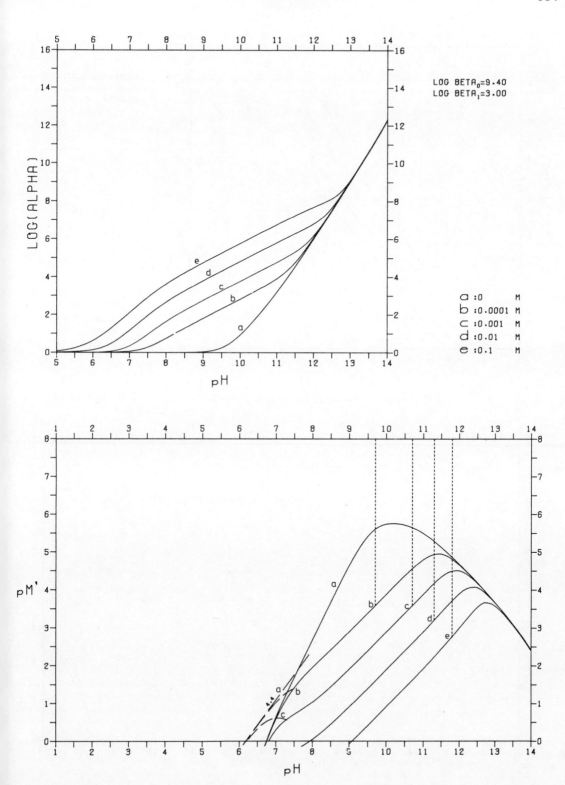

Nickel(II)-Trien [Ch. 27]

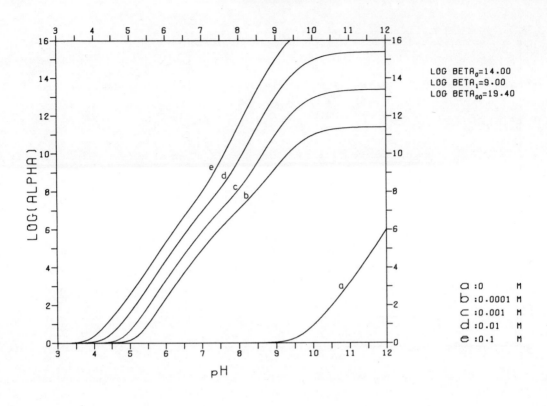

LOG BETA₀=14.00
LOG BETA₁=9.00
LOG BETA₀₀=19.40

a : 0   M
b : 0.0001 M
c : 0.001   M
d : 0.01   M
e : 0.1   M

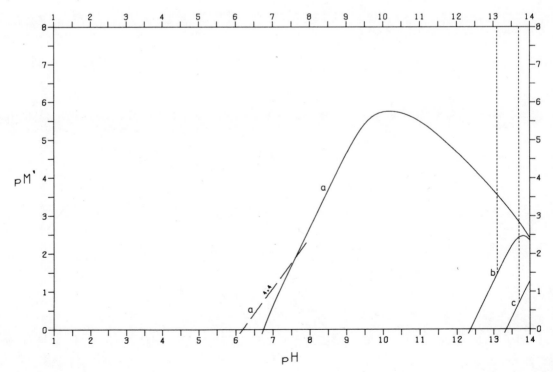

CHAPTER 28

Lead(II)   Pb

Extensive investigations of the hydrolysis of Pb^{2+} ions have resulted in consistent identification of the species and accurate knowledge of the stability constants.

Three distinct solids may precipitate when the pH is raised, but their solubility products are very similar. The stable red form is easily formed, so the following values can be regarded as accurate.

$$\log *\beta_1 = -7.86 \qquad \log *\beta_{43} = -23.95$$
$$\log *\beta_2 = -17.26 \qquad \log *\beta_{44} = -20.30$$
$$\log *\beta_3 = -28.0 \qquad \log *\beta_{86} = -43.3$$
$$\log *\beta_{12} = -6.16 \qquad \log *K_{s0} = 13.0$$

Lead carbonate is less soluble than lead hydroxide (see pages 555–574).

In theory, Pb^{2+} can only be in equilibrium with PbO(s) (in the absence of complexing ligands) between limits given by $E = (E_0 - 59.1\text{ pH})$ mV, where $277 < E_0 < 885$ mV. Below 277 mV, metallic lead is deposited; in air-saturated solutions (i.e. above 885 mV) the solid phase PbO_2, with apparent $\log *K_{s0} = 7.5$, should be formed. Formation of PbO_2 from Pb^{2+} proceeds only slowly in practice, but some inexplicable phenomena encountered may be due to the deposition of PbO_2. For example, addition of reducing agents has been found to improve remarkably the precision of complexometric tritrations at low concentrations.

Lead(II)

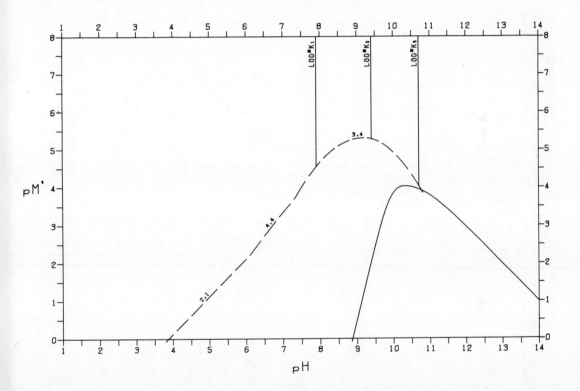

Lead(II)–Acetate

[Ch. 28

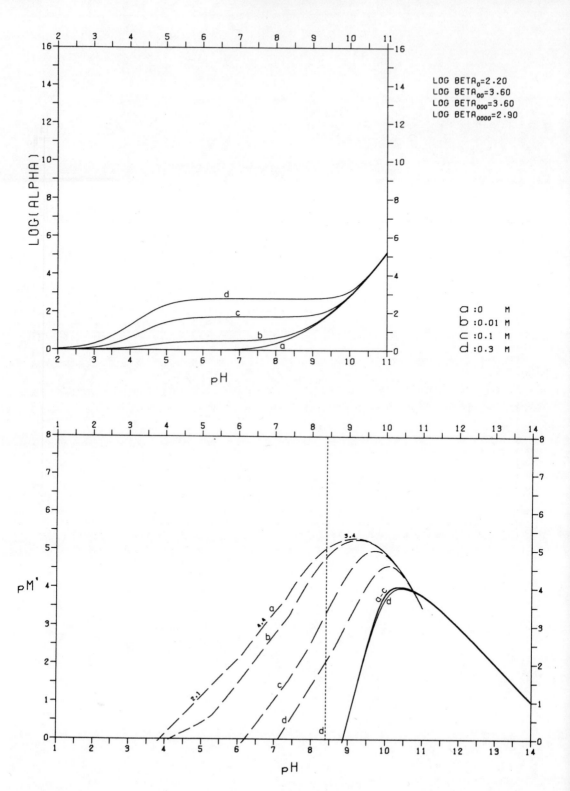

LOG BETA$_0$=2.20
LOG BETA$_{00}$=3.60
LOG BETA$_{000}$=3.60
LOG BETA$_{0000}$=2.90

a : 0   M
b : 0.01 M
c : 0.1 M
d : 0.3 M

Lead(II)- Bromide

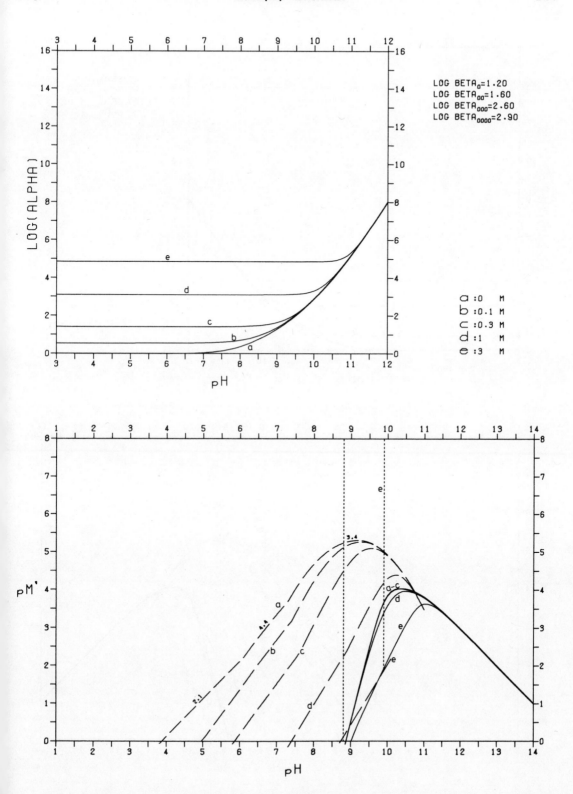

LOG BETA₀=1.20
LOG BETA₀₀=1.60
LOG BETA₀₀₀=2.60
LOG BETA₀₀₀₀=2.90

a : 0 M
b : 0.1 M
c : 0.3 M
d : 1 M
e : 3 M

Lead(II)-Chloride [Ch. 28

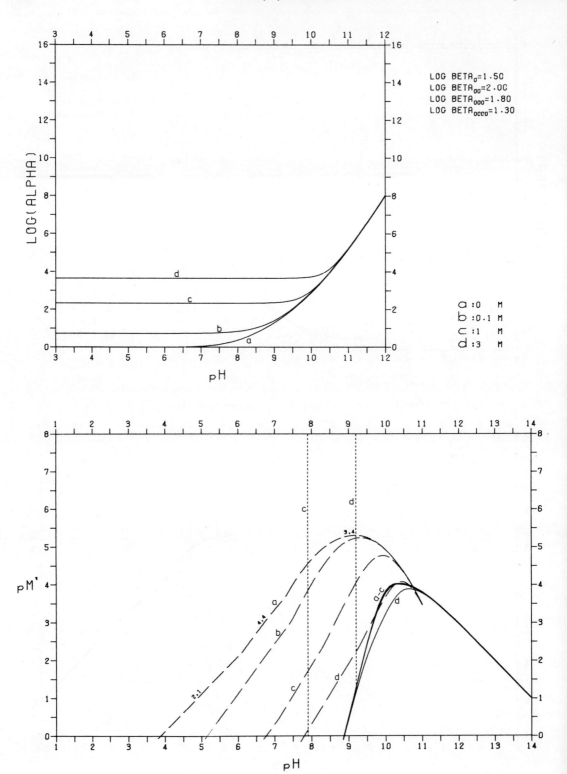

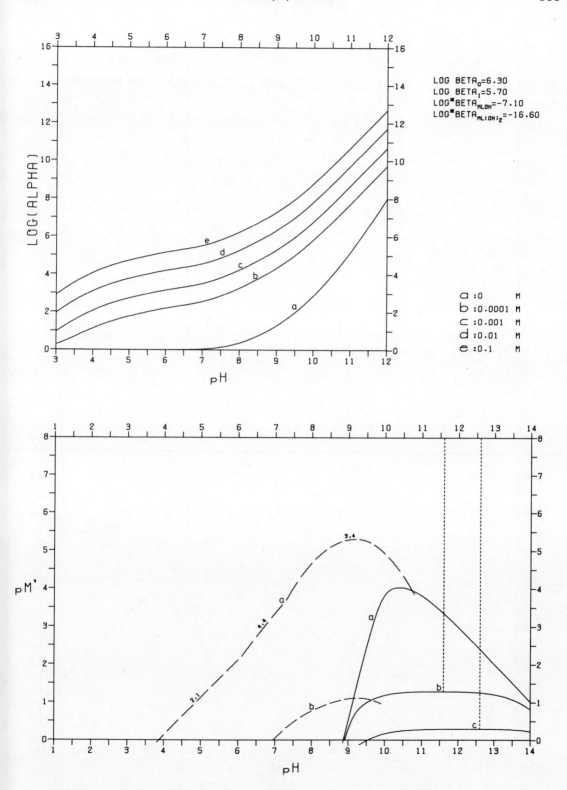

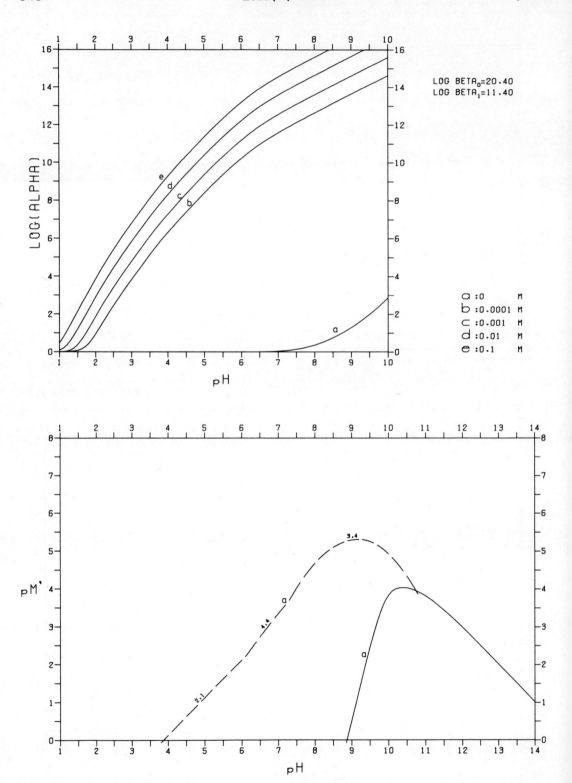

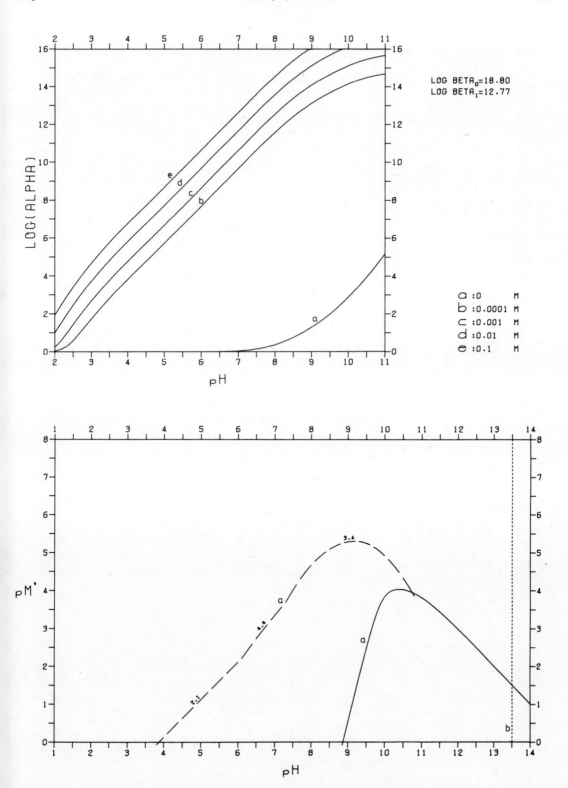

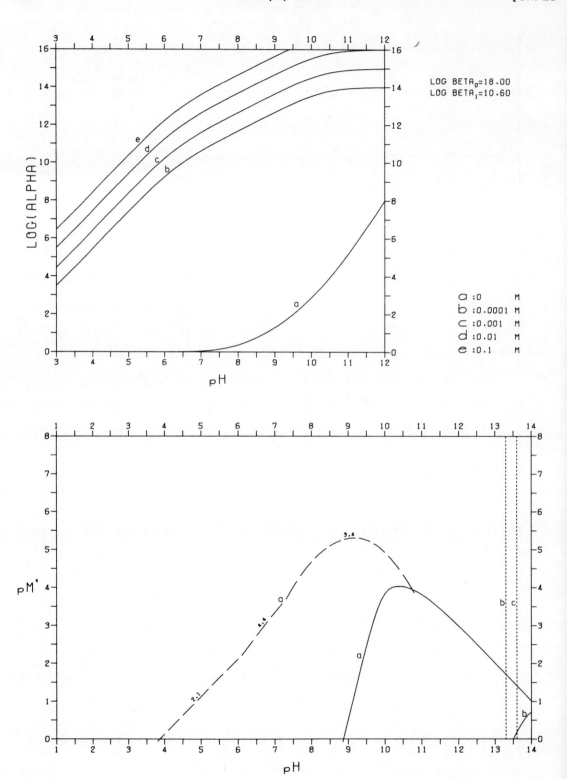

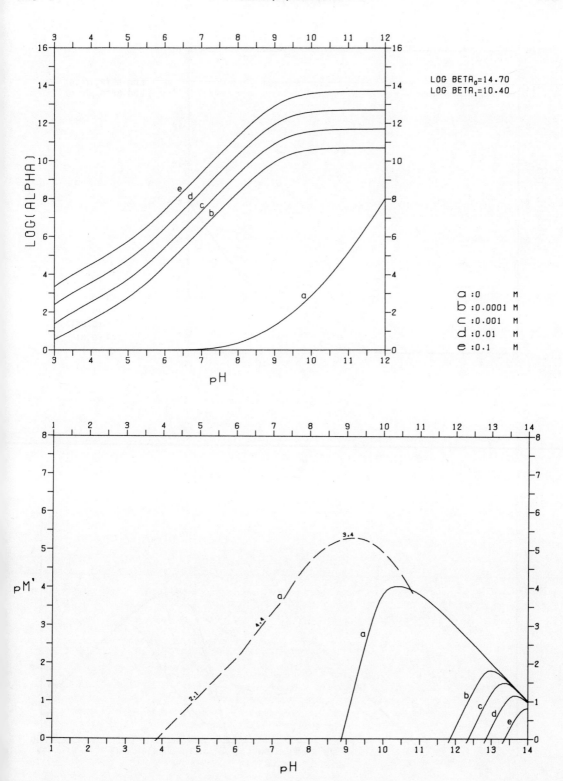

544 Lead(II)-Ethylenediamine [Ch. 28

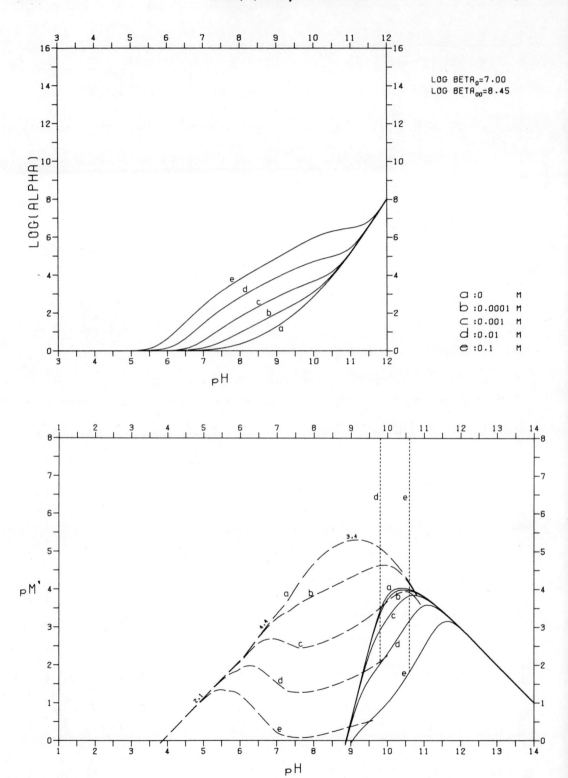

Lead(II)-Glycine

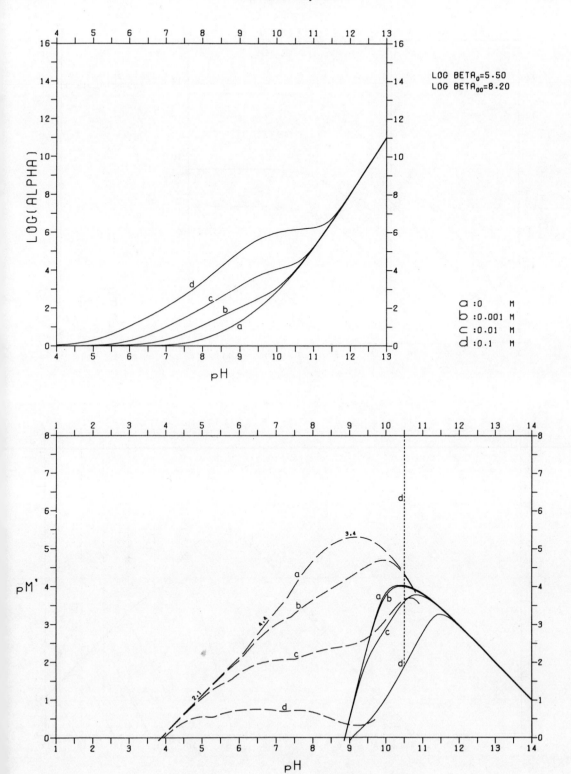

Lead(II)-(OH)-Quinolinesulphonate [Ch. 28]

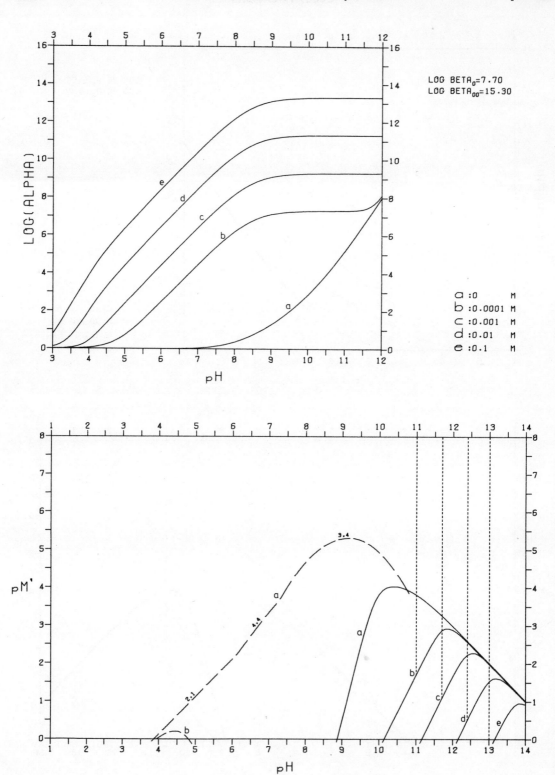

Lead(II)-Iminodiacetate

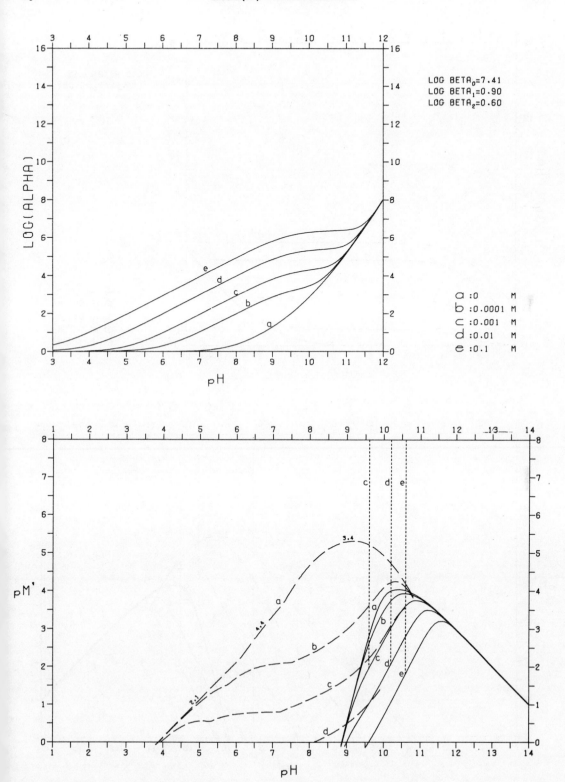

LOG BETA$_0$=7.41
LOG BETA$_1$=0.90
LOG BETA$_2$=0.60

a : 0 M
b : 0.0001 M
c : 0.001 M
d : 0.01 M
e : 0.1 M

Lead(II)-Oxalate [Ch. 28]

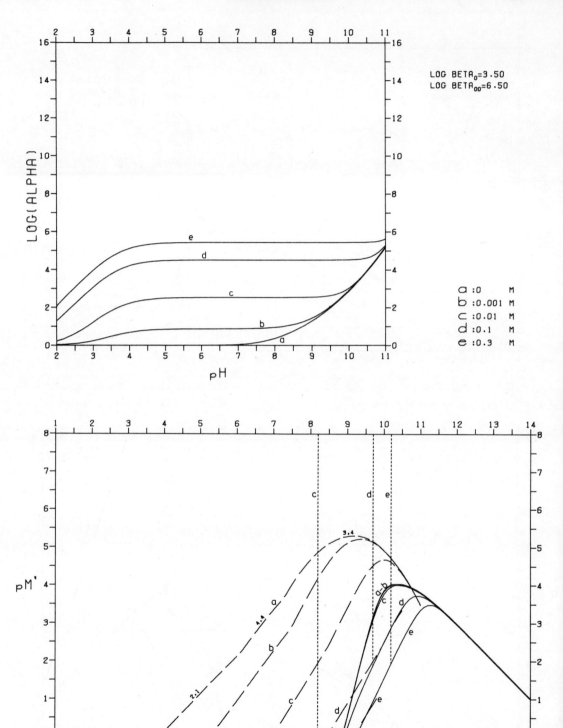

Lead(II)-Pyridine-2,6-dicarboxylate

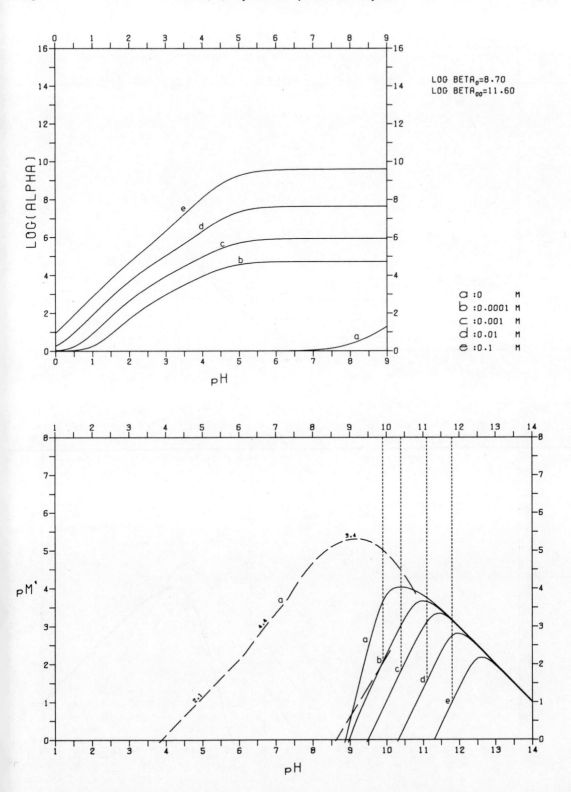

Lead(II)-1,10-Phenanthroline

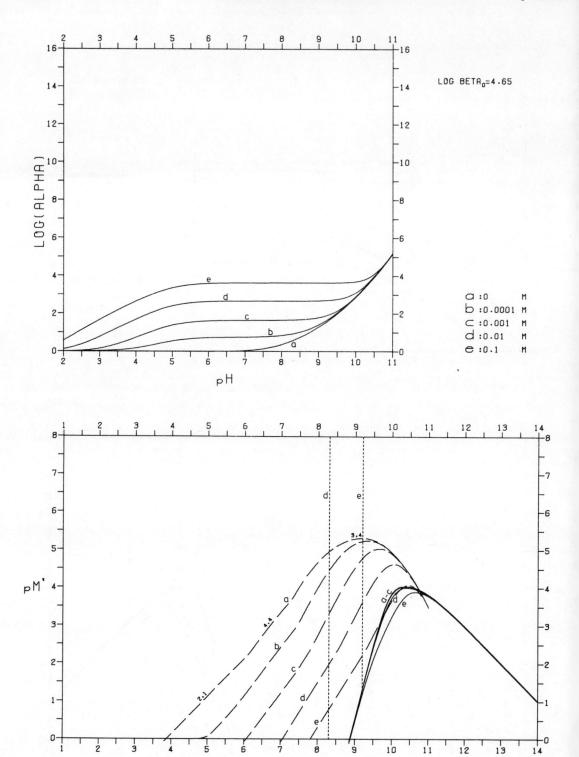

Lead(II)-Tartrate

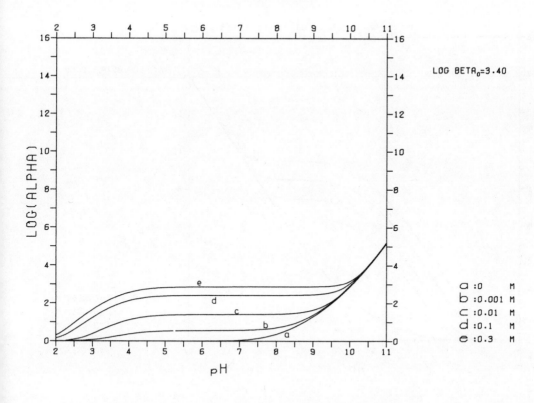

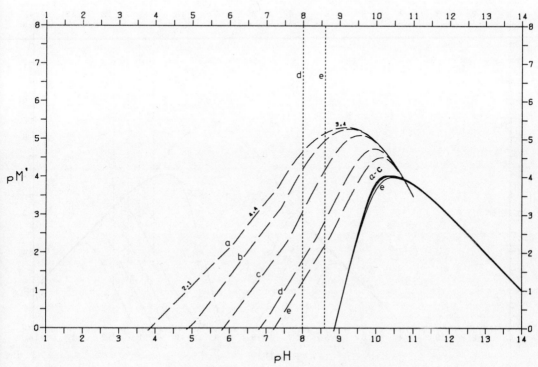

Lead(II)-Tetren

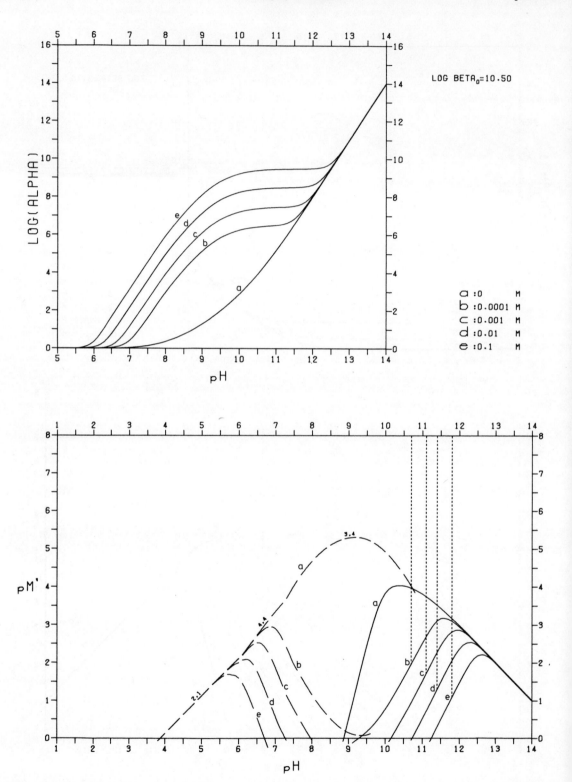

Lead(II)-Tiron

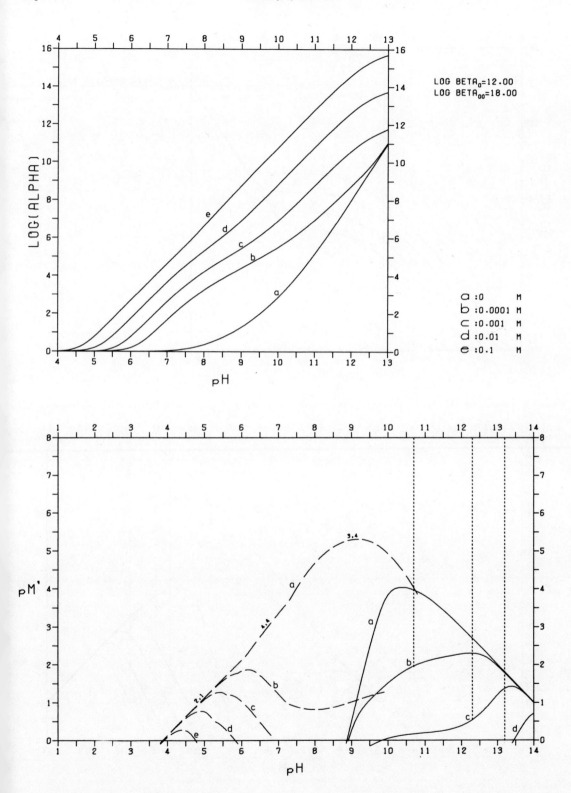

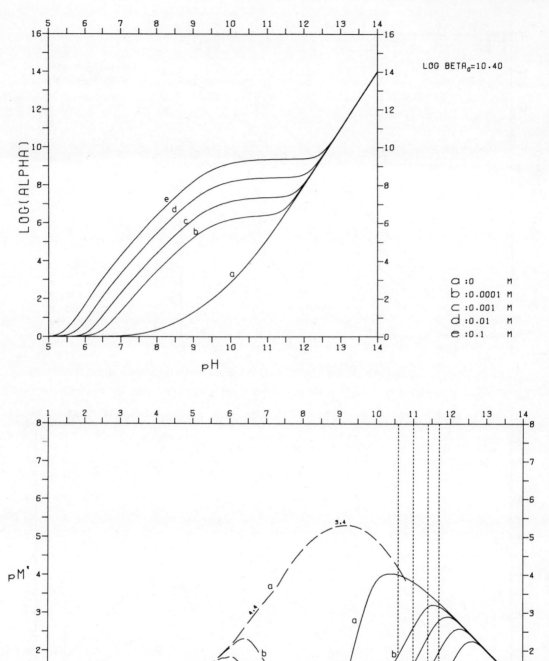

Pb — carbonate system

Lead forms some carbonates which are less soluble than lead hydroxide. The solubility product of $PbCO_3$ (log $*K_{s0}$ = −13.1) leads to an apparent constant log K_{carb} = 8.6 for air-saturated solutions. According to the literature, basic carbonates are also formed, but their stability constants are not accurately known, and their contributions to the plots are of minor importance, so they have been neglected here. The plots were constructed with the following data.

$$\log *\beta_1 = -7.86$$
$$\log *\beta_2 = -17.26$$
$$\log *\beta_3 = -28.0$$
$$\log *\beta_{12} = -6.61$$
$$\log *\beta_{43} = -23.95$$
$$\log *\beta_{44} = -20.30$$
$$\log *\beta_{86} = -43.3$$
$$\log K_{carb} = 8.6$$

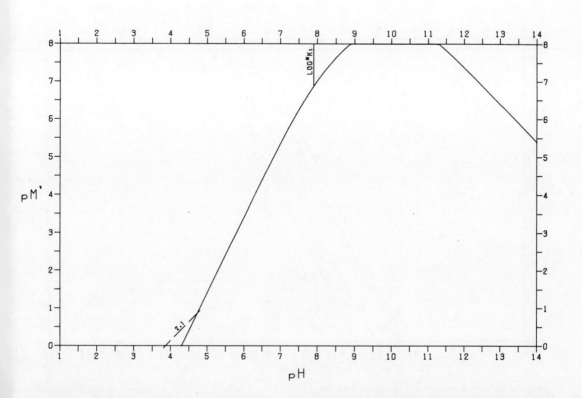

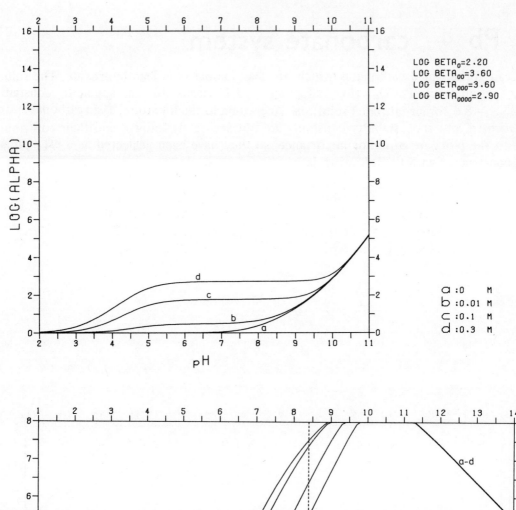

Pb-Carbonate-Bromide

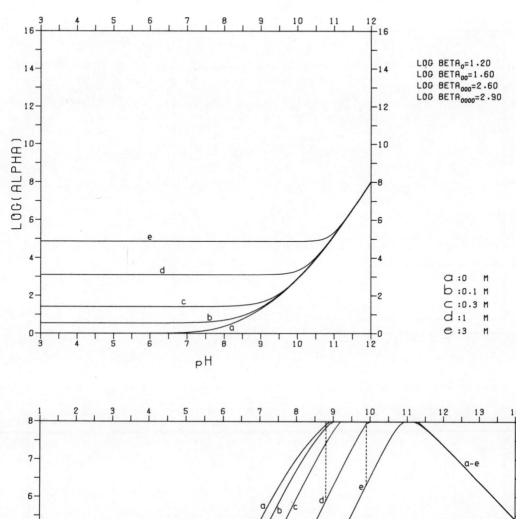

LOG BETA$_0$=1.20
LOG BETA$_{00}$=1.60
LOG BETA$_{000}$=2.60
LOG BETA$_{0000}$=2.90

a : 0 M
b : 0.1 M
c : 0.3 M
d : 1 M
e : 3 M

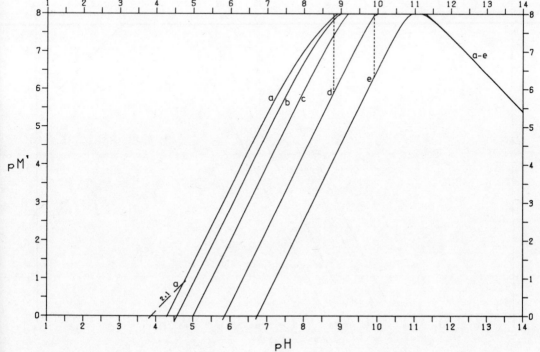

Pb-Carbonate-Chloride [Ch. 28]

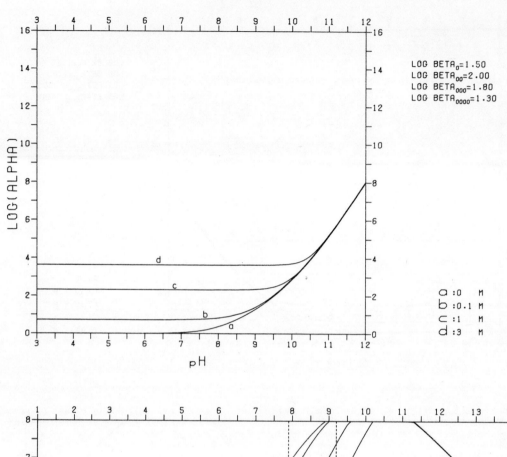

LOG BETA$_0$=1.50
LOG BETA$_{00}$=2.00
LOG BETA$_{000}$=1.80
LOG BETA$_{0000}$=1.30

a : 0 M
b : 0.1 M
c : 1 M
d : 3 M

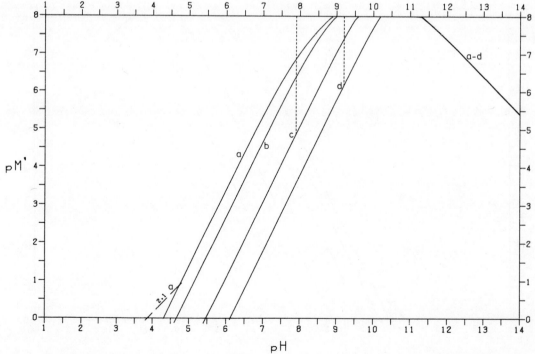

Pb-Carbonate-Citrate

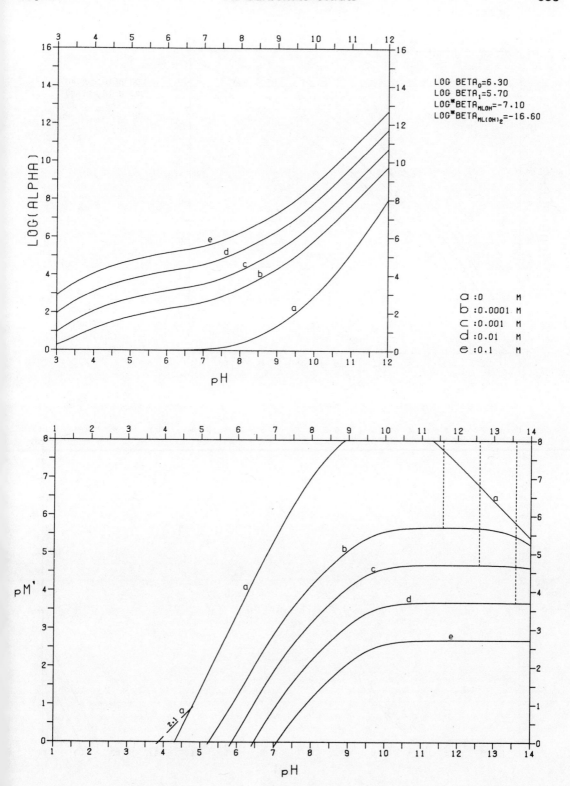

Pb-Carbonate-DCTA [Ch. 28]

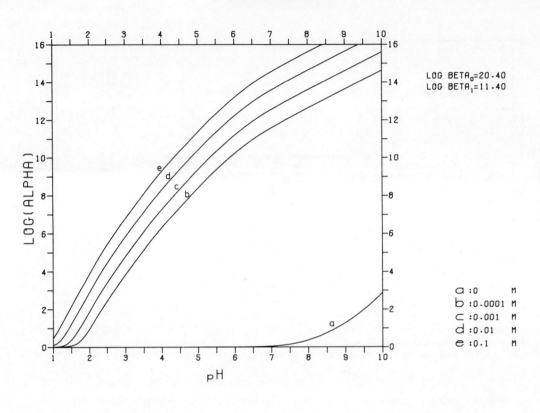

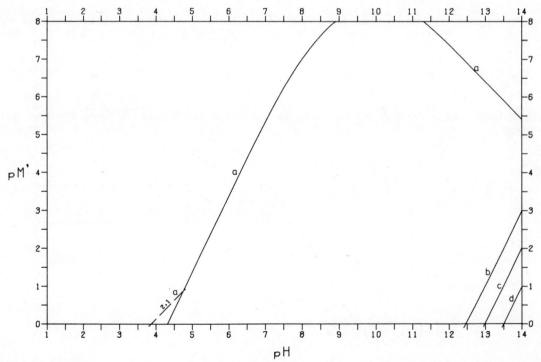

Pb-Carbonate-DTPA

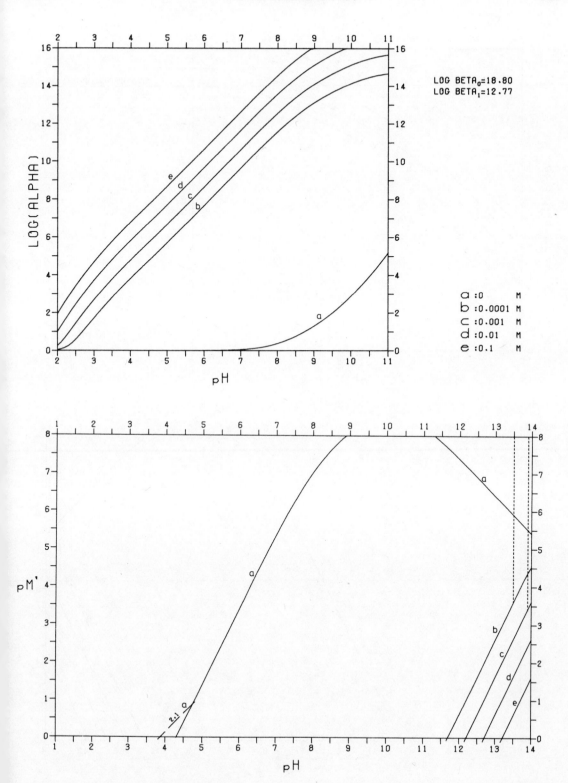

Pb-Carbonate-EDTA [Ch. 28]

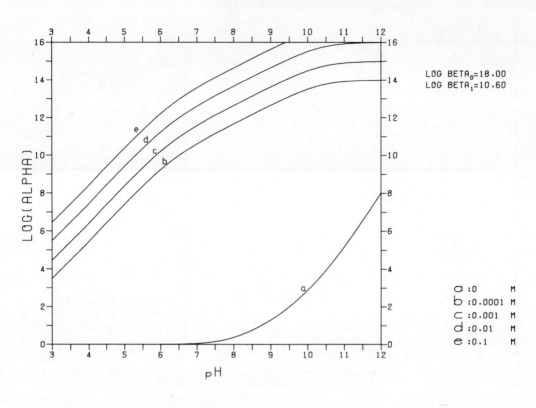

LOG BETA$_0$=18.00
LOG BETA$_1$=10.60

a : 0   M
b : 0.0001 M
c : 0.001 M
d : 0.01   M
e : 0.1   M

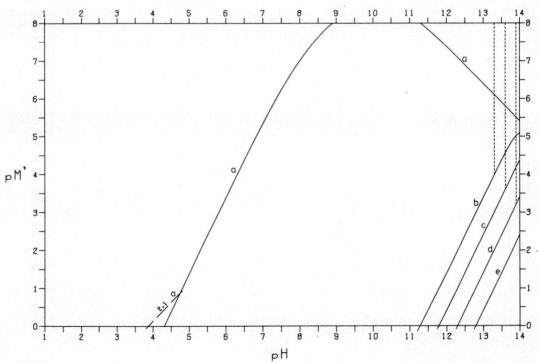

Pb-Carbonate-EGTA

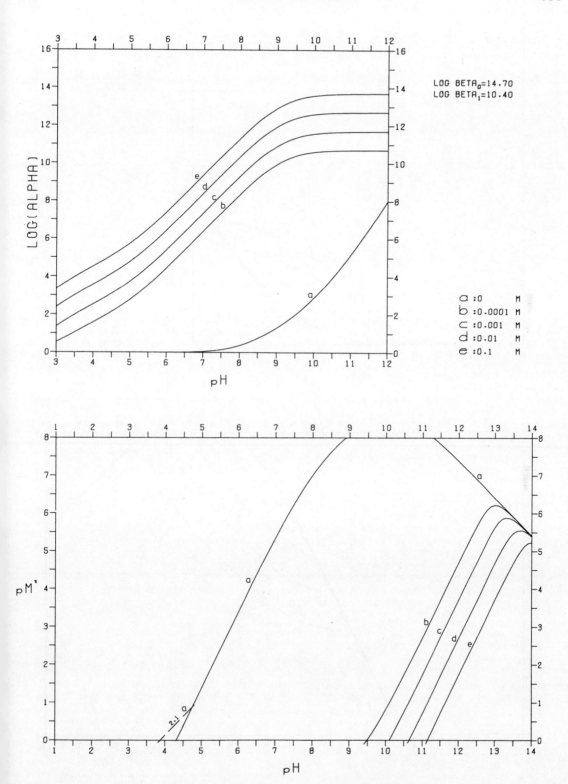

Pb–Carbonate–Ethylenediamine

[Ch. 28

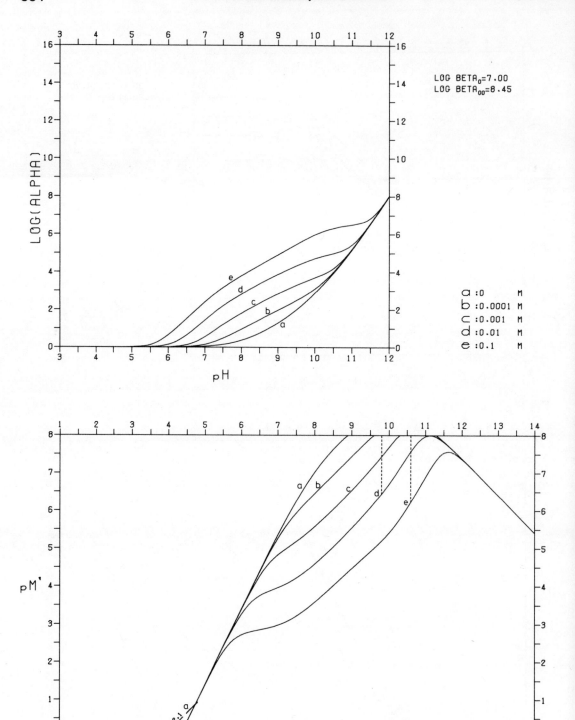

Pb-Carbonate-Glycine

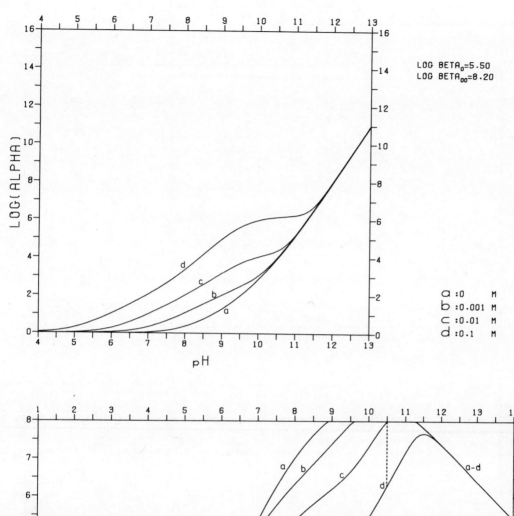

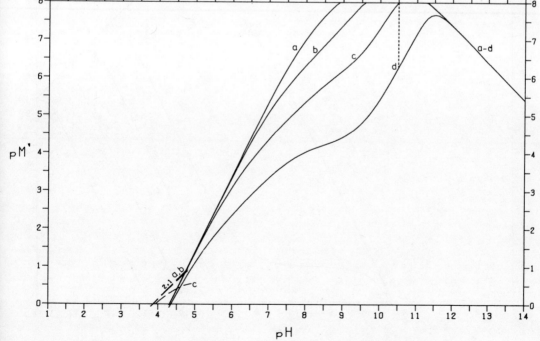

566 Pb-Carbonate-(OH)-Quinolinesulphonate [Ch. 28

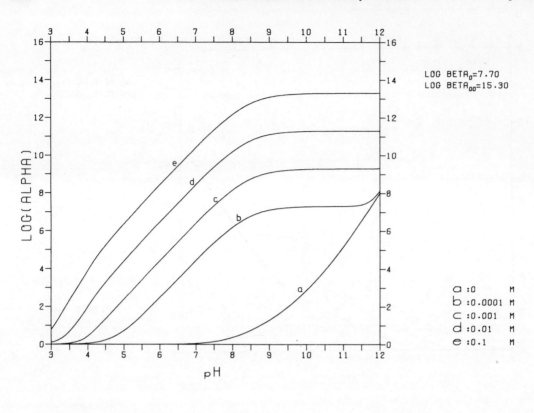

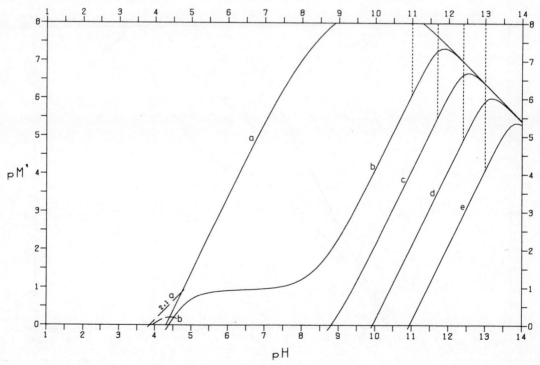

Pb-Carbonate-Iminodiacetate

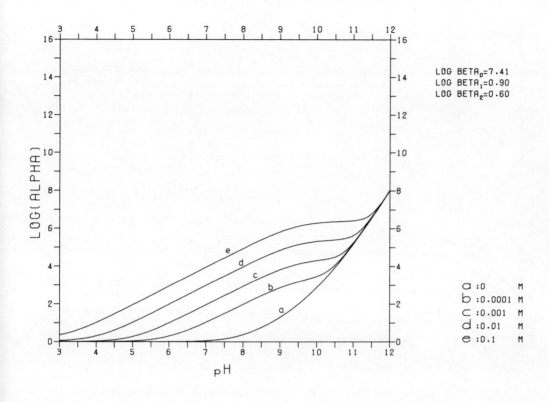

LOG BETA$_0$ = 7.41
LOG BETA$_1$ = 0.90
LOG BETA$_2$ = 0.60

a : 0   M
b : 0.0001   M
c : 0.001   M
d : 0.01   M
e : 0.1   M

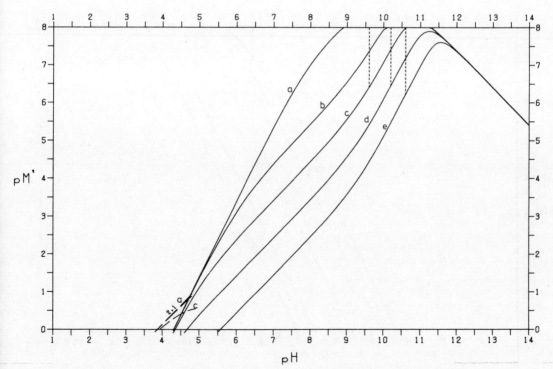

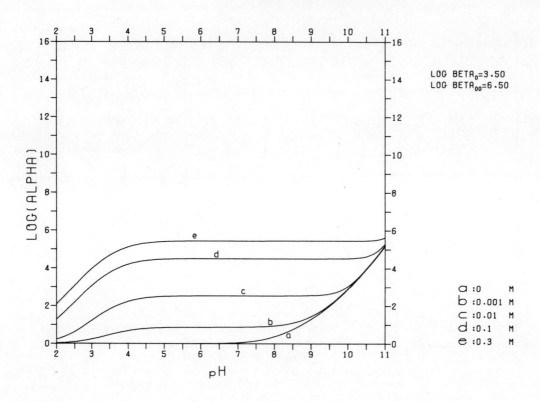
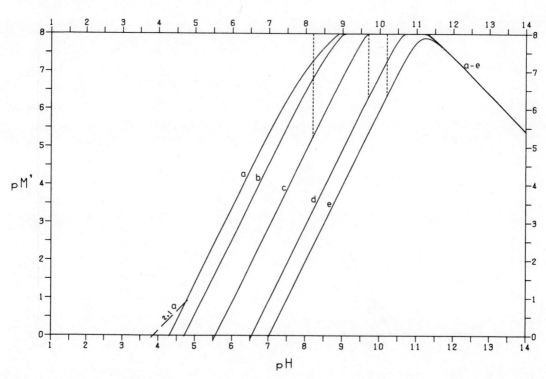

Pb-Carbonate-Pyridine-2,6-dicarboxylate

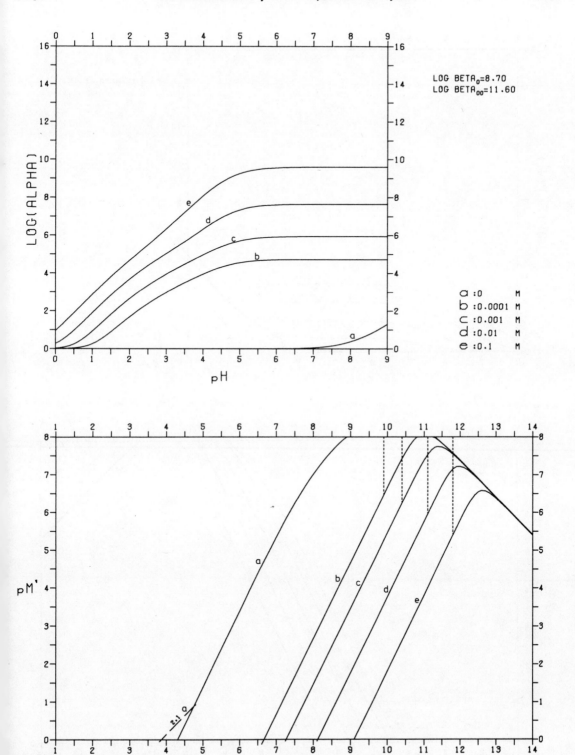

Pb-Carbonate-1,10-Phenanthroline [Ch. 28]

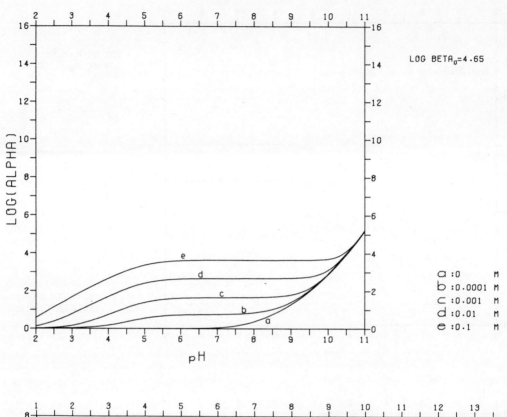

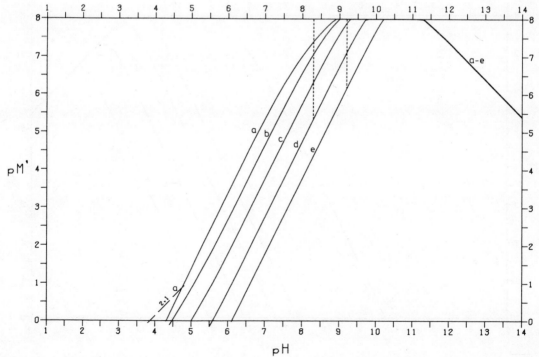

Pb-Carbonate-Tartrate

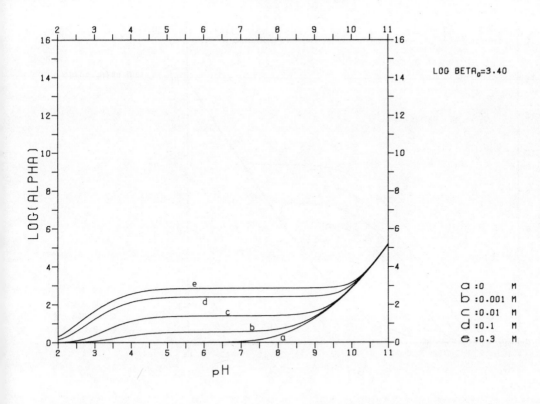

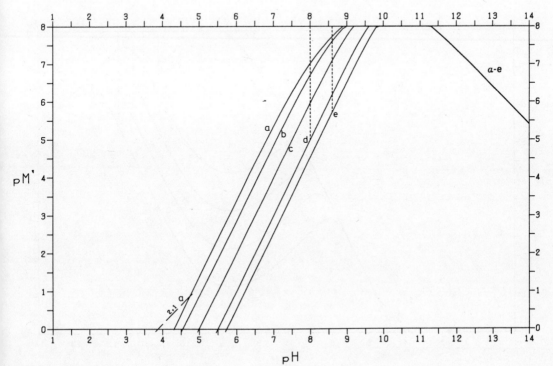

Pb-Carbonate-Tetren [Ch. 28]

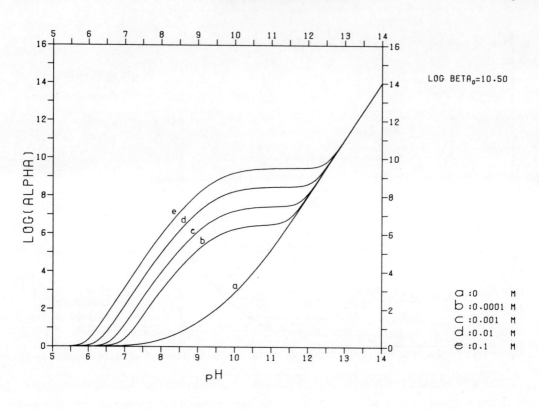

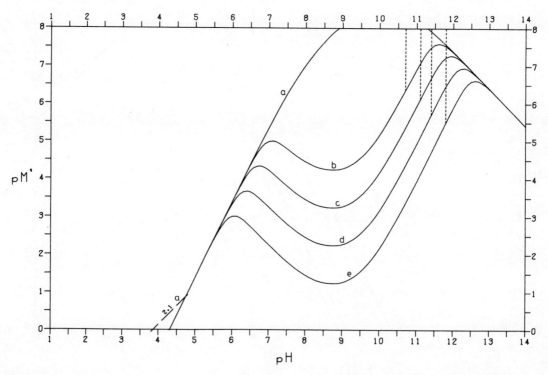

Pb-Carbonate-Tiron

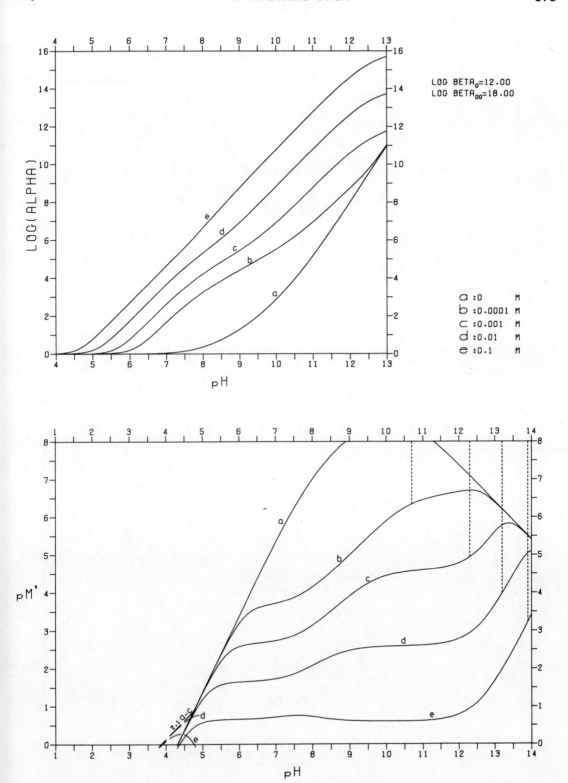

574   **Pb-Carbonate-Trien**   [Ch. 28

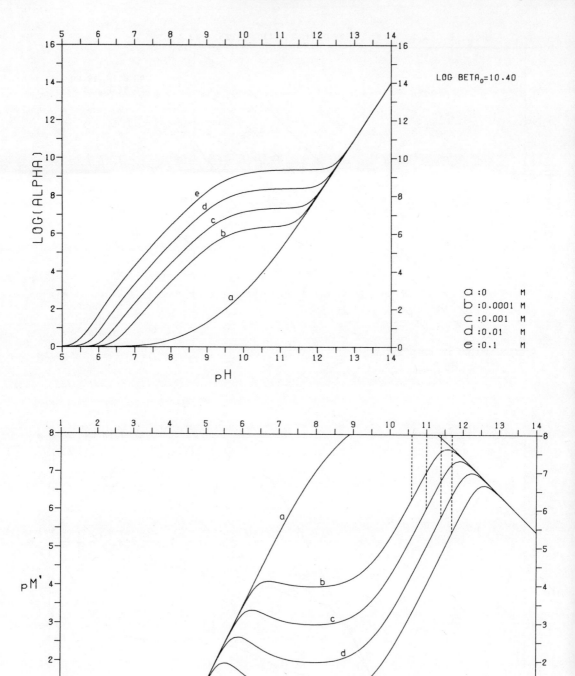

CHAPTER 29

Palladium (II)   Pd

A wide range of values is found in the literature for the hydrolysis constants of Pd^{2+}. The reason is that every trace of complexing ligand must be eliminated from a test solution to enable accurate measurements of hydrolysis equilibria to be made, because Pd^{2+} forms so many strong complexes. The values for log $*\beta_1$ and log $*\beta_2$ selected by Baes and Mesmer [1] seem to be reliable. Their values for log $*\beta_3$ and log $*\beta_4$ have such a large uncertainty that they will not be used; this means that the plots are less certain above pH 9.

Although thermodynamic calculations indicate that PdO is the stable solid phase, Pd^{2+} invariably precipitates as $Pd(OH)_2$ (log $*K_{s0} = -0.6$).

The plots were constructed with the following data.

$$\log *\beta_1 = -2.3 \qquad \log *K_{s0} = -0.6$$
$$\log *\beta_2 = -4.8$$

Pd^{2+} is stable in air-saturated solution. At potentials below E = (897 −59.1 pH) mV, $Pd(OH)_2(s)$ is reduced to metallic Pd. Pd^{2+} is reduced at potentials below E = (987 + 29.5 log $[Pd^{2+}]$) mV.

[1] C. F. Baes and R. E. Mesmer, *The Hydrolysis of Cations*, Wiley, New York, 1976.

Palladium(II)

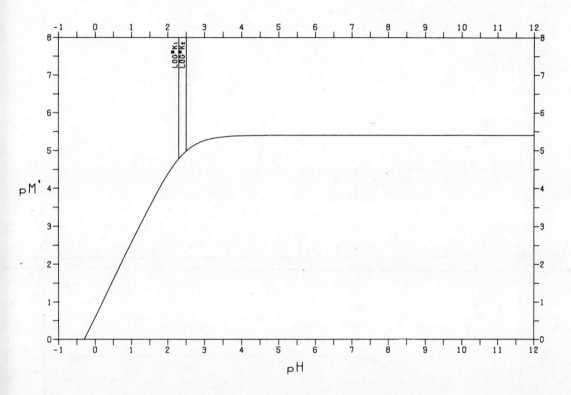

Palladium(II)-Acetylacetone [Ch. 29]

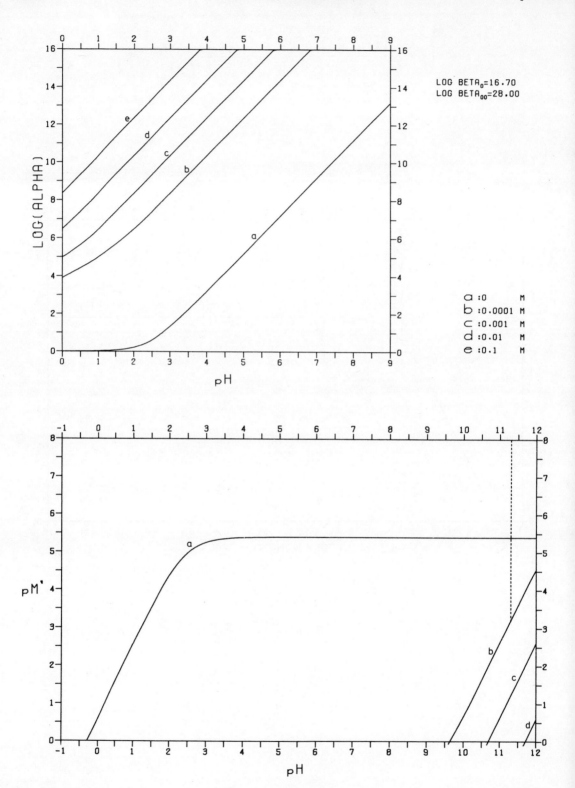

LOG BETA$_0$=16.70
LOG BETA$_{00}$=28.00

a : 0 M
b : 0.0001 M
c : 0.001 M
d : 0.01 M
e : 0.1 M

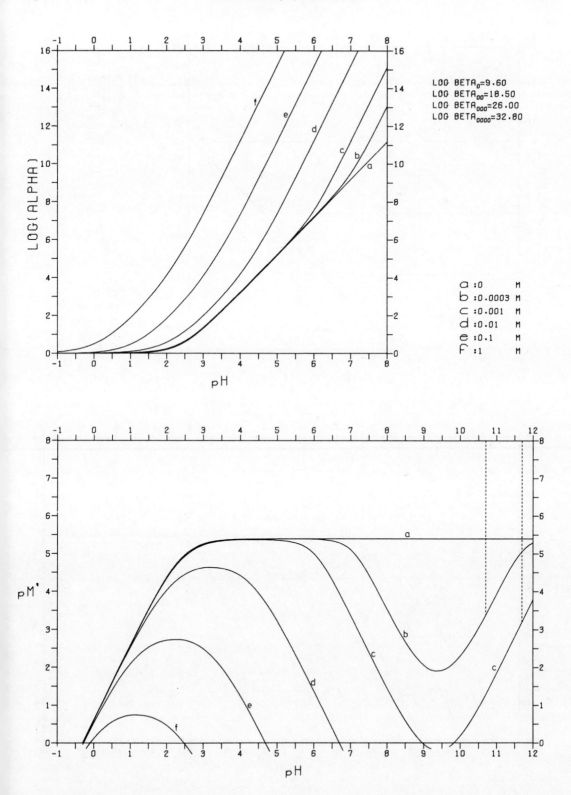

Palladium(II)-Bromide

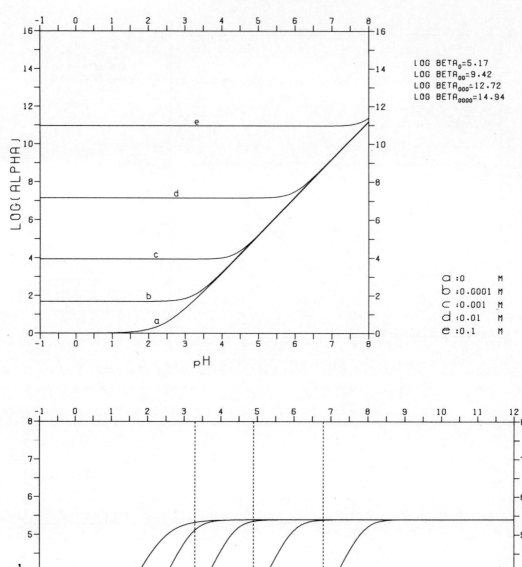

Palladium(II)-Chloride

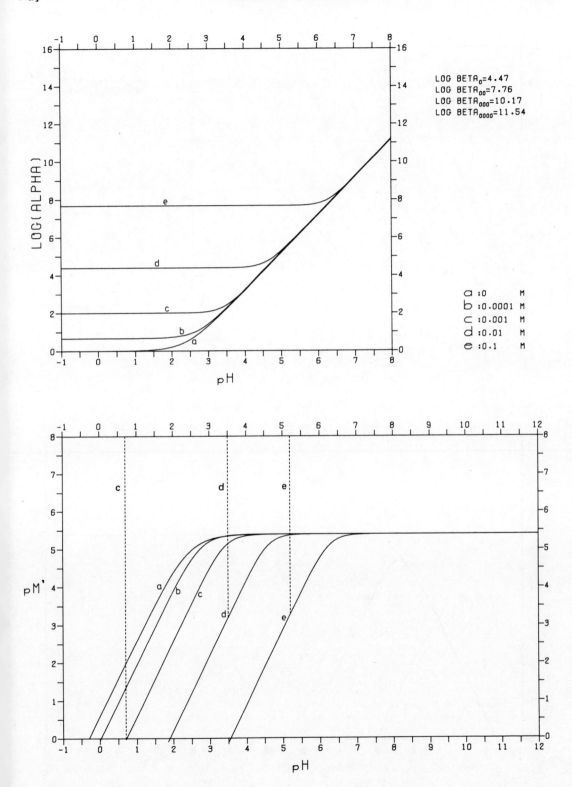

LOG BETA₀=4.47
LOG BETA₀₀=7.76
LOG BETA₀₀₀=10.17
LOG BETA₀₀₀₀=11.54

a : 0   M
b : 0.0001 M
c : 0.001   M
d : 0.01   M
e : 0.1   M

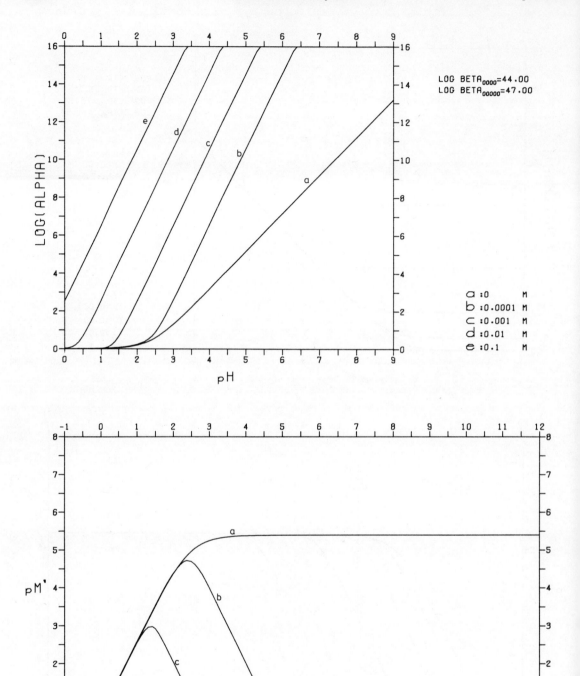

Palladium(II)–EDTA

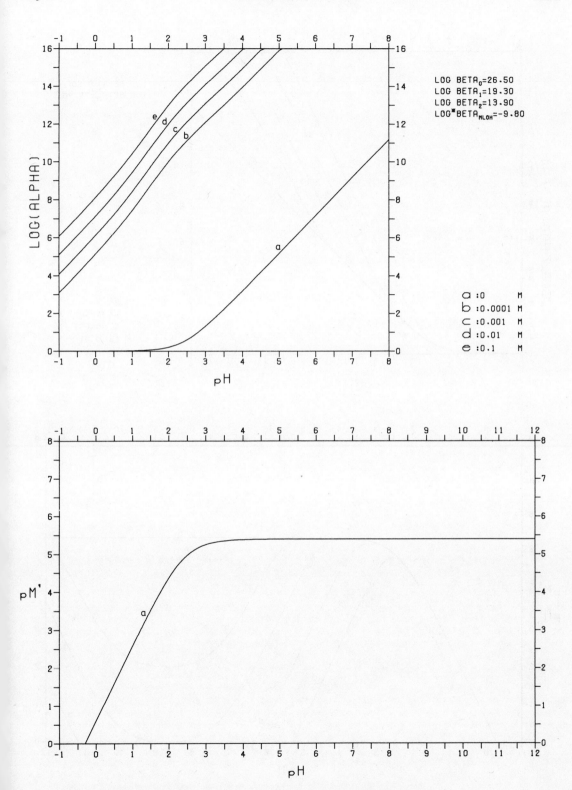

LOG BETA$_0$=26.50
LOG BETA$_1$=19.30
LOG BETA$_2$=13.90
LOG*BETA$_{MLOH}$=-9.80

a : 0   M
b : 0.0001 M
c : 0.001 M
d : 0.01   M
e : 0.1   M

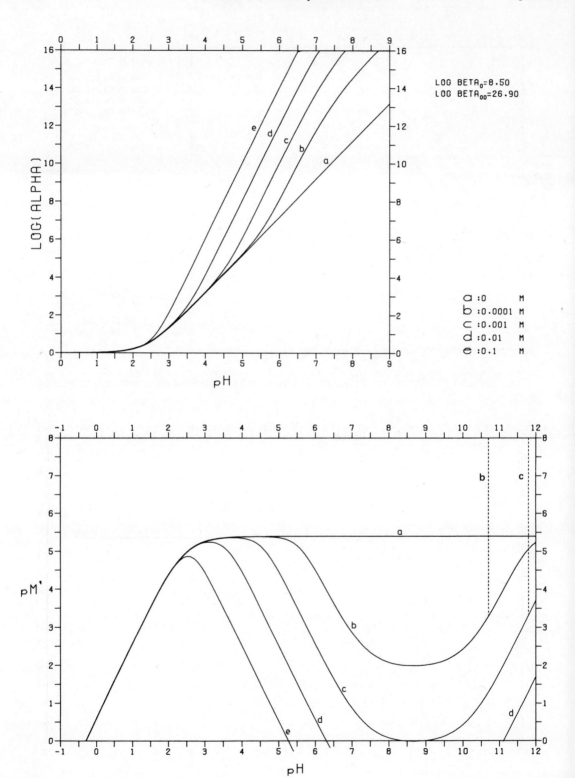

Palladium(II)-Glycine

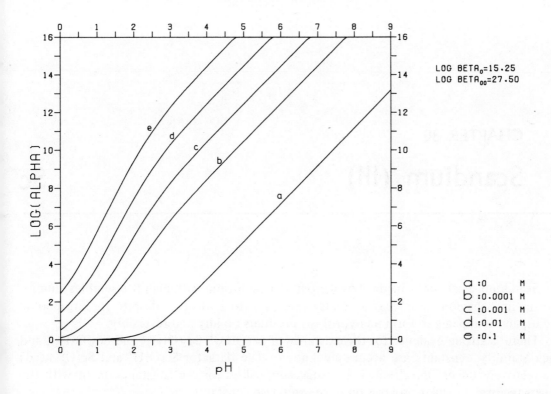

LOG BETA$_0$=15.25
LOG BETA$_{00}$=27.50

a : 0   M
b : 0.0001 M
c : 0.001 M
d : 0.01   M
e : 0.1   M

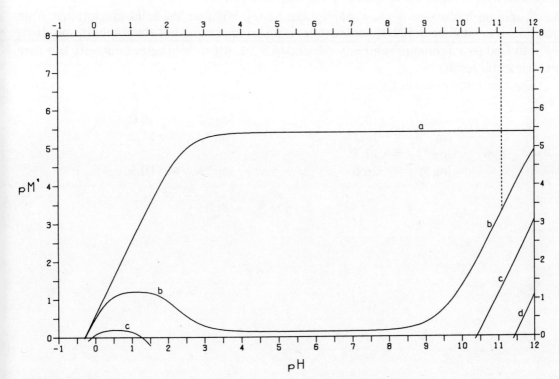

CHAPTER 30

Scandium (III)   Sc

Scandium exists only in the tervalent state in solution. Although it resembles aluminium in its behaviour, Sc^{3+} has a greater tendency to hydrolyse despite its larger ionic radius, and it appears to form its hydrolysis products rapidly and reversibly.

There is strong evidence for the existence of $Sc(OH)^{2+}$, $Sc_2(OH)_2^{4+}$ and $Sc_3(OH)_5^{4+}$, and their stability constants are accurately known. The data for $Sc(OH)_2^+$ and $Sc(OH)_3(aq)$ are relatively uncertain, although the suggested values for the constants do fit with the others to form a regular progression in the stepwise constants.

Scandium hydroxide is more stable than Sc_2O_3. Whether the solid precipitated from solution is crystalline $Sc(OH)_3$ or $ScO(OH)$ remains uncertain, but the solubility products of both lead to a minimum solubility of about 10^{-7}M, which is in agreement with our own experimental results.

The constants used for the plots are:

$\log {}^*\beta_1 = -4.9$ $\log {}^*\beta_{22} = -6.0$
$\log {}^*\beta_2 = -10.7$ $\log {}^*\beta_{53} = -17.2$
$\log {}^*\beta_3 = -17.3$
$\log {}^*\beta_4 = -26.6$ $\log {}^*K_{s0} = 10.5$

Scandium(III)

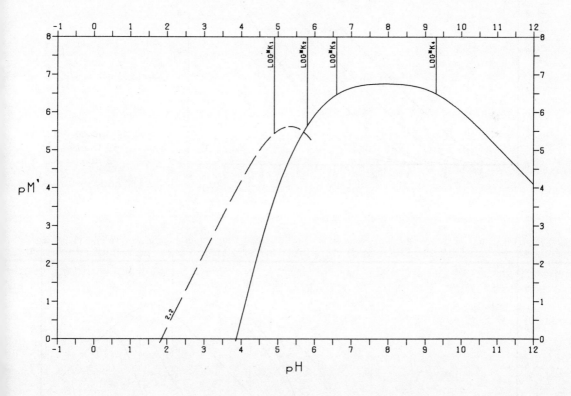

588   **Scandium(III)-Acetylacetone**   [Ch. 30

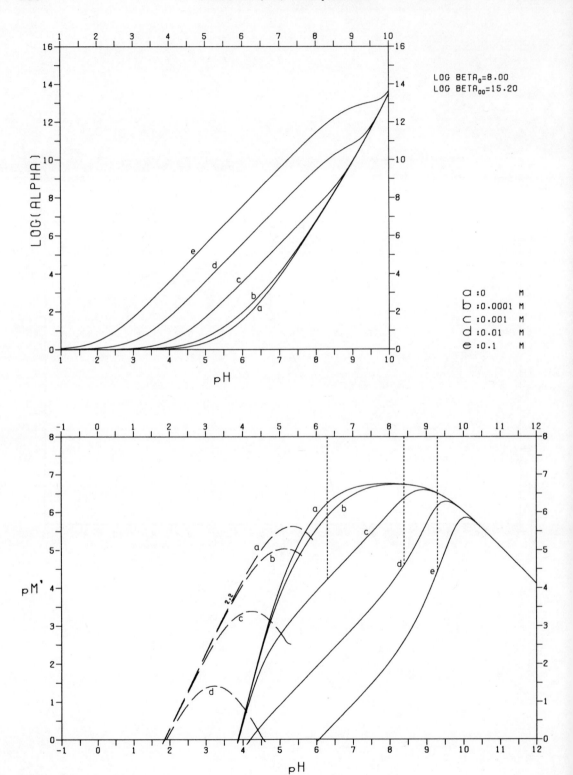

Scandium(III)-Citrate

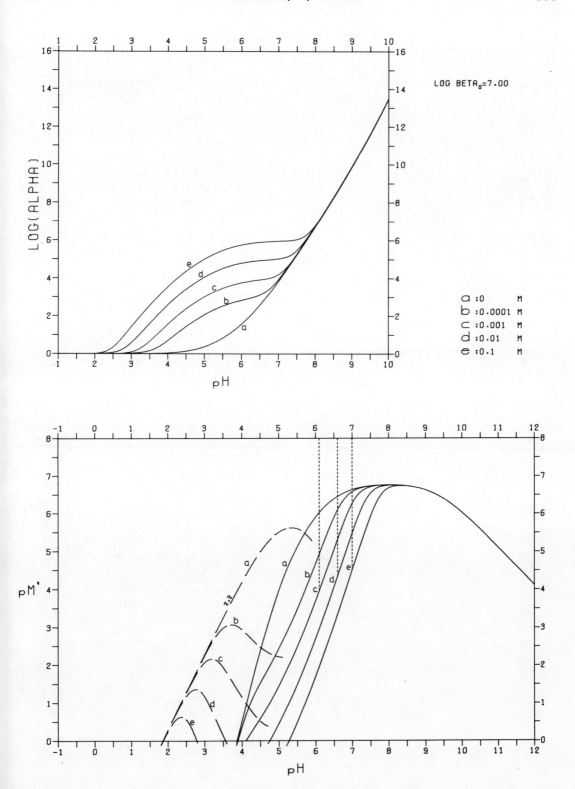

Scandium(III)–DCTA

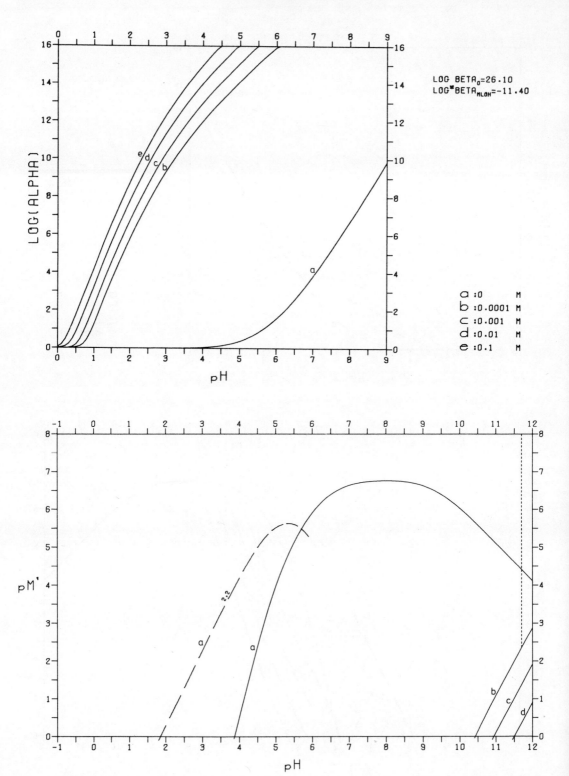

Scandium(III)-DTPA

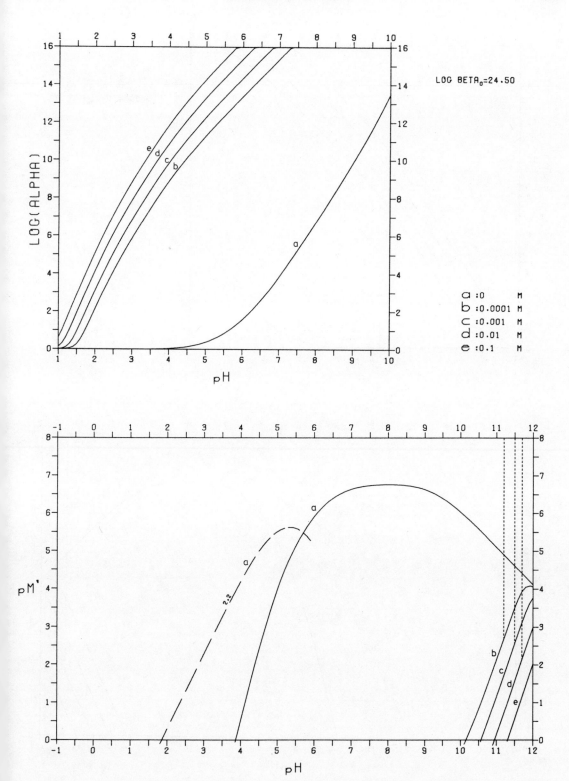

LOG BETA$_0$ = 24.50

a : 0 M
b : 0.0001 M
c : 0.001 M
d : 0.01 M
e : 0.1 M

Scandium(III)-EDTA [Ch. 30]

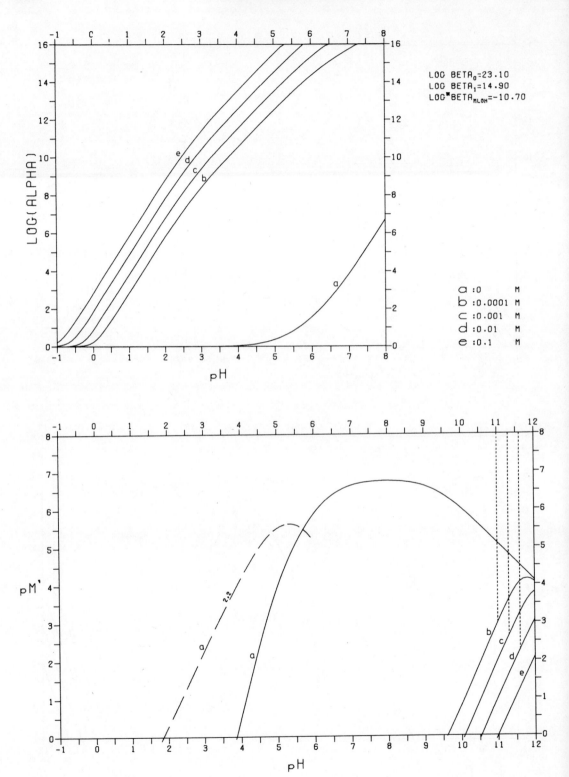

Scandium(III)-Fluoride

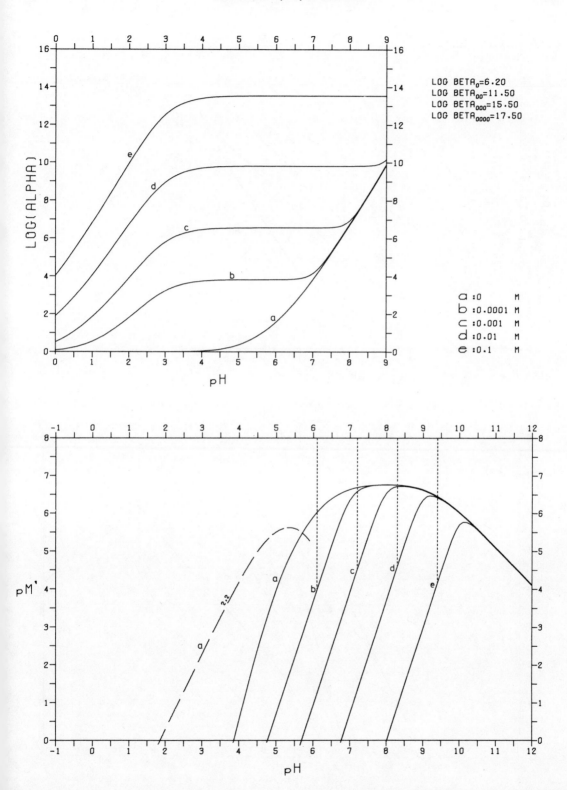

LOG BETA$_0$=6.20
LOG BETA$_{00}$=11.50
LOG BETA$_{000}$=15.50
LOG BETA$_{0000}$=17.50

a : 0 M
b : 0.0001 M
c : 0.001 M
d : 0.01 M
e : 0.1 M

Scandium(III)-Iminodiacetate [Ch. 30]

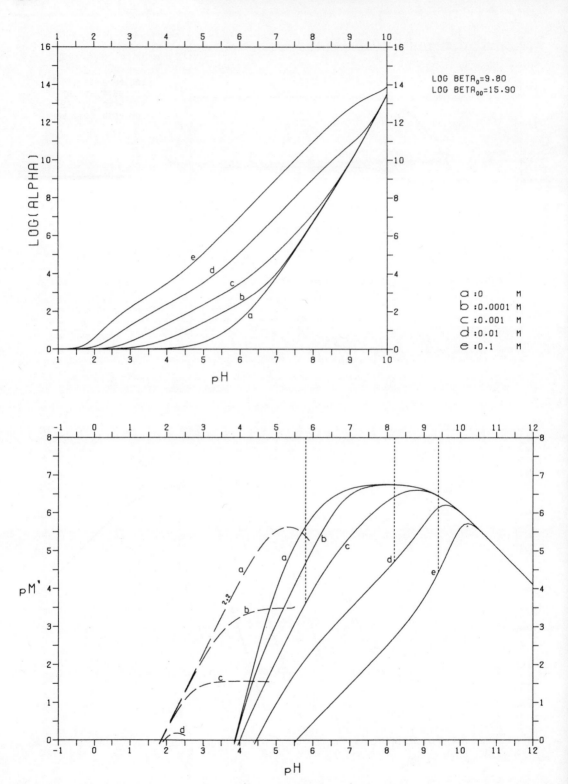

Scandium(III)-Oxalate

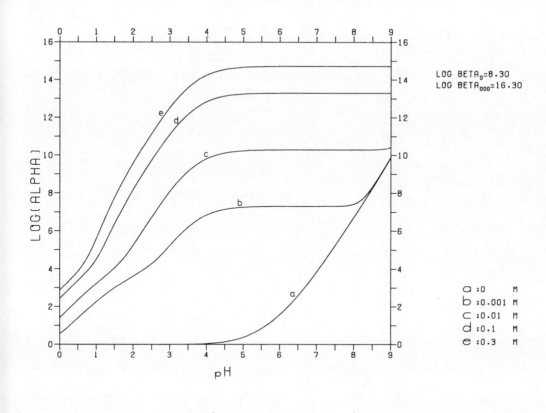

LOG BETA$_0$=8.30
LOG BETA$_{000}$=16.30

a : 0   M
b : 0.001 M
c : 0.01 M
d : 0.1   M
e : 0.3   M

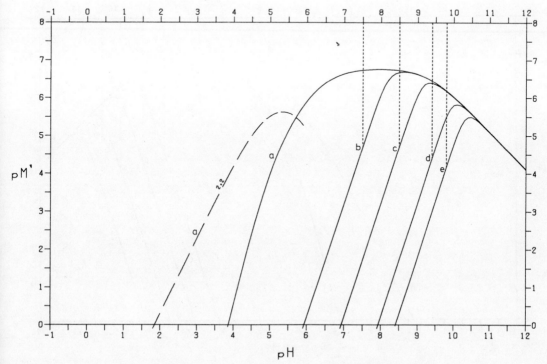

Scandium(III)-Pyridine-2,6-dicarboxylate

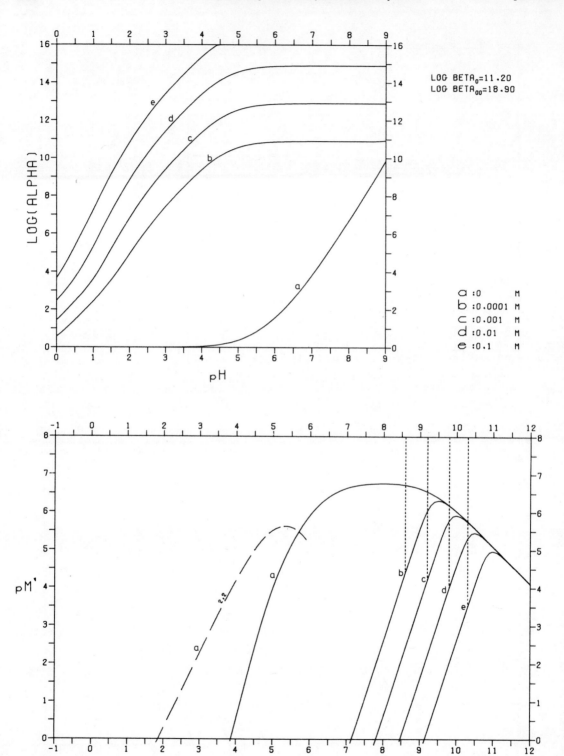

Scandium(III)-Sulphate

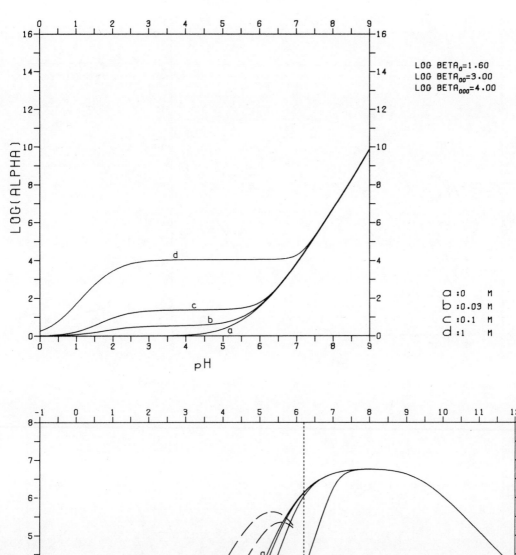

LOG BETA$_0$=1.60
LOG BETA$_{00}$=3.00
LOG BETA$_{000}$=4.00

a : 0   M
b : 0.03   M
c : 0.1   M
d : 1   M

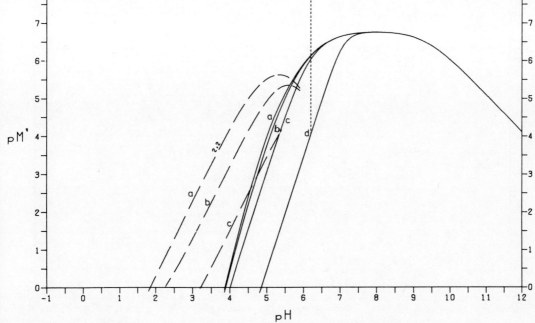

Scandium(III)-Tartrate

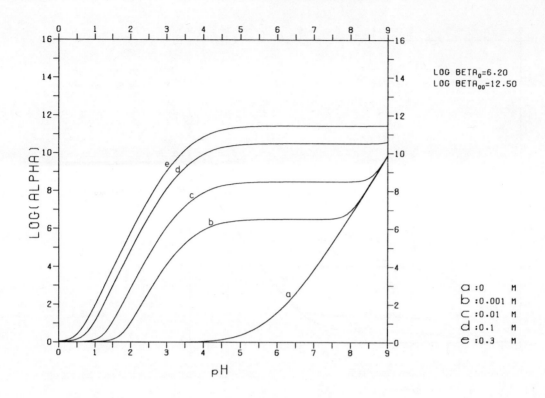

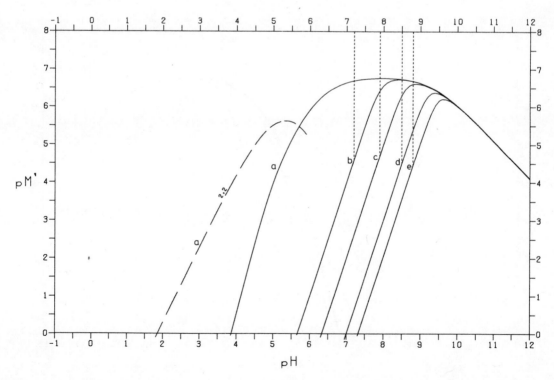

CHAPTER 31

Samarium (III)   Sm

Very little information about the hydrolysis of Sm^{3+} ions is available, but there is a little evidence for polycomplexation. The data used for construction of the plots were estimated by comparison with other rare earth elements.

$$\log {}^*\beta_1 = -\ 7.9 \qquad\qquad \log {}^*\beta_{22} = -13.7$$
$$\log {}^*\beta_2 = -16.5$$
$$\log {}^*\beta_3 = -25$$
$$\log {}^*\beta_4 = -37 \qquad\qquad \log {}^*K_{s0} = \ 18.5$$

From the value of the solubility product of samarium carbonate it is apparent that the solubility of the carbonate in air-saturated solution is equal to that of the hydroxide. This means that the pM' – pH plots for the carbonate system are approximately the same as for the hydroxide system.

Samarium(III)

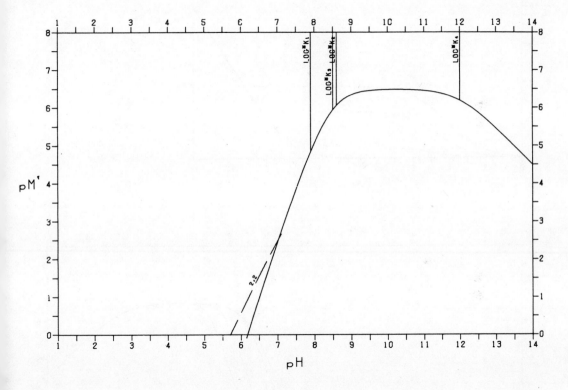

Samarium(III)-Acetate

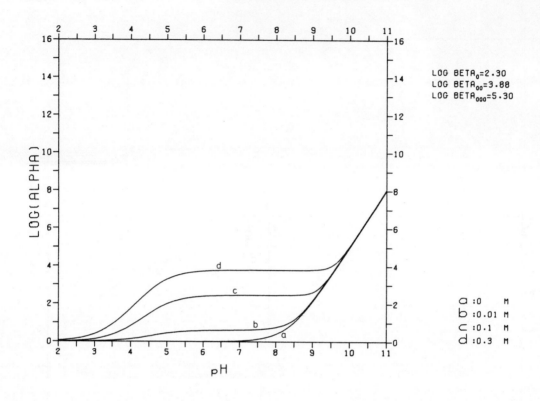

LOG BETA$_0$=2.30
LOG BETA$_{00}$=3.88
LOG BETA$_{000}$=5.30

a : 0   M
b : 0.01   M
c : 0.1   M
d : 0.3   M

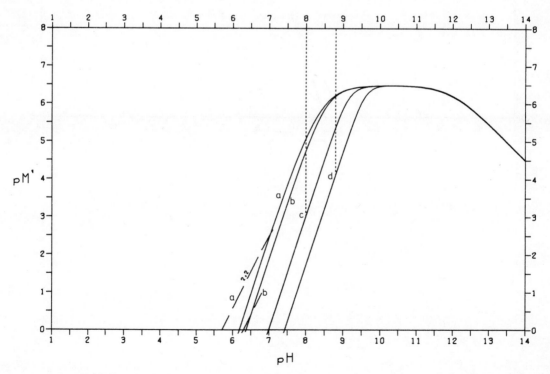

Samarium(III)-Acetylacetone

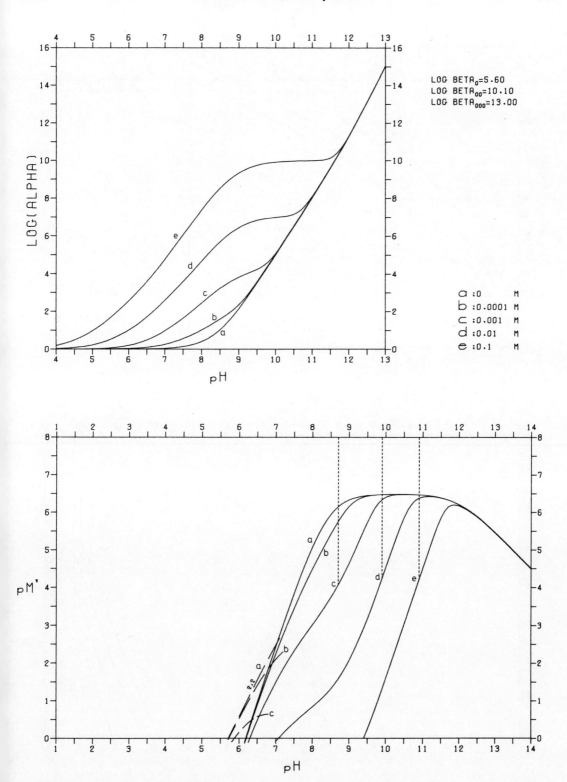

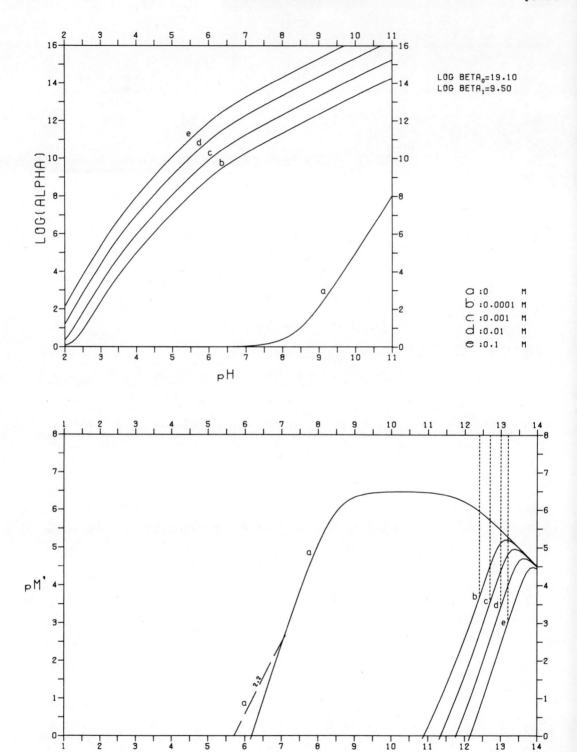

Samarium(III)-DTPA

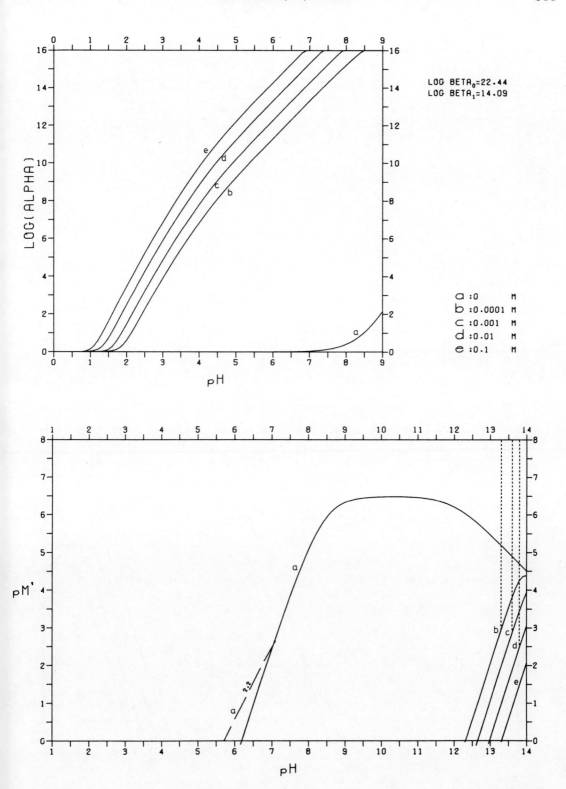

LOG BETA$_0$ = 22.44
LOG BETA$_1$ = 14.09

a : 0   M
b : 0.0001 M
c : 0.001   M
d : 0.01   M
e : 0.1   M

Samarium(III)-EDTA

[Ch. 31

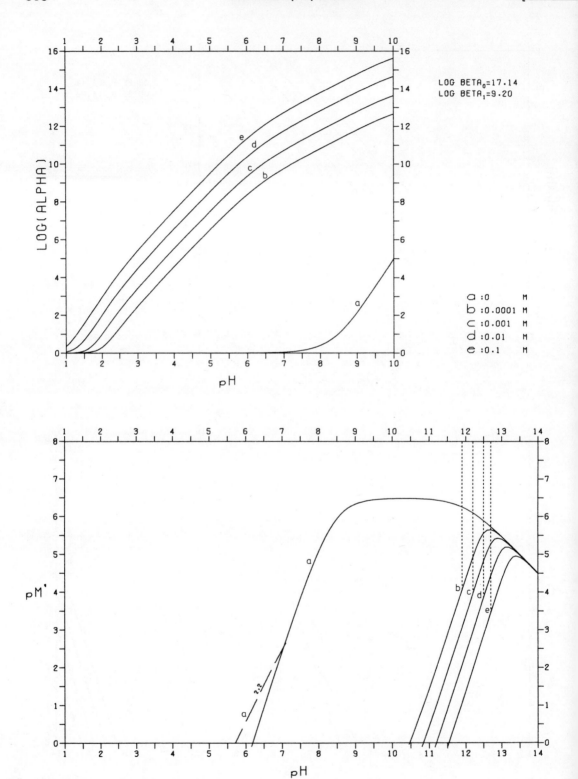

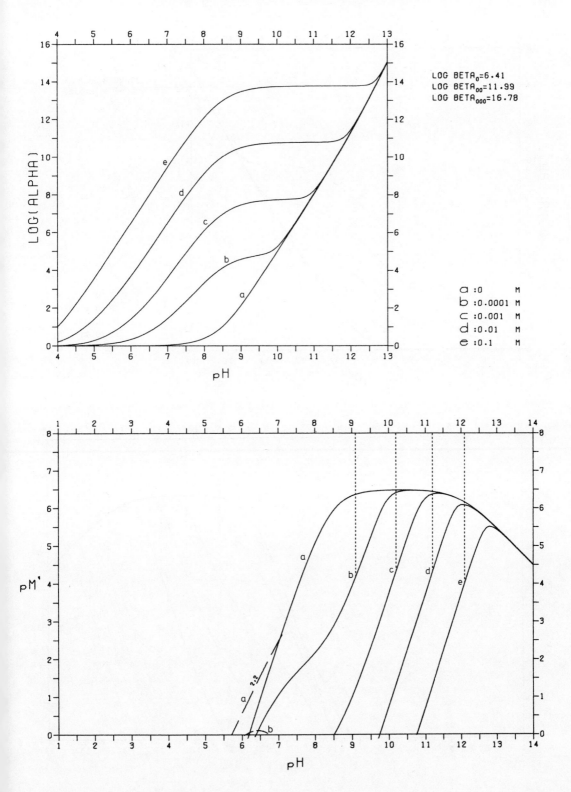

Samarium(III)-Iminodiacetate

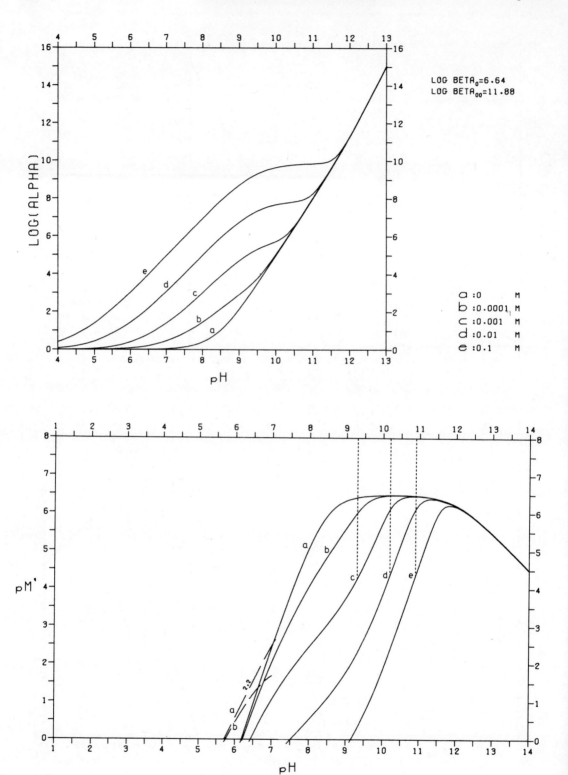

Samarium(III)-Pyridine-2,6-dicarboxylate

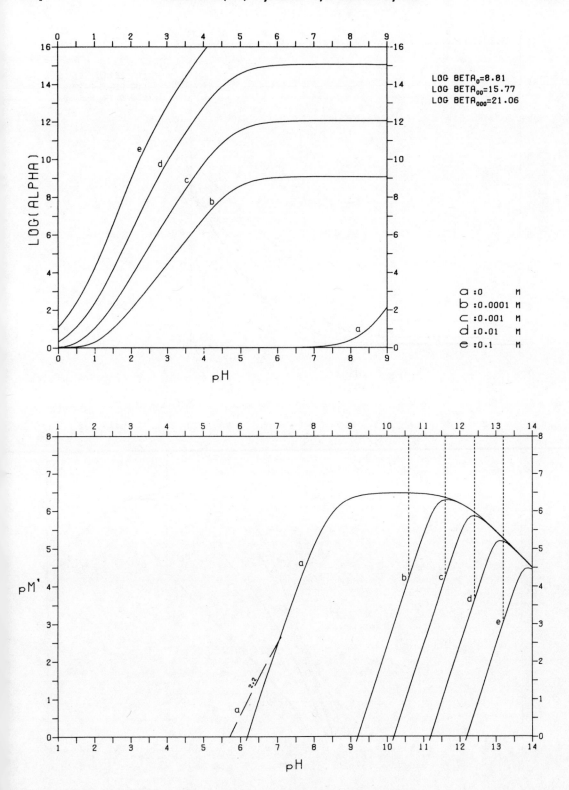

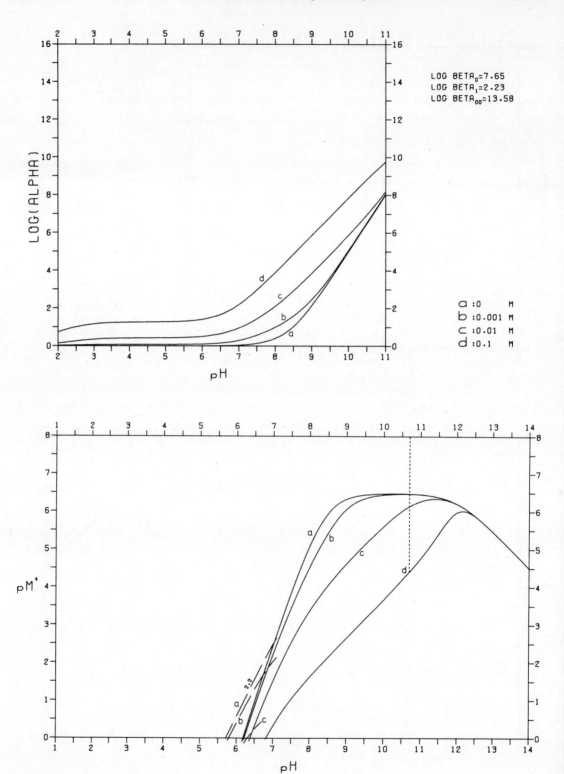

Samarium(III)-Tartrate

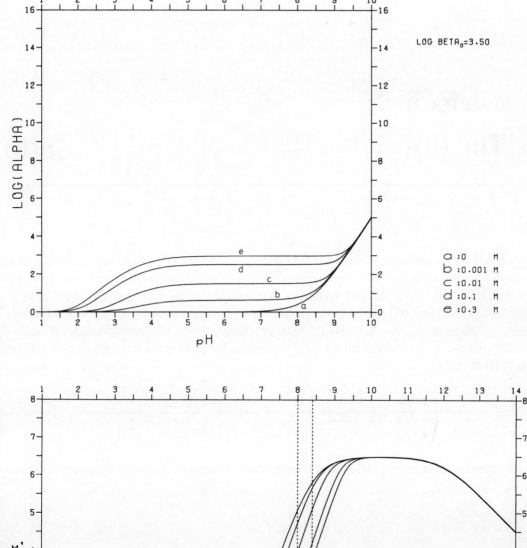

CHAPTER 32
Tin (II)   Sn(II)

The stability constants of $Sn(OH)^+$, $Sn_2(OH)_2^{2+}$ and $Sn_3(OH)_4^{2+}$ are accurately known. The polycomplexes have been investigated in solutions which were supersaturated with respect to SnO(s); they do not occur in normal solutions. The stability constants of $Sn(OH)_2(aq)$ and $Sn(OH)_3^-$ have been determined from measurements of the solubility of SnO in dilute solutions. $Sn(OH)_2(s)$ may precipitate from supersaturated solution, but generally the more stable solid phase SnO separates. The data used for construction of the plots refer to SnO.

$$\log {}^*\beta_1 = -\ 3.64 \qquad \log {}^*\beta_{22} = -\ 4.99$$
$$\log {}^*\beta_2 = -\ 7.31 \qquad \log {}^*\beta_{43} = -\ 7.33$$
$$\log {}^*\beta_3 = -16.62 \qquad \log {}^*K_{s0} = \ \ \ 2.0$$

In aqueous solution, Sn(II) is stable only when oxygen is absent; it is readily oxidized by air.

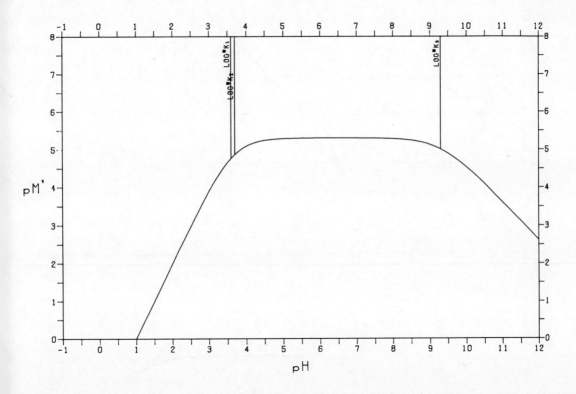

Tin(II)-Bromide

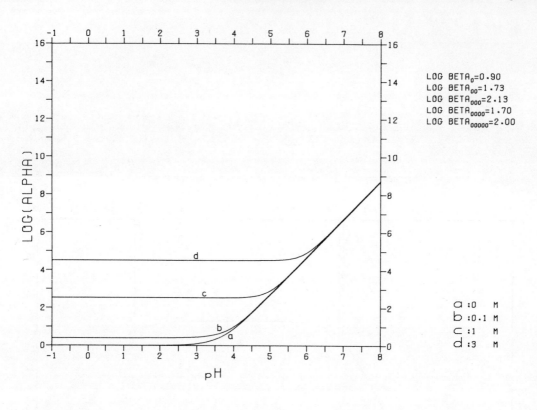

LOG BETA$_0$=0.90
LOG BETA$_{00}$=1.73
LOG BETA$_{000}$=2.13
LOG BETA$_{0000}$=1.70
LOG BETA$_{00000}$=2.00

a : 0 M
b : 0.1 M
c : 1 M
d : 3 M

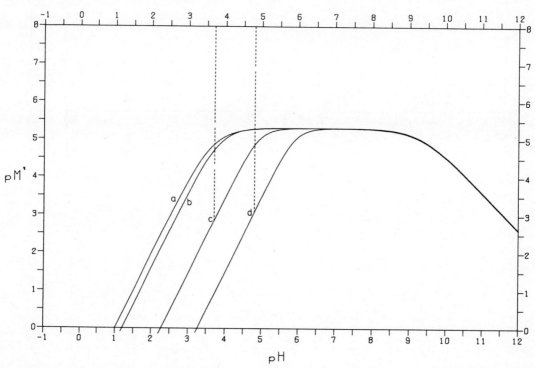

Sn(II) — Tin(II)-Chloride

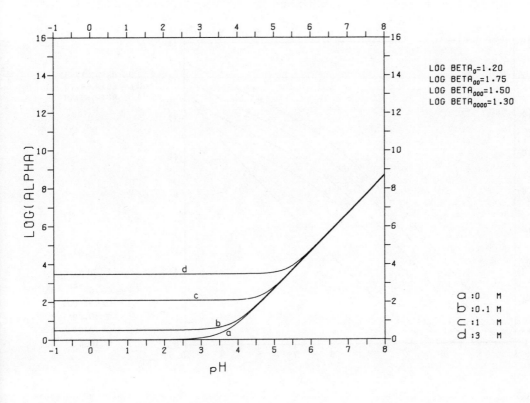

LOG BETA$_0$=1.20
LOG BETA$_{00}$=1.75
LOG BETA$_{000}$=1.50
LOG BETA$_{0000}$=1.30

a : 0   M
b : 0.1 M
c : 1   M
d : 3   M

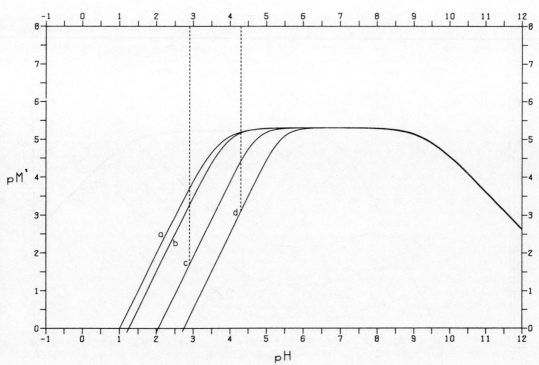

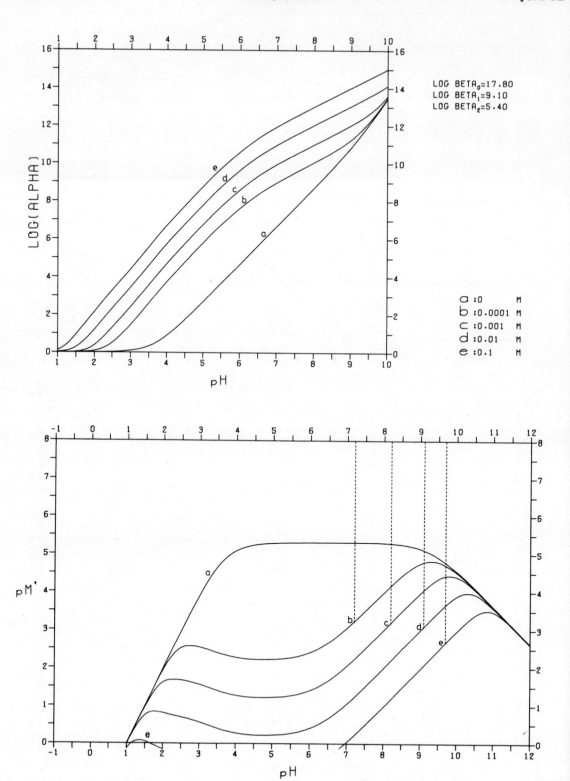

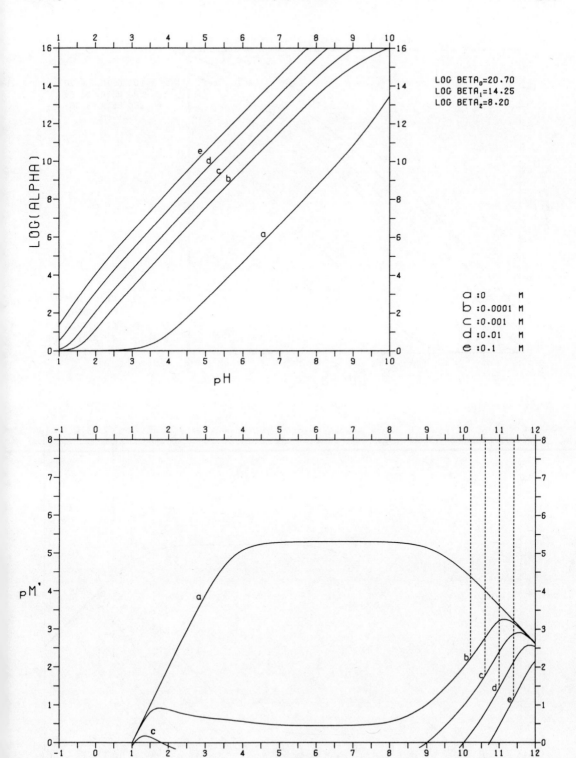

Tin(II)-EDTA [Ch. 32]

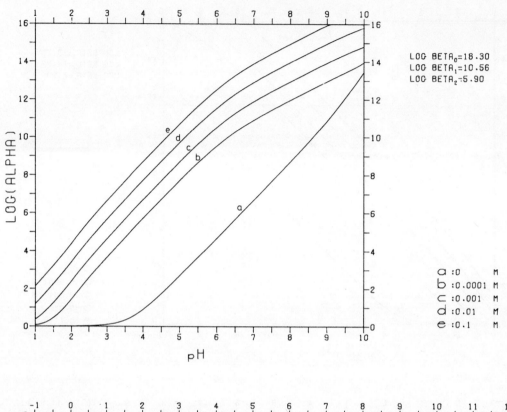

LOG BETA$_0$=18.30
LOG BETA$_1$=10.56
LOG BETA$_2$=5.90

a : 0   M
b : 0.0001 M
c : 0.001 M
d : 0.01   M
e : 0.1   M

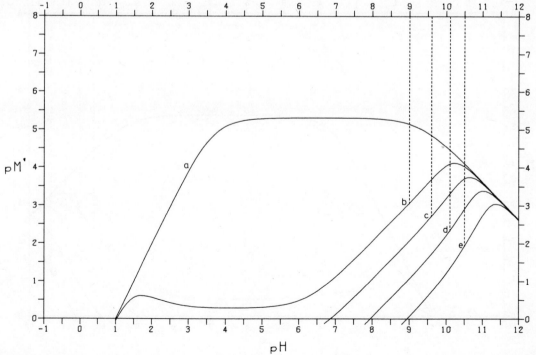

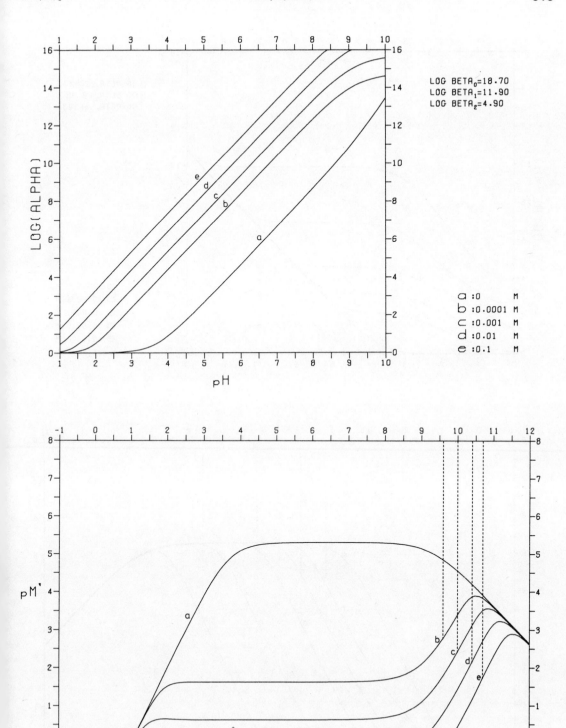

Tin(II)-Fluoride [Ch. 32]

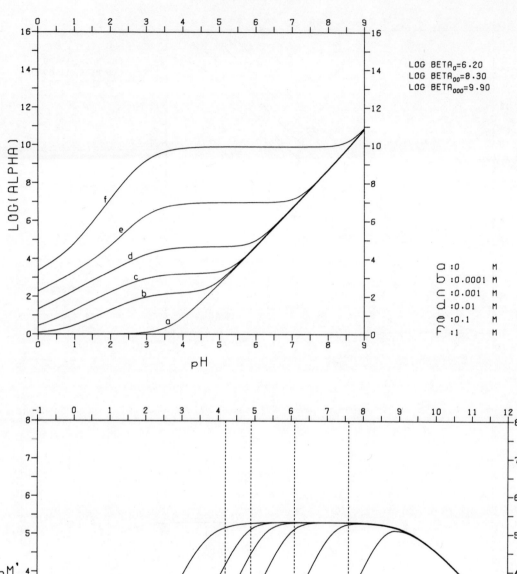

Tin(II)-Tartrate

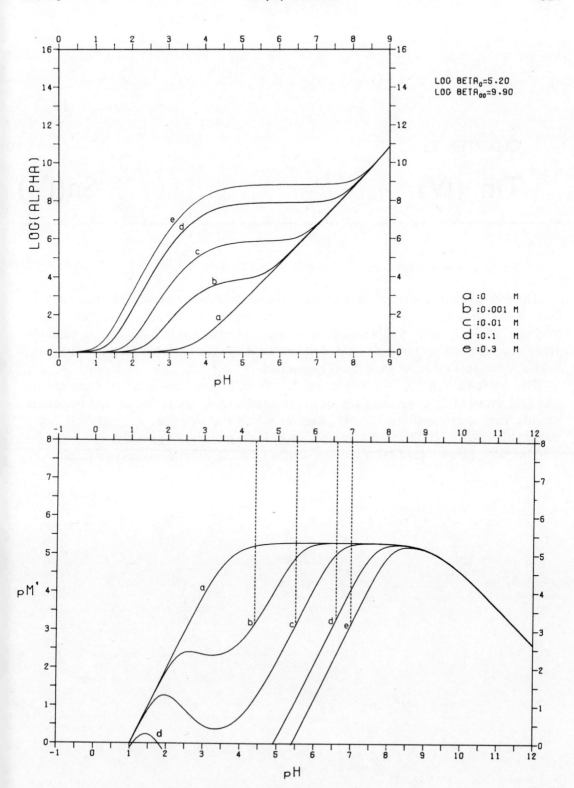

CHAPTER 33
Tin (IV)   Sn(IV)

The +IV oxidation state of tin is particularly stable in natural environments, and solid SnO_2 is the main source of the element. Little quantitative information is available on the solubility of SnO_2. It is soluble only in strong acid and strong base, and its solubility depends greatly on its physical state. For practical applications, log $*K_{s0}$ = −3 appears to be a reasonable value for the solubility product.

The constants log $*\beta_1$ = −1 and log $*\beta_6$ = −22 have been deduced from electrochemical data [1]. The missing data for the constants from $*\beta_2$ to $*\beta_5$ are less important, because they correspond to the pH range in which the solubility of Sn(IV) is out of measurable range (<10^{-9}M), and construction of log α plots is not possible.

The pM' − pH plot for EDTA was constructed directly from experimental results [1].

[1] J. Kragten, *Talanta*, 1975, 22, 505.

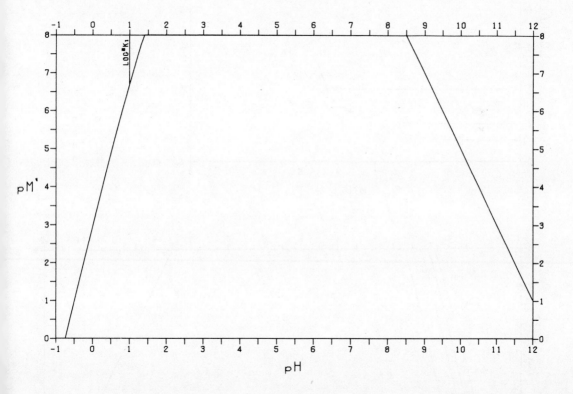

Tin(IV)-Chloride

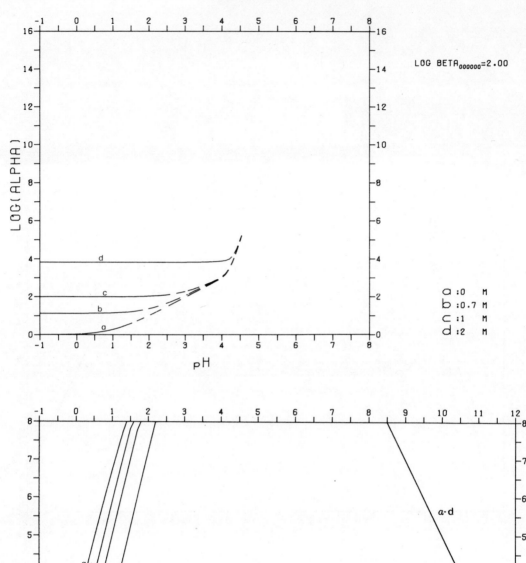

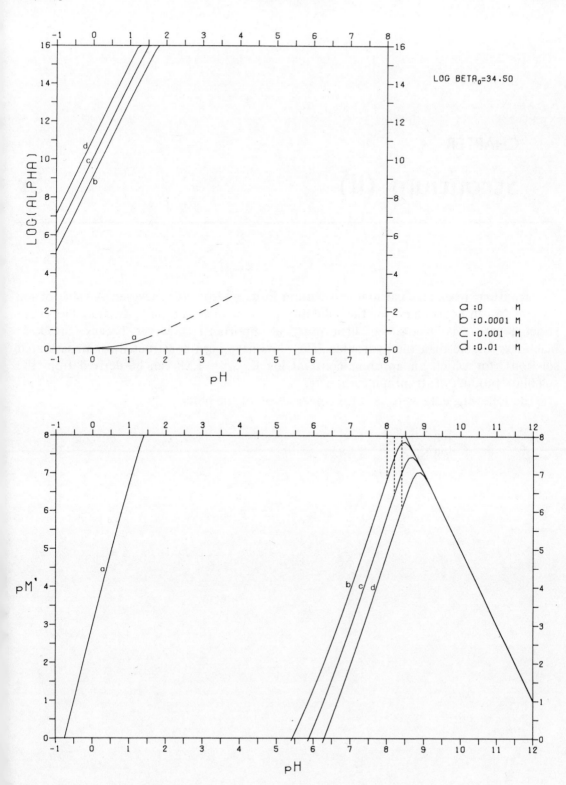

CHAPTER 34

Strontium (II)   Sr

Strontium resembles calcium and barium in its hydrolysis behaviour. A value of log $*K_{s0} = 24$ can be assumed for the solubility product of strontium hydroxide. Plots constructed with this value are of little practical importance, however, because $SrCO_3$ is much less soluble than the hydroxide. The plots have been constructed for air-saturated solutions, for which an apparent constant log $K_{carb} = 12.8$ can be derived from the solubility product of strontium carbonate.

The following data were used for construction of the plots.

$$\log *\beta_1 = -13.4$$
$$\log K_{carb} = 12.8$$

Sr–Carbonate

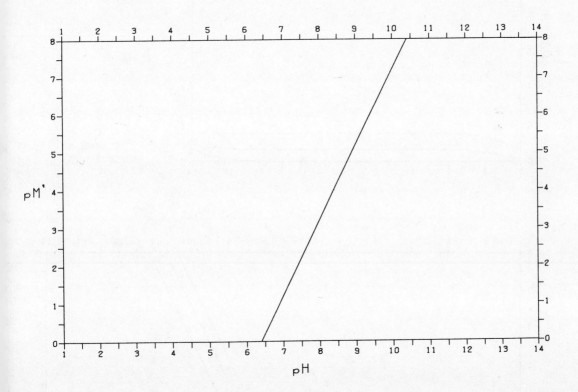

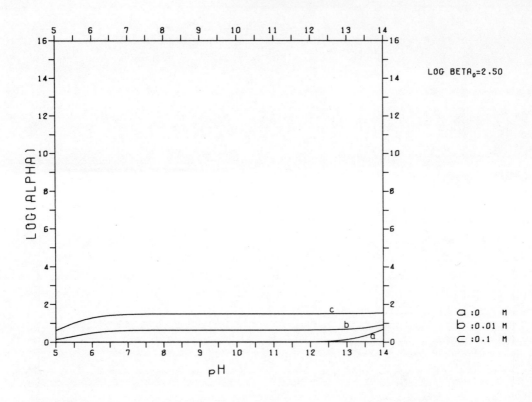
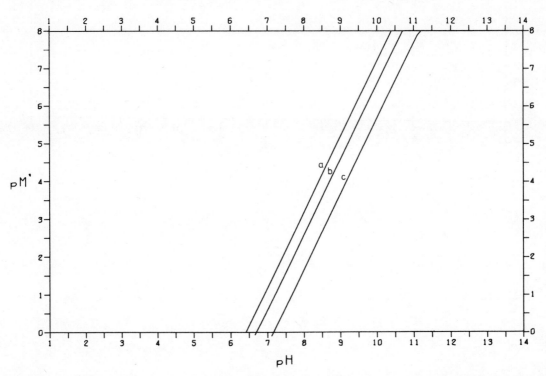

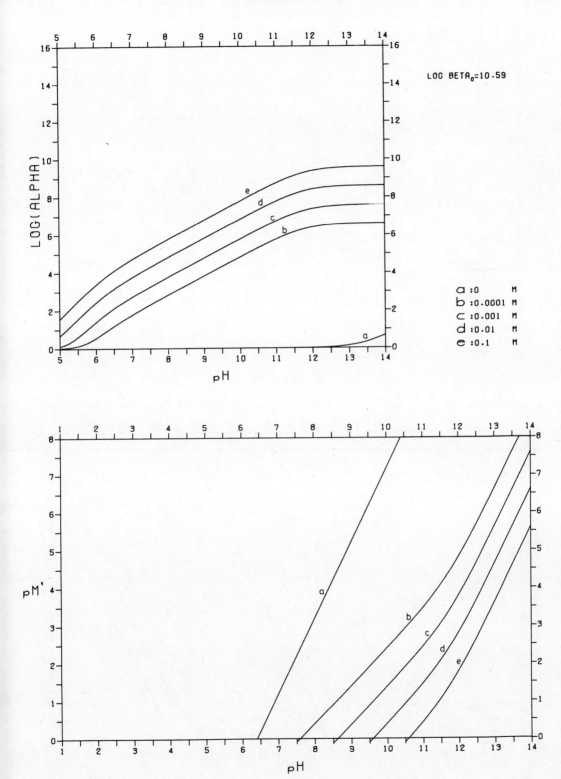

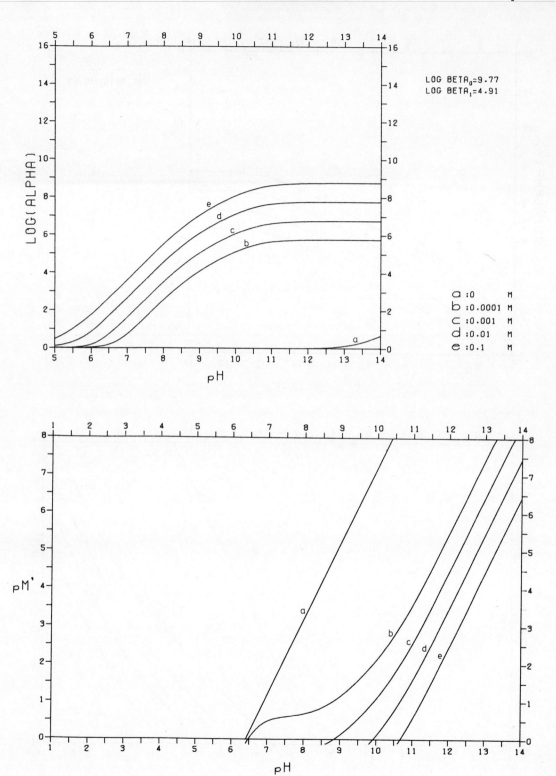

Sr-Carbonate-EDTA

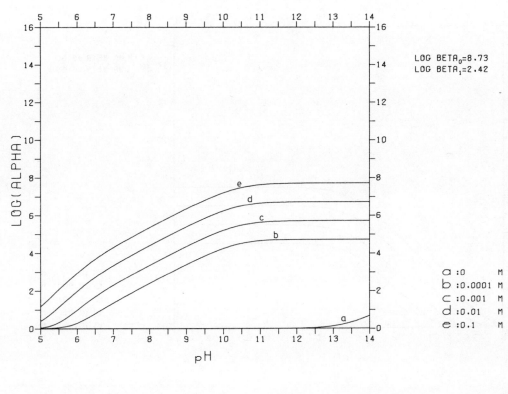

LOG BETA$_0$=8.73
LOG BETA$_1$=2.42

a : 0   M
b : 0.0001 M
c : 0.001   M
d : 0.01   M
e : 0.1   M

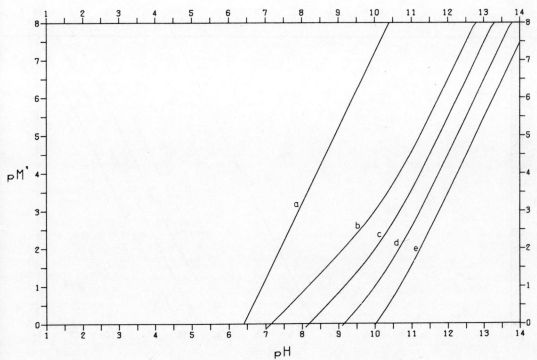

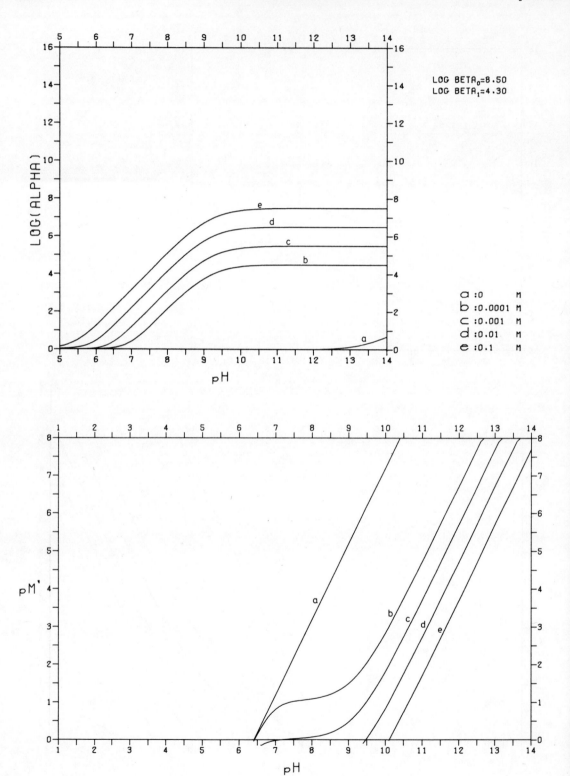

Sr-Carbonate-Iminodiacetate

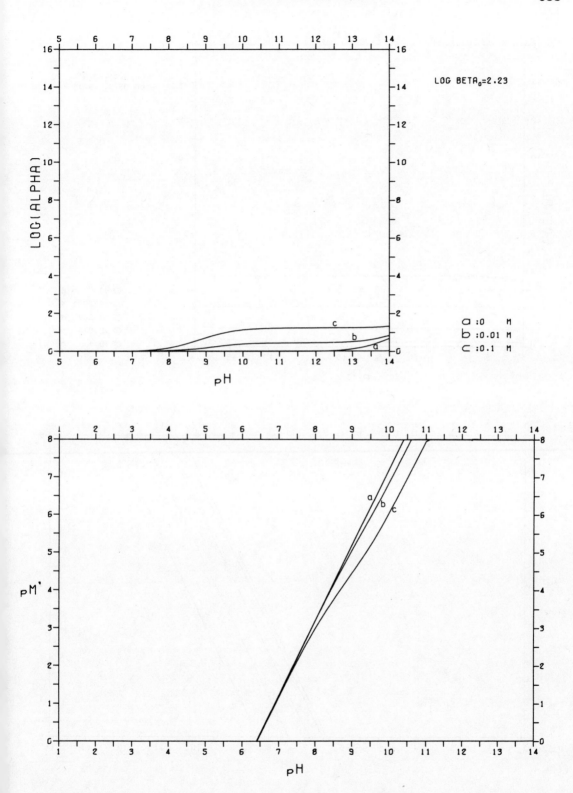

634 Sr-Carbonate-Pyridine-2,6-dicarboxylate [Ch. 34

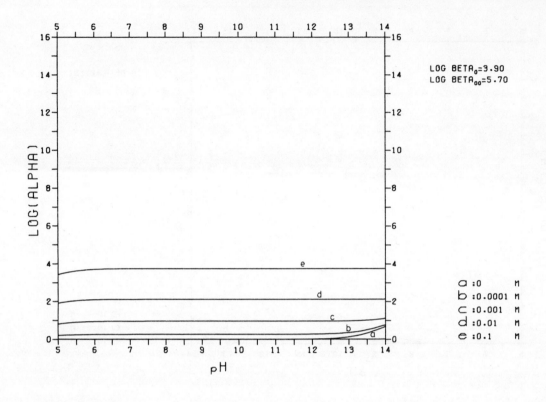

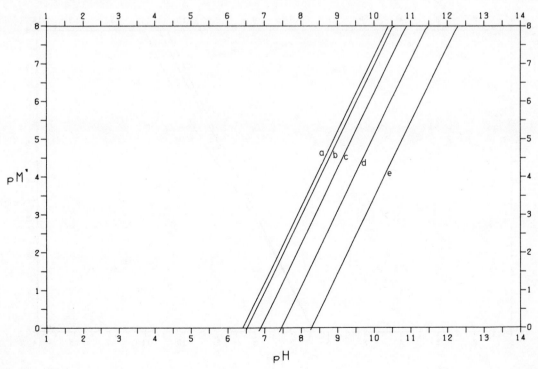

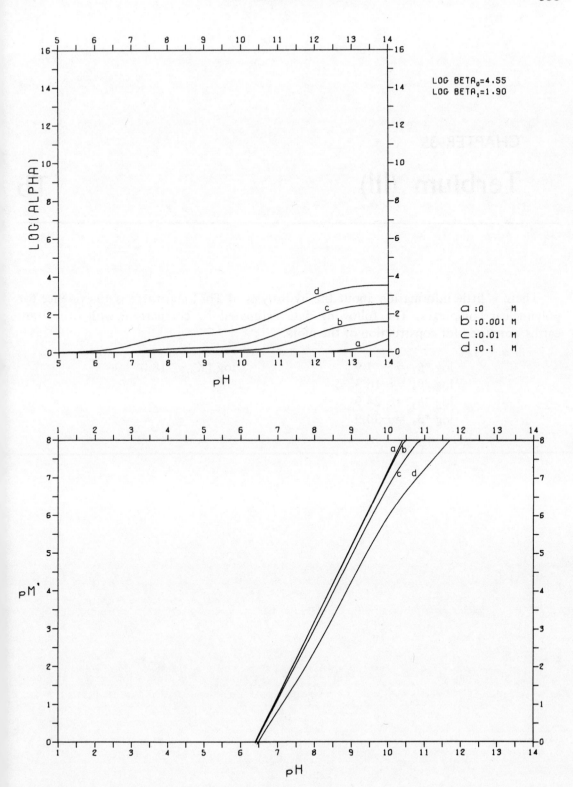

CHAPTER 35
Terbium (III)   Tb

There is little information about the hydrolysis of Tb^{3+}, and there is no evidence for polynuclear complexes. The following data estimated by comparison with other rare earths were used for construction of the plots.

$$\log {}^*\beta_1 = -7.9 \qquad \log {}^*K_{s0} = 18.0$$
$$\log {}^*\beta_2 = -16.3$$
$$\log {}^*\beta_3 = -24.9$$
$$\log {}^*\beta_4 = -34.9$$

Terbium(III)

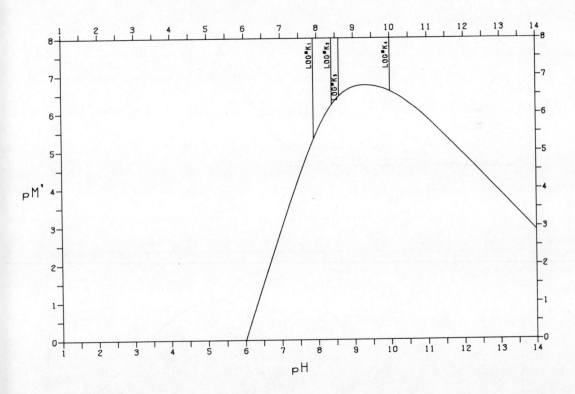

Terbium(III)-Acetate [Ch. 35

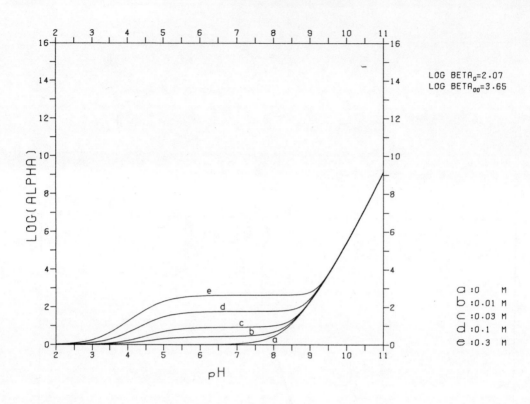

LOG BETA$_0$=2.07
LOG BETA$_{00}$=3.65

a : 0 M
b : 0.01 M
c : 0.03 M
d : 0.1 M
e : 0.3 M

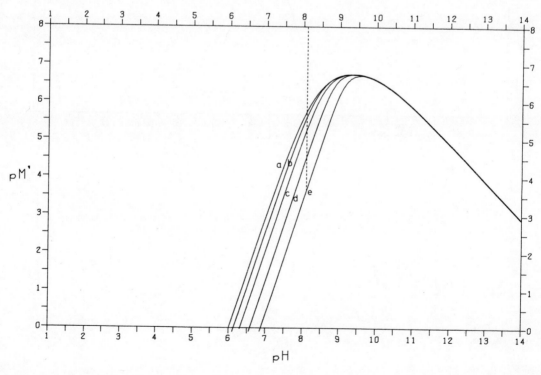

Terbium(III)-Acetylacetone

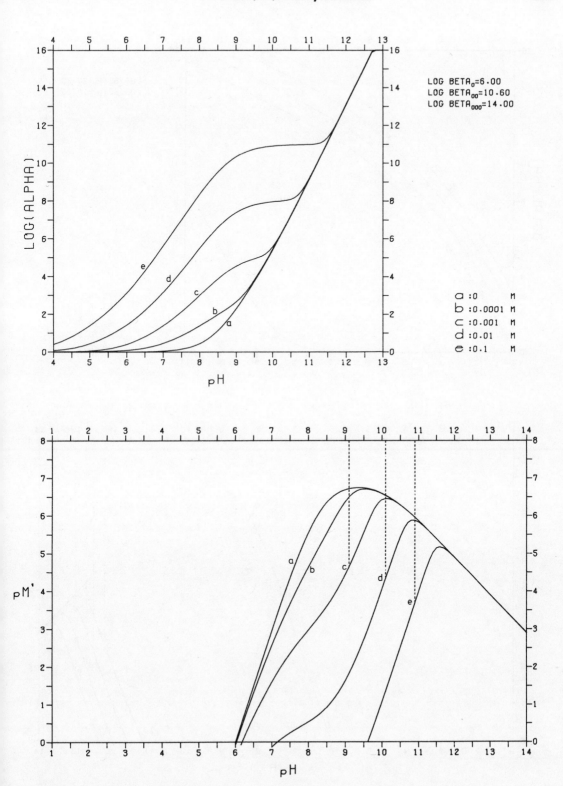

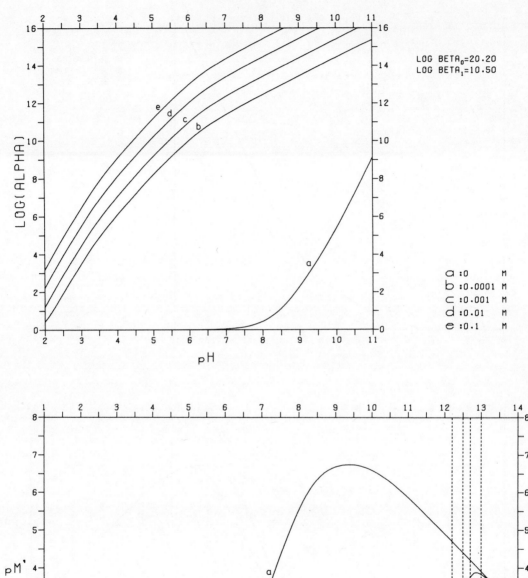

Terbium(III)-DTPA

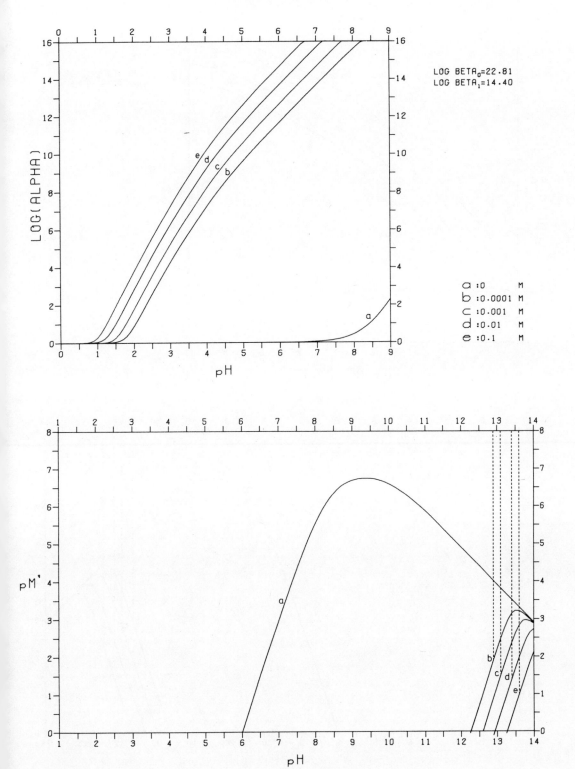

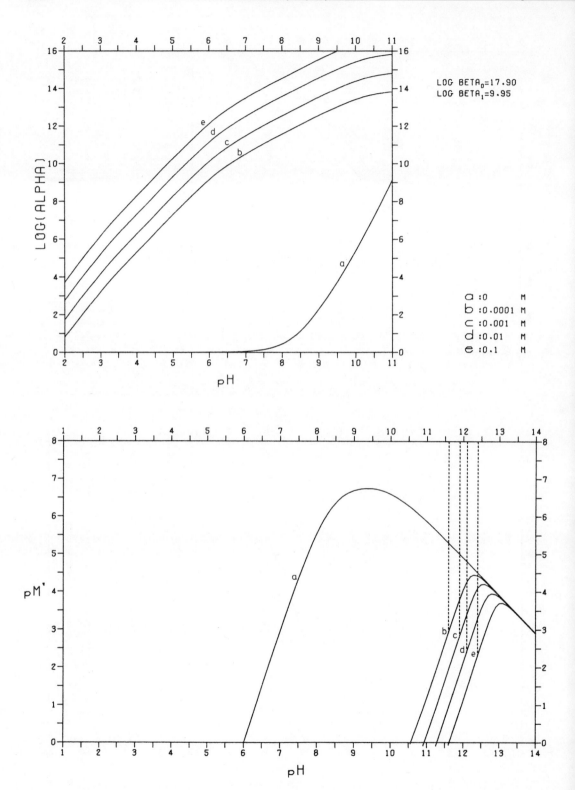

Terbium(III)-Iminodiacetate 643

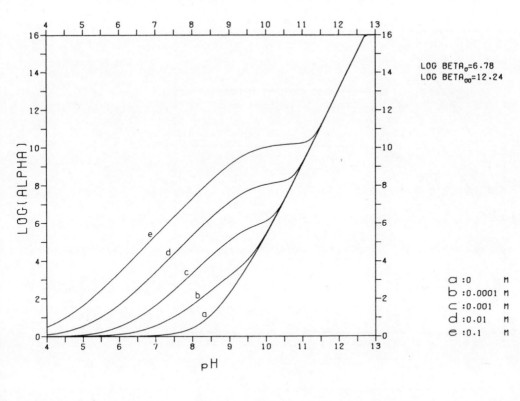

LOG BETA$_0$=6.78
LOG BETA$_{00}$=12.24

a : 0 M
b : 0.0001 M
c : 0.001 M
d : 0.01 M
e : 0.1 M

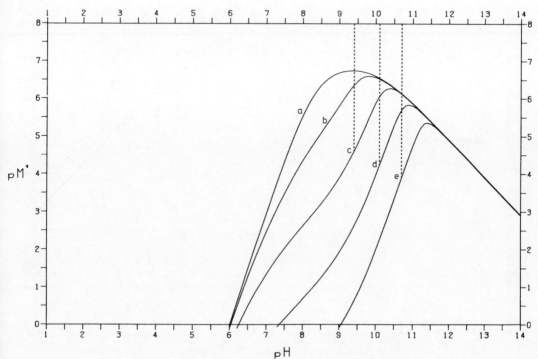

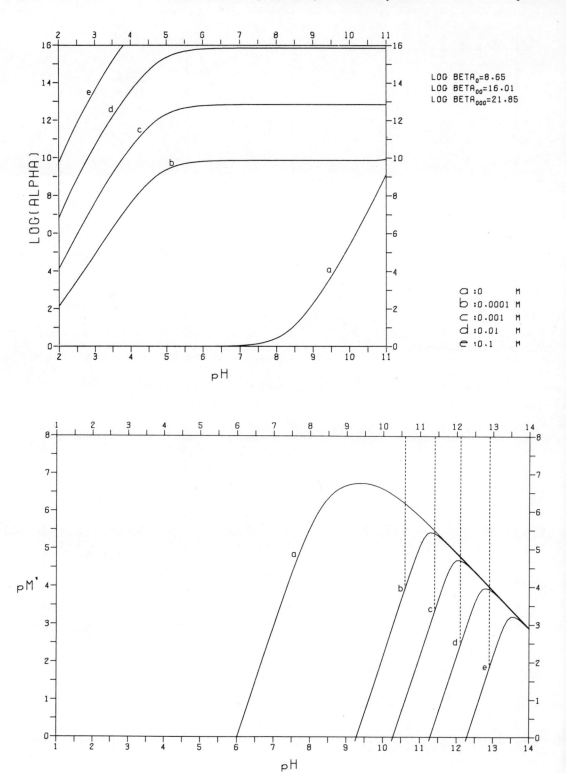

Terbium(III)-5-Sulphosalicylate

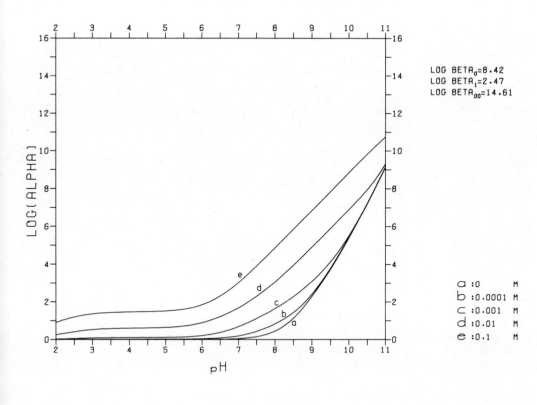

LOG BETA$_0$=8.42
LOG BETA$_1$=2.47
LOG BETA$_{00}$=14.61

a : 0 M
b : 0.0001 M
c : 0.001 M
d : 0.01 M
e : 0.1 M

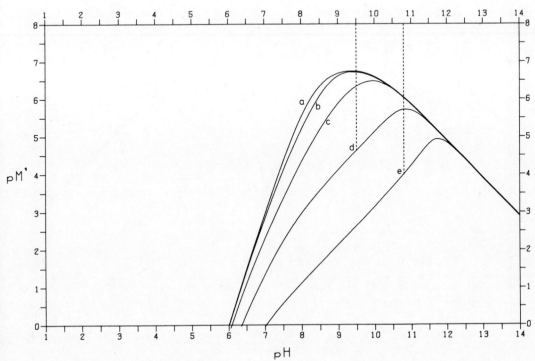

Terbium(III)–Tartrate [Ch. 35

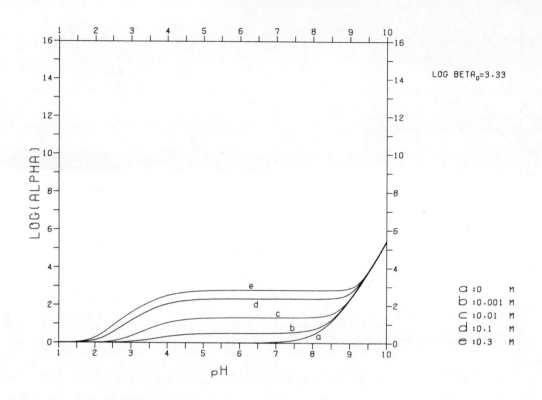

LOG BETA₀=3.33

a : 0   M
b : 0.001 M
c : 0.01   M
d : 0.1   M
e : 0.3   M

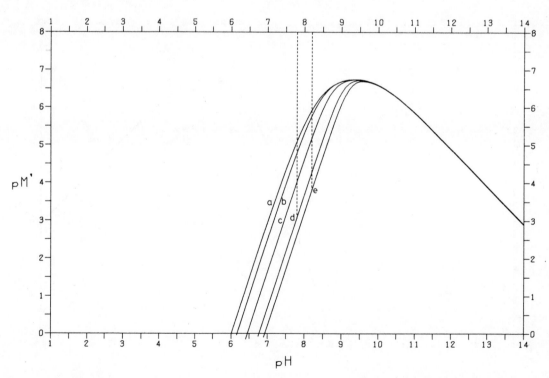

CHAPTER 36

Thorium (IV)   Th

The Th^{4+} ion is the most resistant to hydrolysis of all the quadrivalent ions because it is the largest. $Th(OH)^{3+}$ and $Th(OH)_2^{2+}$ become detectable at pH 2-3; their stability constants have been determined fairly accurately. Because polynuclear complex formation interferes in measurements of the pH-dependence of the solubility of active hydrated ThO_2, the value of log $*\beta_4$ is less accurately known. There is no evidence for the formation of $Th(OH)_3^+$ at any pH; log $*\beta_3$ has here been assigned a value equal to the maximum value it could take and still remain only a minor component of the system.

Solutions of thorium can be very supersaturated with respect to the precipitation of the oxide. From solubility measurements, a value of log $*K_{s0} = 11.4$ has been estimated for the solubility product of the active hydrated oxide. This unstable solid phase is ill-defined, but it is of practical importance in solution chemistry.

Polynuclear complex formation occurs extensively. The species $Th_2(OH)_2^{6+}$ and $Th_6(OH)_{15}^{9+}$ seem to be the most certain. There is little agreement about the identity of other products because anions appear to be involved in their constitution.

Since thorium is usually used in solutions of high ionic strength, the data used for construction of the plots refer to an ionic strength of 2.

$$\log *\beta_1 = -4.2$$
$$\log *\beta_2 = -7.7$$
$$\log *\beta_3 = -12.4$$
$$\log *\beta_4 = -17.7$$
$$\log *K_{s0} = 11.4$$

$$\log *\beta_{22} = -4.7$$
$$\log *\beta_{15,6} = -40.0$$
$$\log *\beta_{84} = -19$$
$$\log *\beta_{32} = -8.8$$
$$\log *\beta_{14,6} = -36.5$$
$$\log *\beta_{12} = -2.9$$

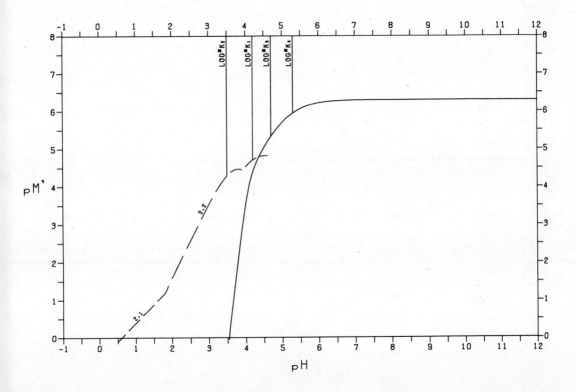

Thorium(IV)-Acetate

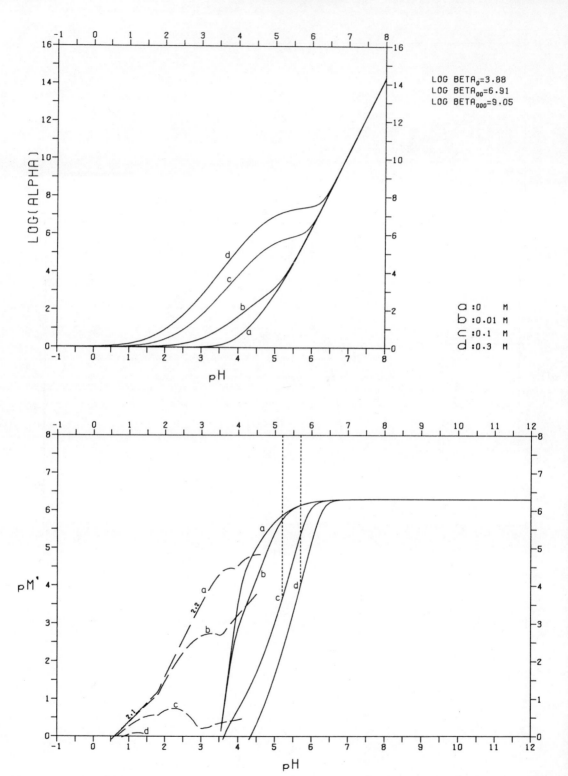

Thorium(IV)-Acetylacetone

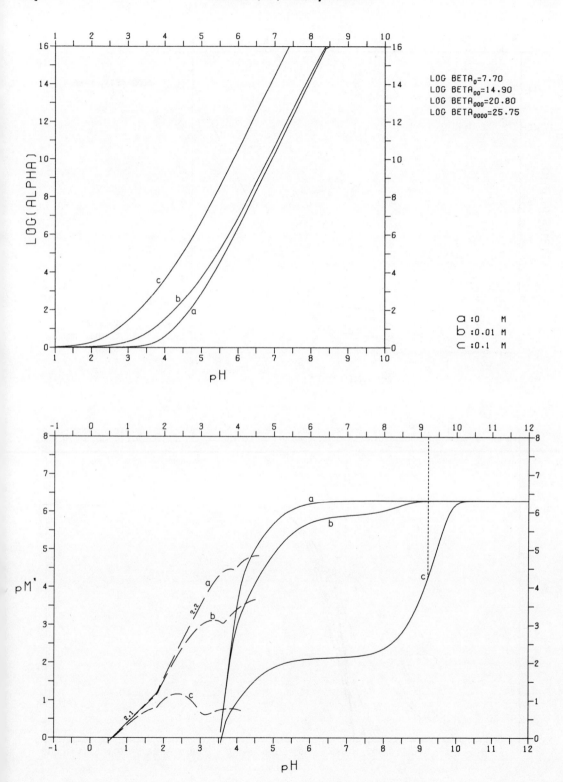

LOG BETA$_0$=7.70
LOG BETA$_{00}$=14.90
LOG BETA$_{000}$=20.80
LOG BETA$_{0000}$=25.75

a :0   M
b :0.01 M
c :0.1 M

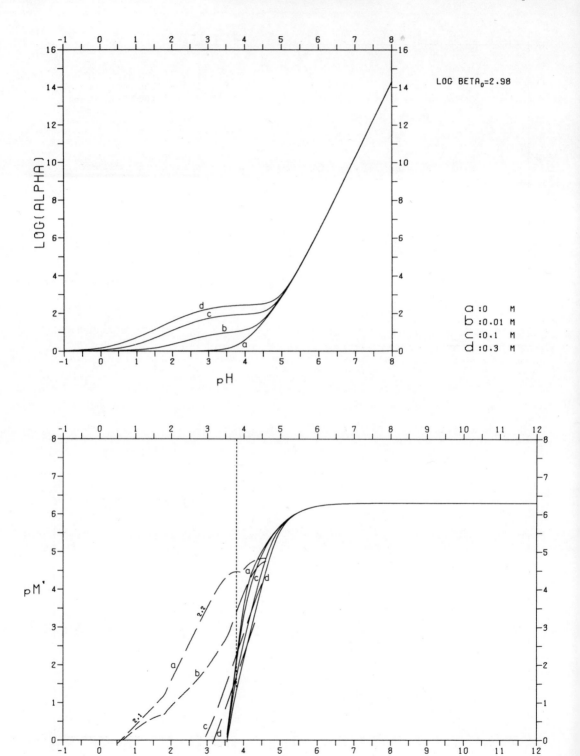

Thorium(IV)-Citrate

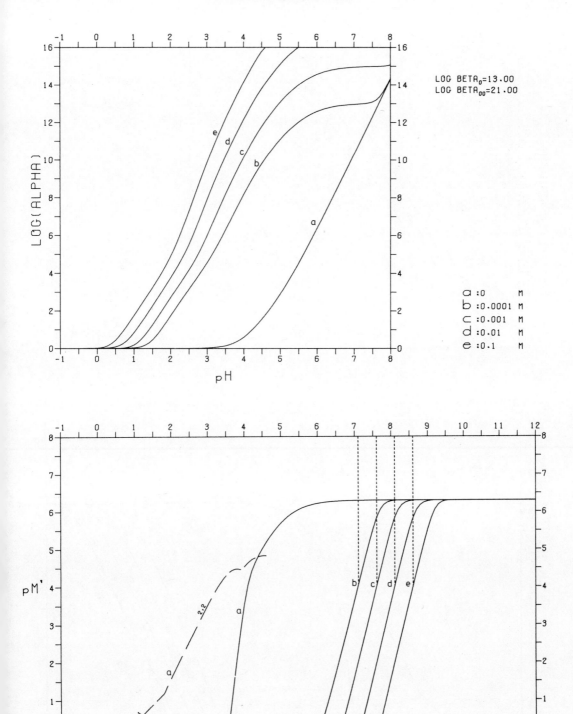

Thorium(IV)-DCTA [Ch. 36

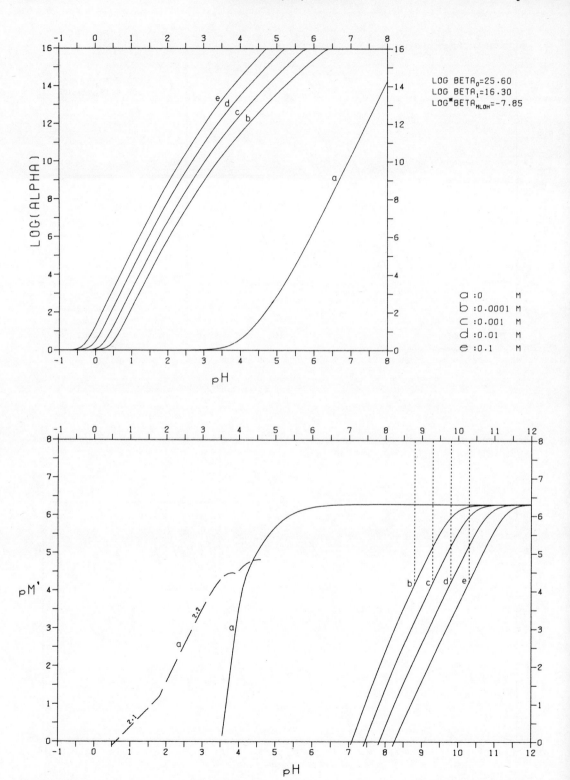

Thorium(IV)-DTPA   655

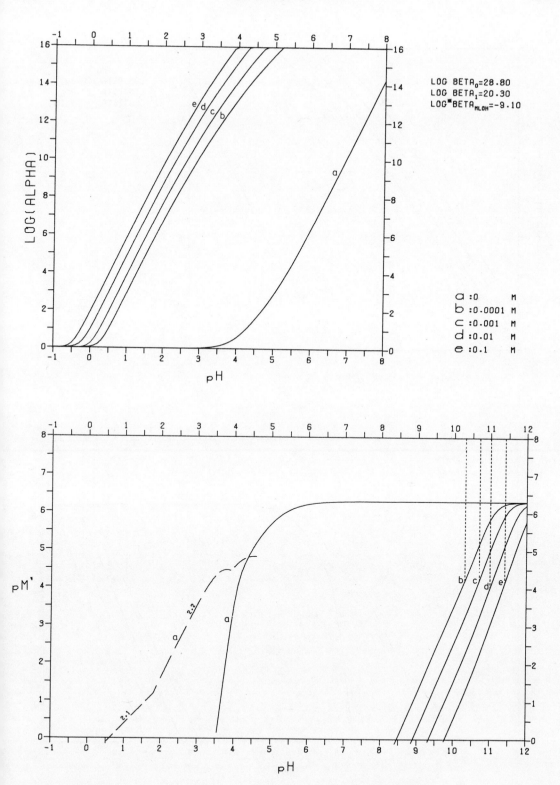

LOG BETA$_0$=28.80
LOG BETA$_1$=20.30
LOG*BETA$_{MLOH}$=-9.10

a : 0   M
b : 0.0001 M
c : 0.001 M
d : 0.01   M
e : 0.1   M

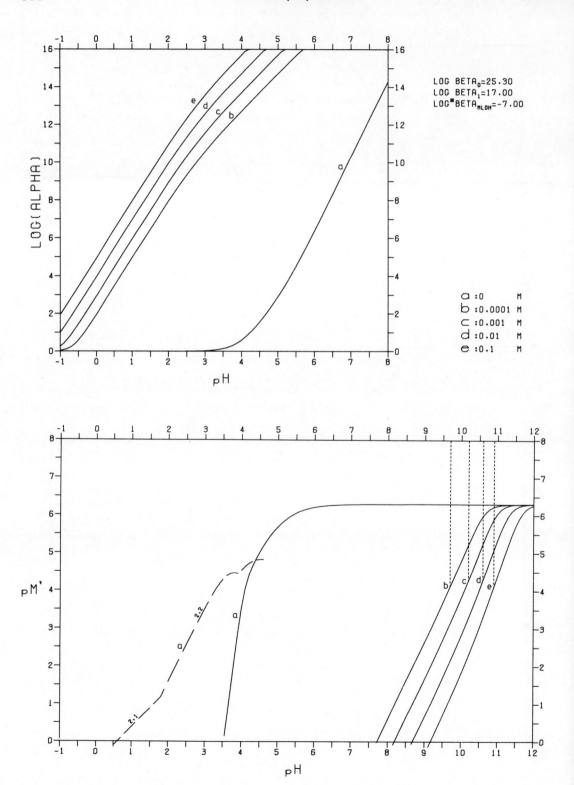

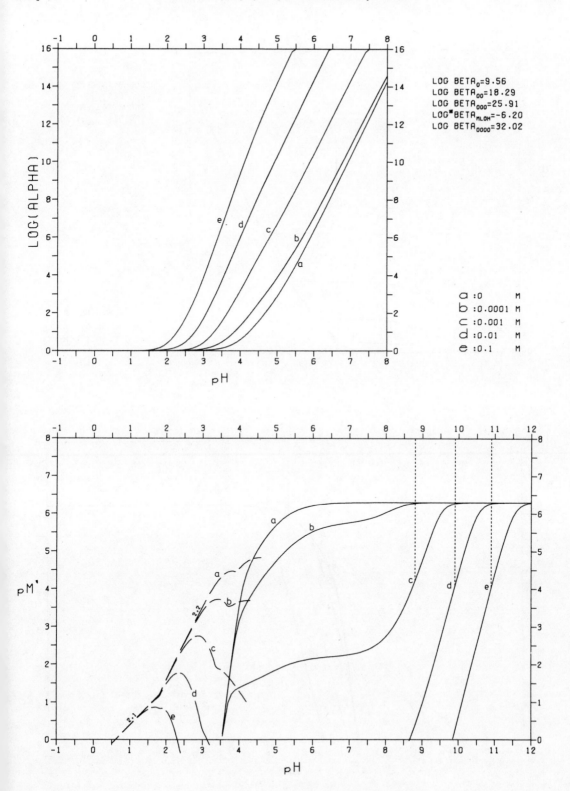

Thorium(IV) Sulphate [Ch. 36]

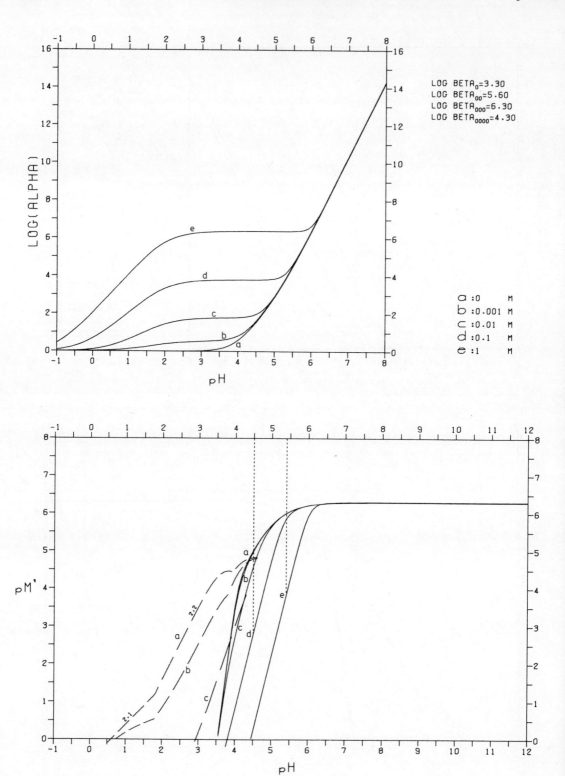

Thorium(IV)-5-Sulphosalicylate

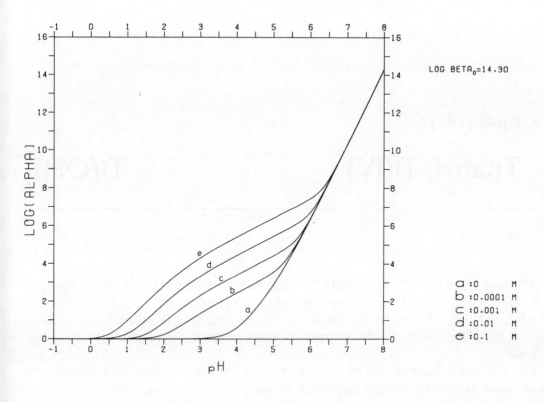

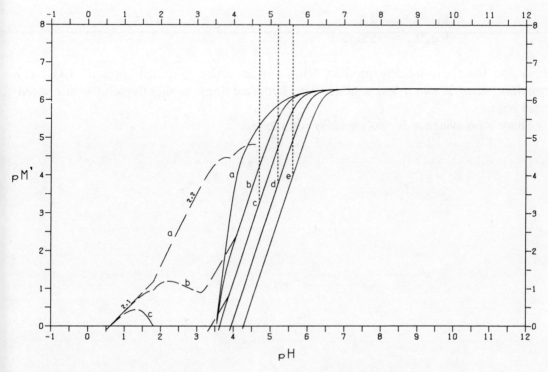

CHAPTER 37
Titanyl, Ti(IV) $Ti(OH)_2^{2+}$

The hydrolysis of Ti^{4+} to $Ti(OH)_2^{2+}$ is complete in 1M $HClO_4$; no values have been reported for the relevant formation constants. The titanyl ion hydrolyses further to $Ti(OH)_3^+$ and $Ti(OH)_4(aq)$, as indicated by the increase in the solubility of hydrated TiO_2 as the pH is increased. The stepwise stability constants were estimated from solubility data.

For the calculations of the plots, the "titanyl" ion was treated as a separate "bivalent" metal with the following stability constants.

$$\log {}^*\beta_1 = -2.5 \qquad \log {}^*K_{s0} = -0.5$$
$$\log {}^*\beta_2 = -5.0$$

The value for the solubility product holds for an active hydrated form of TiO_2. The crystalline form is much less soluble. The hydrated form is slightly soluble in concentrated base.

There is no evidence for polynuclear complexes.

Titanyl

Ti(OH)$_2^{2+}$]

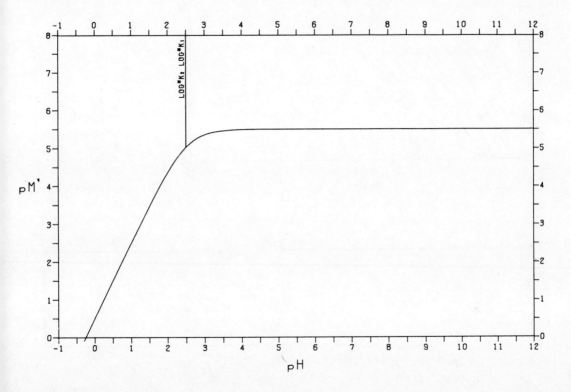

Titanyl–EDTA [Ch. 37]

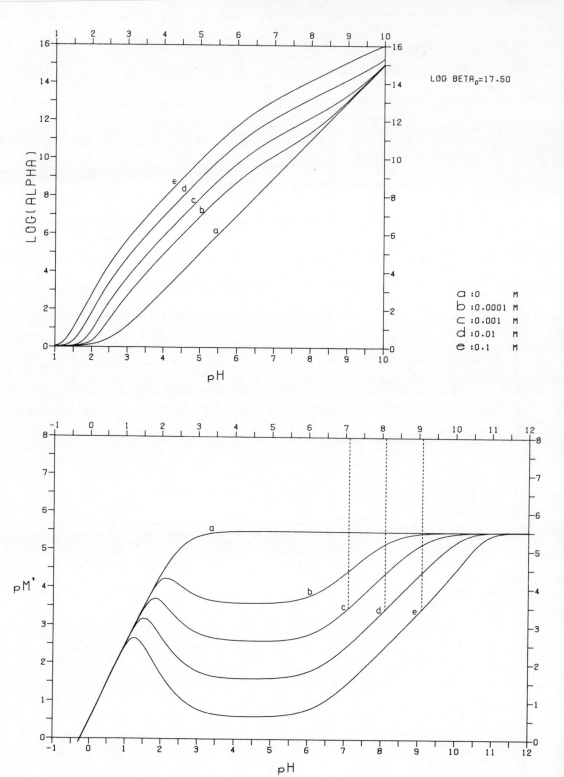

Ti(OH)$_2^{2+}$] Titanyl-Fluoride 663

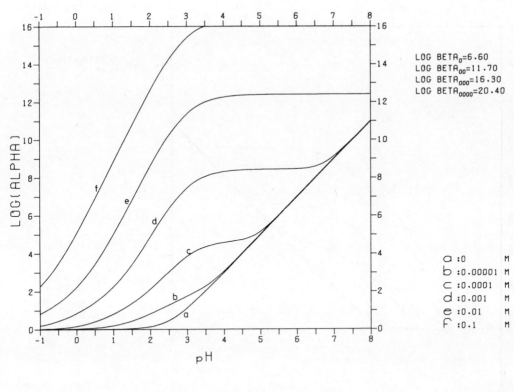

LOG BETA$_0$=6.60
LOG BETA$_{00}$=11.70
LOG BETA$_{000}$=16.30
LOG BETA$_{0000}$=20.40

a : 0 M
b : 0.00001 M
c : 0.0001 M
d : 0.001 M
e : 0.01 M
f : 0.1 M

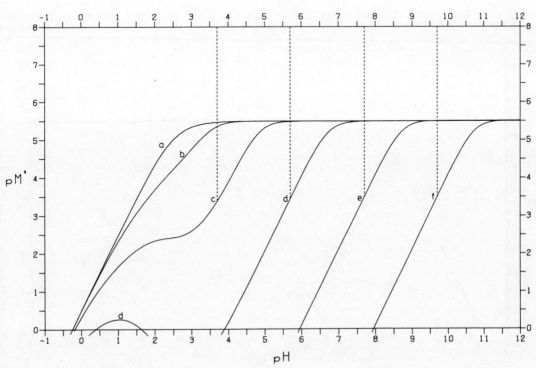

Titanyl-Oxalate [Ch. 37]

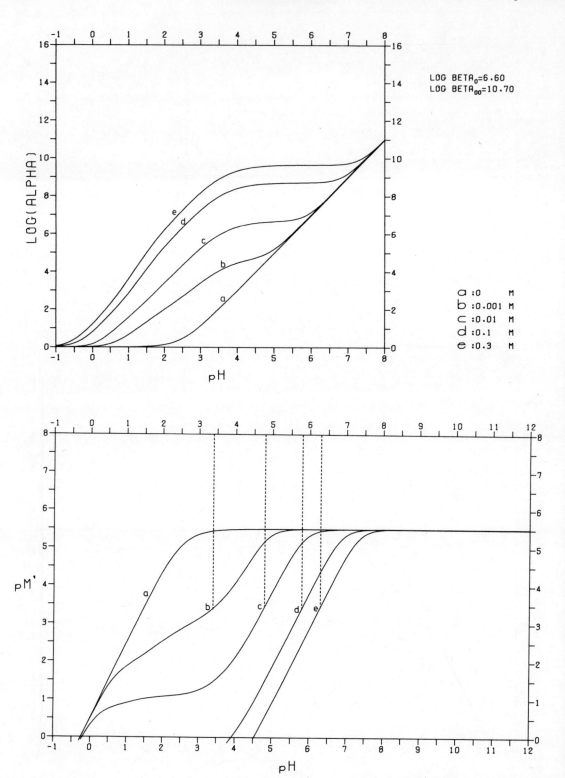

LOG BETA$_0$=6.60
LOG BETA$_{00}$=10.70

a : 0   M
b : 0.001 M
c : 0.01 M
d : 0.1   M
e : 0.3   M

Titanyl-Salicylate

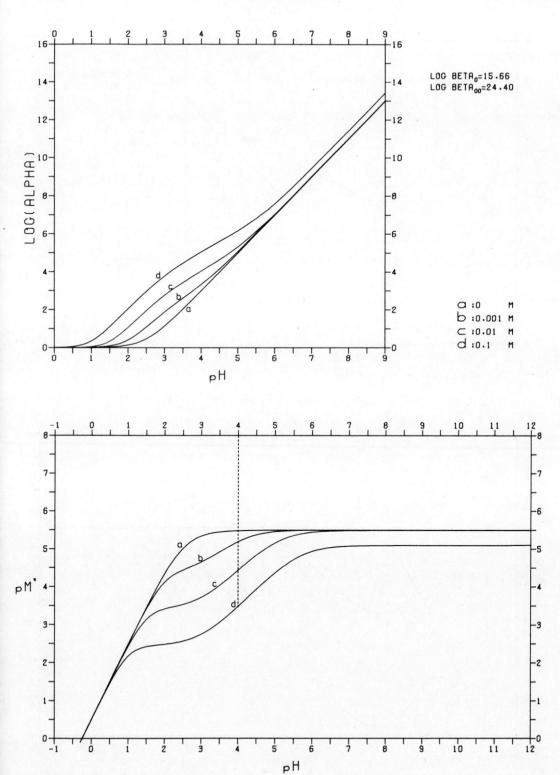

LOG BETA$_0$=15.66
LOG BETA$_{00}$=24.40

a : 0 M
b : 0.001 M
c : 0.01 M
d : 0.1 M

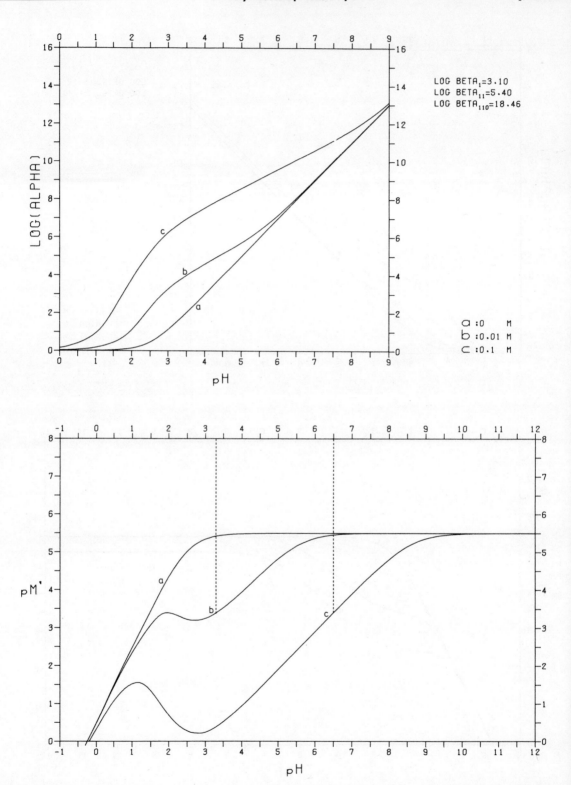

Titanyl-Tartrate

$[Ti(OH)_2^{2+}]$

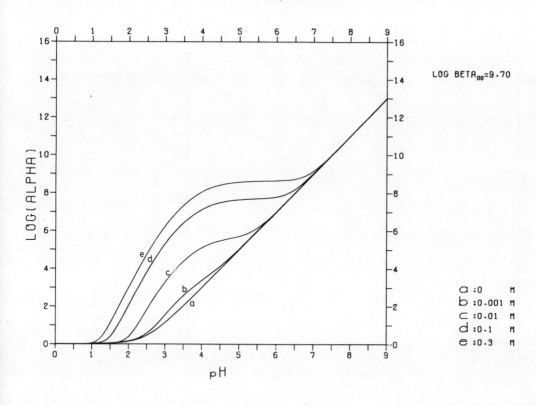

LOG BETA$_{00}$ = 9.70

a : 0 M
b : 0.001 M
c : 0.01 M
d : 0.1 M
e : 0.3 M

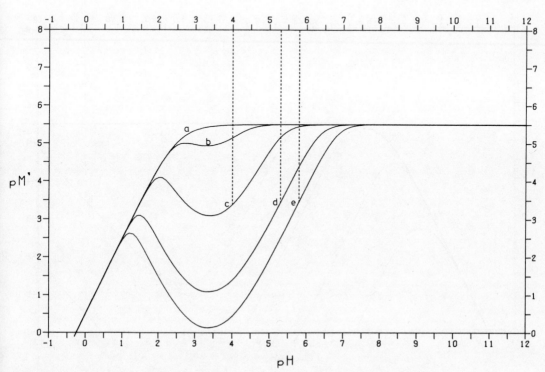

Titanyl-Tiron [Ch. 37]

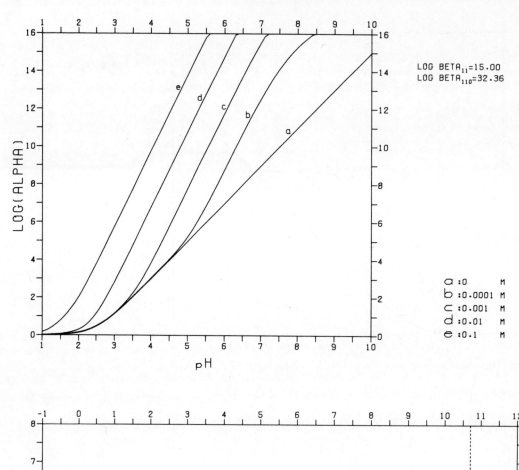

LOG BETA$_{11}$ = 15.00
LOG BETA$_{110}$ = 32.36

a : 0   M
b : 0.0001 M
c : 0.001   M
d : 0.01   M
e : 0.1   M

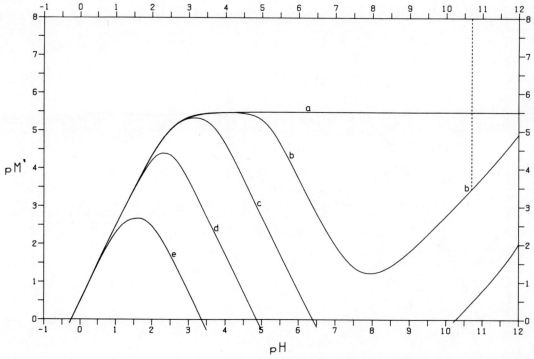

CHAPTER 38

Thallium (III) Tl

Thallium exists as Tl^+ and Tl^{3+} in aqueous solution; the Tl^{3+} ion is easily reduced. In its +III state Tl should oxidize water, but the reaction does not proceed at a measureable rate. The +III state is stabilized by complexing ligands, because Tl^+ forms much weaker complexes than Tl^{3+}.

The stability constants of $Tl(OH)^{2+}$, $Tl(OH)_2^+$ and $Tl(OH)_3(aq)$ have been determined potentiometrically; the first two are known accurately. There is no evidence of polynuclear complex formation. A value for log $*\beta_4$ could be estimated from the dependence of the solubility of Tl_2O_3 on pH.

There is a slight tendency for supersaturation to occur, but the solubility of a precipitate, once it has been formed, changes little on aging. The following data were used for construction of the plots.

$$\log *\beta_1 = -0.94$$
$$\log *\beta_2 = -2.09$$
$$\log *\beta_3 = -3.8$$
$$\log *\beta_4 = -15.3$$
$$\log *K_{s0} = -3.35$$

Thallium(III)

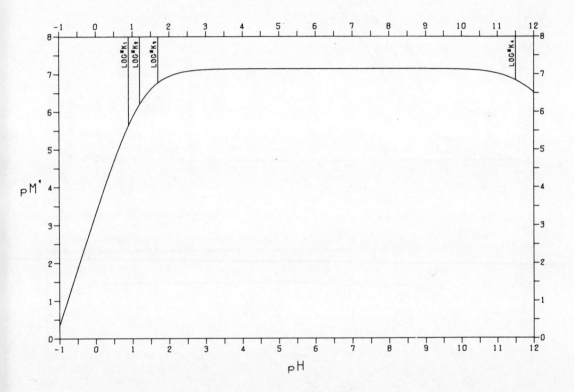

Thallium(III)-Acetate

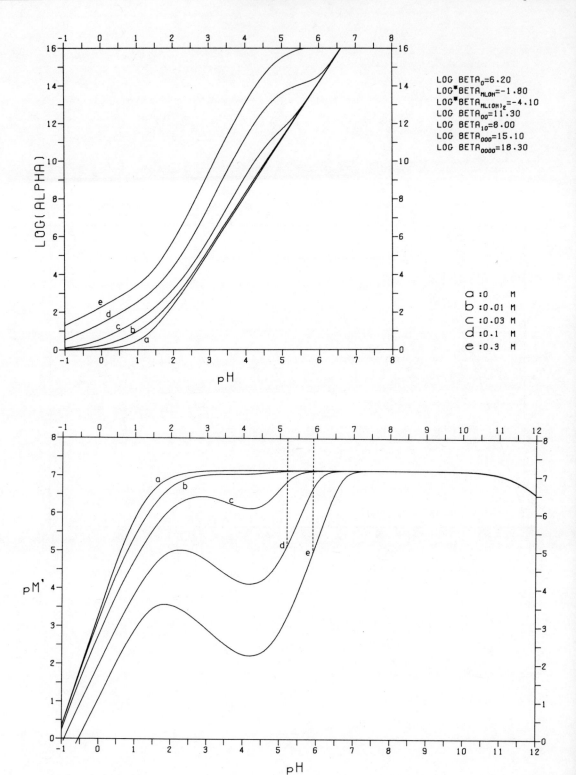

Thallium(III)-Bromide

673

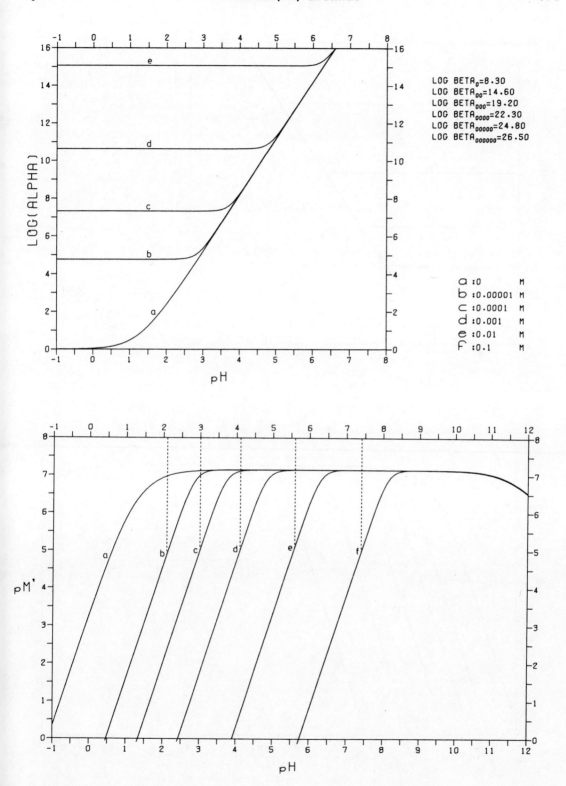

Thallium(III)-Chloride [Ch. 38]

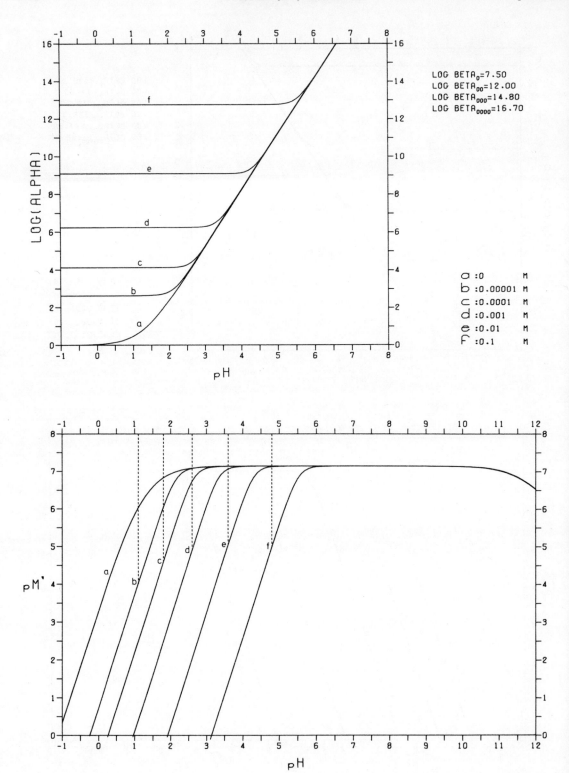

Thallium(III)-DCTA

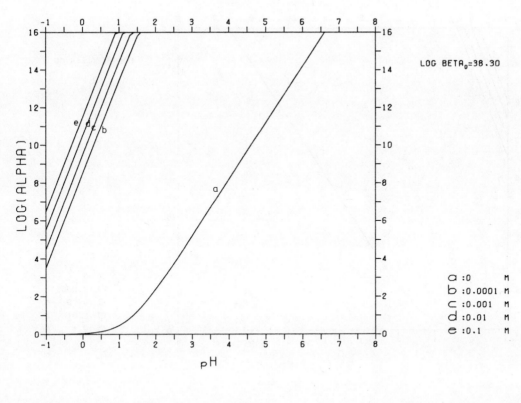

LOG BETA$_0$=38.30

a : 0   M
b : 0.0001 M
c : 0.001   M
d : 0.01   M
e : 0.1   M

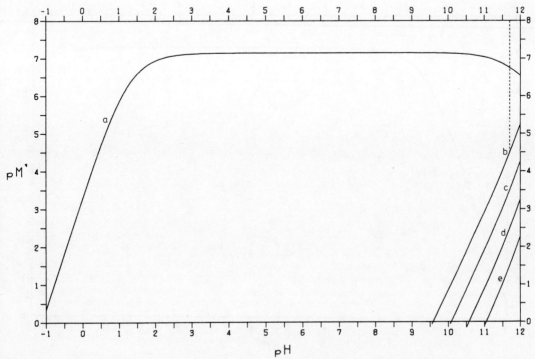

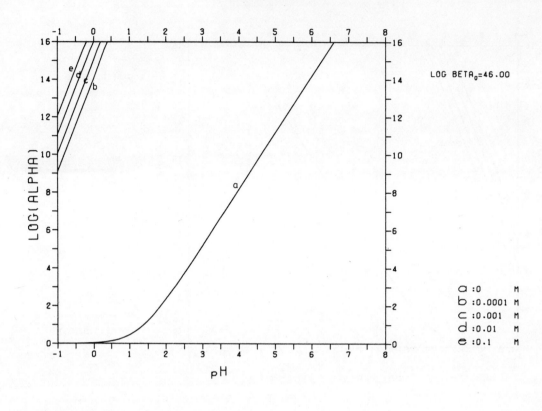
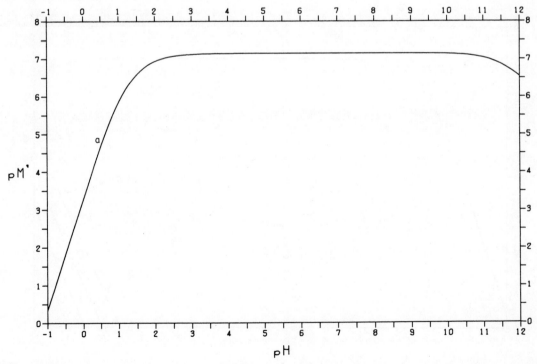

Thallium(III)-EDTA

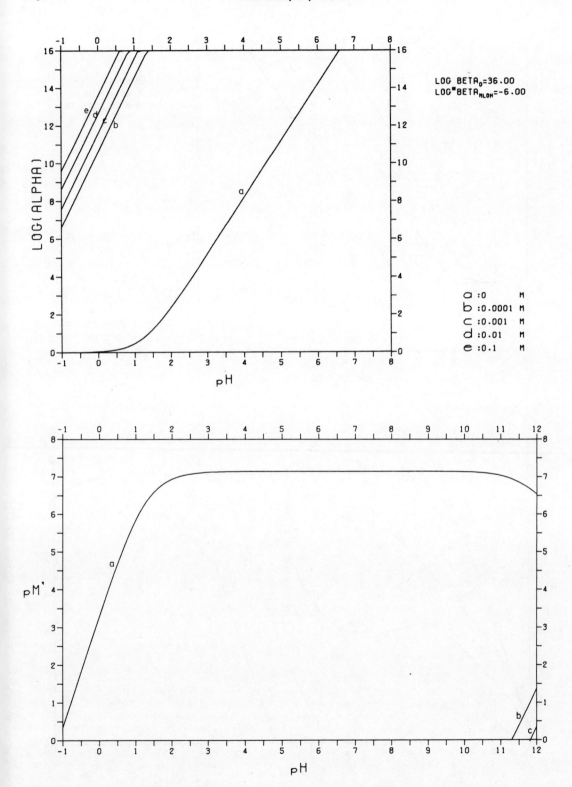

Thallium(III)-1,10-Phenanthroline [Ch. 38]

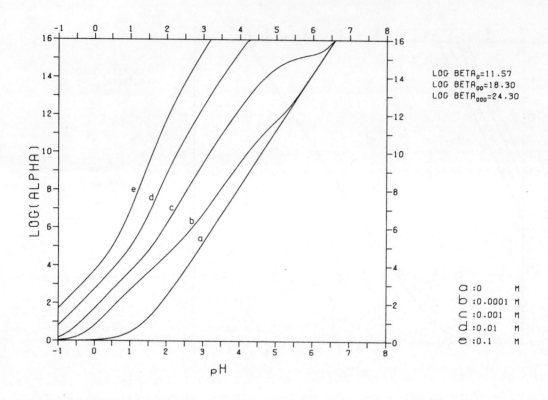

LOG BETA$_0$=11.57
LOG BETA$_{00}$=18.30
LOG BETA$_{000}$=24.30

a : 0   M
b : 0.0001 M
c : 0.001   M
d : 0.01   M
e : 0.1   M

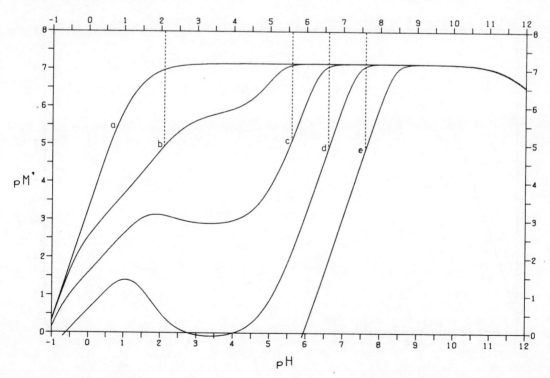

Thallium(III)–Sulphate

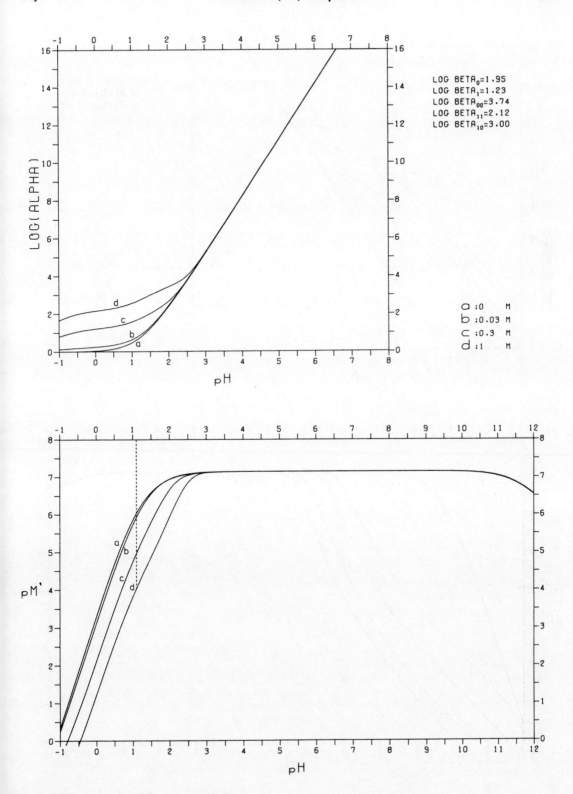

Thallium(III)-Tartrate

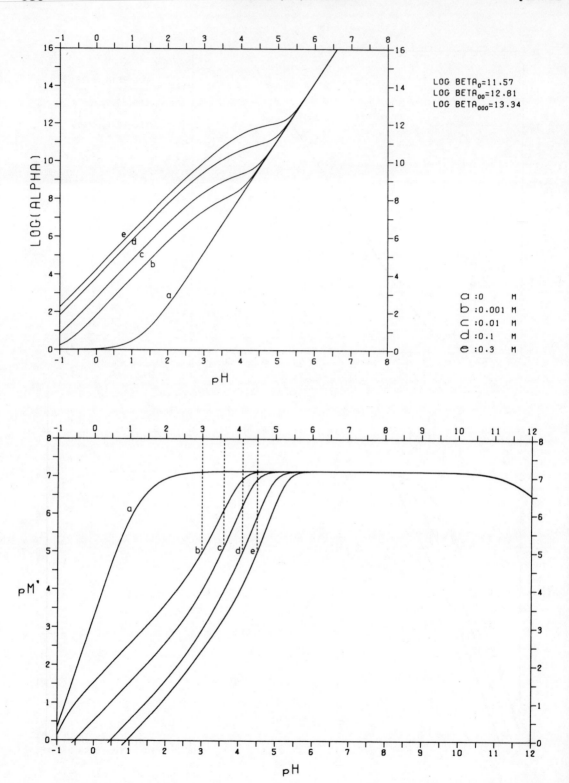

CHAPTER 39

Uranium (IV)   U

In the absence of complexing ligands, quadrivalent uranium can exist in aqueous solutions of redox potentials below E = (270 - 118 pH) mV. Under these conditions, it may exist in equilibrium with UO_3, but at potentials below (204 - 59 pH) mV the trioxide is reduced, and hydrolysed uranium(IV) solutions occur in equilibrium with hydrated UO_2.

Polynuclear complex formation interferes in the study of the mononuclear species. In acid, the hydrolysis of U^{4+} is explained adequately by log $*\beta_1 = -1.6$ and log $*\beta_{15,6} = -17.2$. Log $*\beta_5 = -17.6$ can be estimated from solubility measurements at high pH. The values log $*\beta_2 = -3$, log $*\beta_3 = -7.3$, and log $*\beta_4 = -11.8$ (at ionic strength = 2) can be estimated by assuming that, at infinite dilution, the stepwise constants form a regular progression.

As commonly occurs with multivalent metals, uranium(IV) solutions can be supersaturated with respect to precipitation of $UO_2.xH_2O$. Log $*K_{s0} = 5$ can be estimated for the solubility product of the freshly precipitated hydrated precipitate. This value approaches zero after aging of the precipitate for several hours. The plots refer to the freshly precipitated oxide.

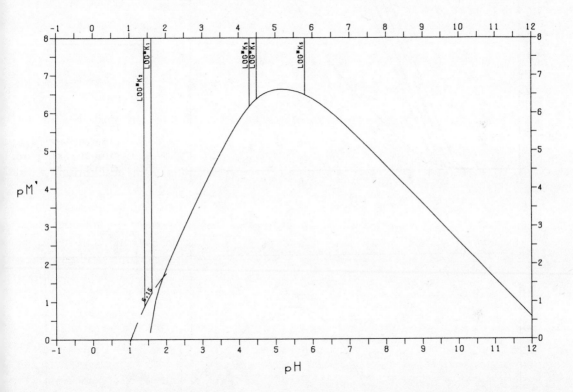

684 Uranium(IV)-Citrate [Ch. 39

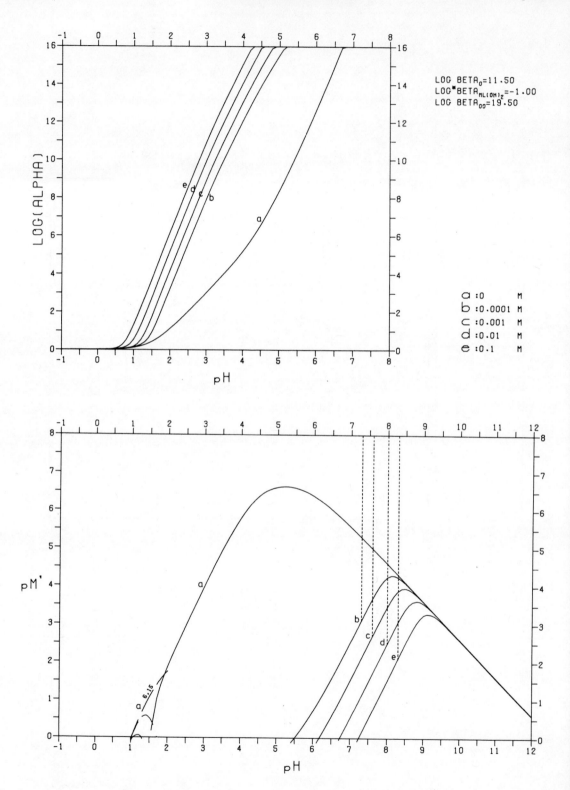

Uranium(IV)-DCTA

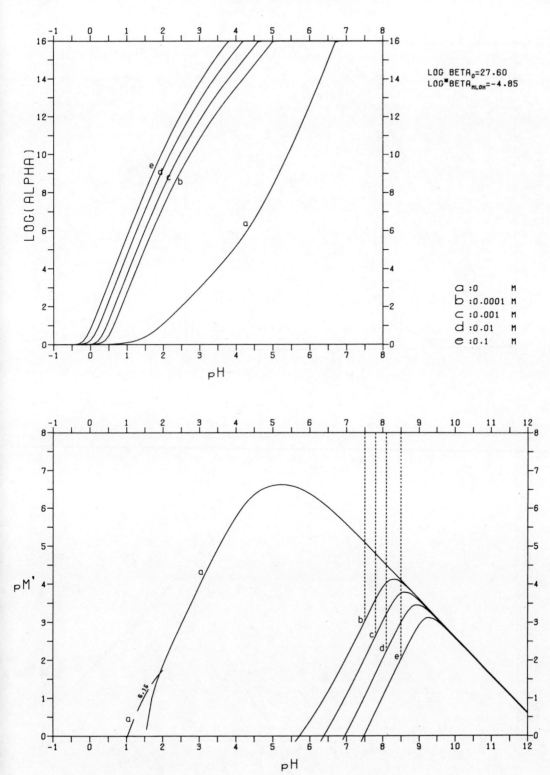

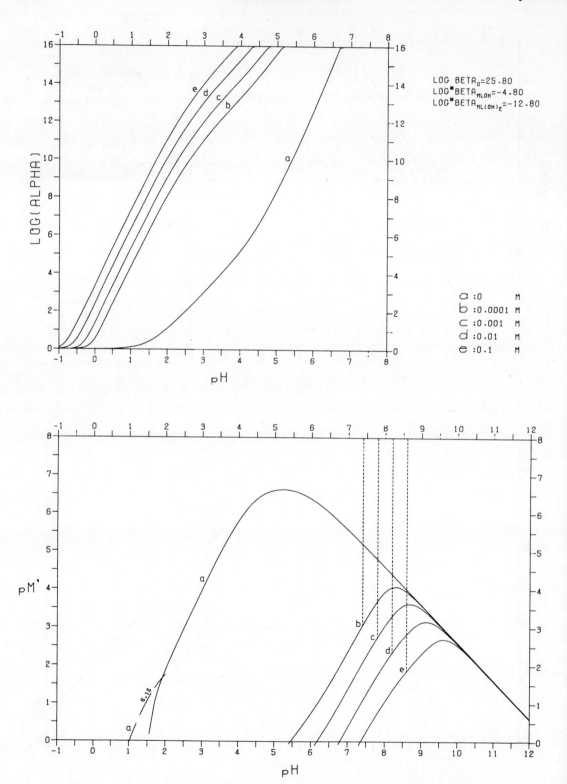

Uranium(IV)-Sulphate

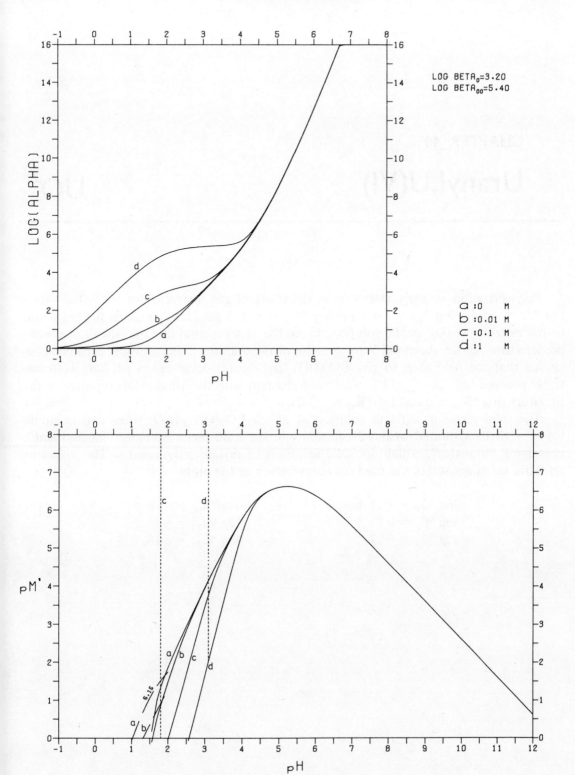

LOG BETA$_0$=3.20
LOG BETA$_{00}$=5.40

a : 0 M
b : 0.01 M
c : 0.1 M
d : 1 M

CHAPTER 40

Uranyl, U(VI) $\qquad$ UO_2^{2+}

Polymerization strongly interferes in the study of the hydrolysis of UO_2^{2+}. The values log $*\beta_1 = -5.4$, $*\log \beta_{22} = -6.0$ and log $*\beta_{53} = -16.3$ adequately explain the behaviour of the metal in weakly acidic solution. From the experimental fact that the (3,5) species predominates above about pH 5, irrespective of the uranium concentration, it can be concluded that the next step to give $UO_2(OH)_2$(aq) cannot occur below pH 5.4; we tentatively propose log $*\beta_2 = -11.2$, which is consistent with the other values reported in the literature (log $*K_{s0} = 6$ and log $*K_{s2} = -5.2$).

The solid phase in aqueous solution at 25° is $UO_2(OH)_2 \cdot H_2O$. When it is in equilibrium with its saturated weakly basic solution, the solid phase becomes "incongruent", containing "uranates", which are solid solutions of various polyuranates. The following tentative set of constants was used for construction of the plots.

$$\log *\beta_1 = -5.4 \qquad \log *K_{s0} = 6$$
$$\log *\beta_2 = -11.2 \qquad \log *\beta_{22} = -6.0$$
$$\log *\beta_3 = -23.7 \qquad \log *\beta_{53} = -16.3$$

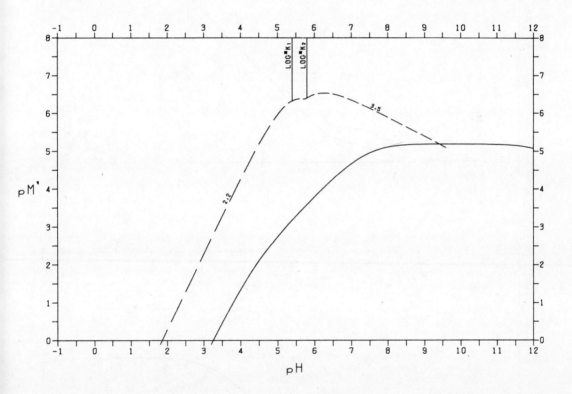

690 Uranyl-Acetate [Ch. 40

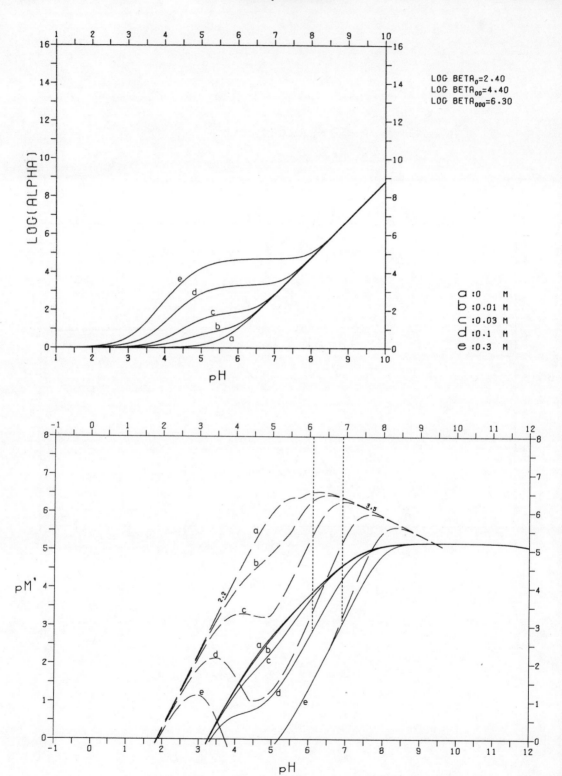

Uranyl-Acetylacetone

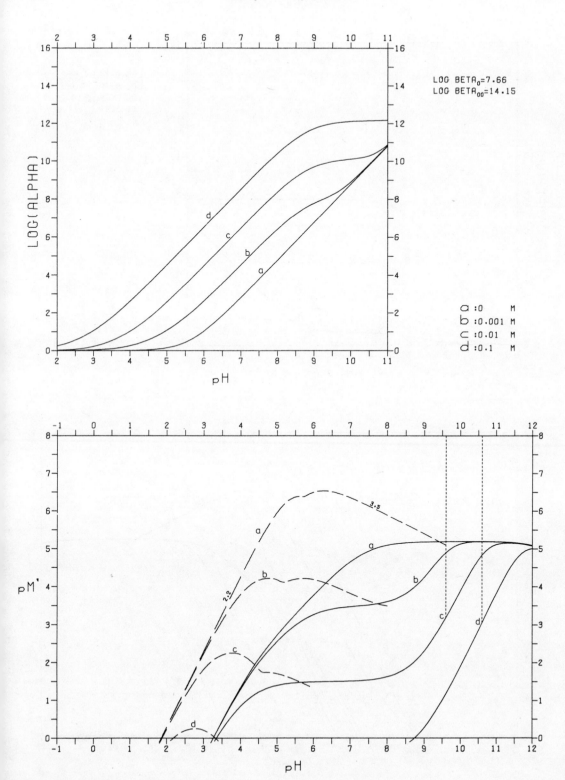

Uranyl-Fluoride [Ch. 40

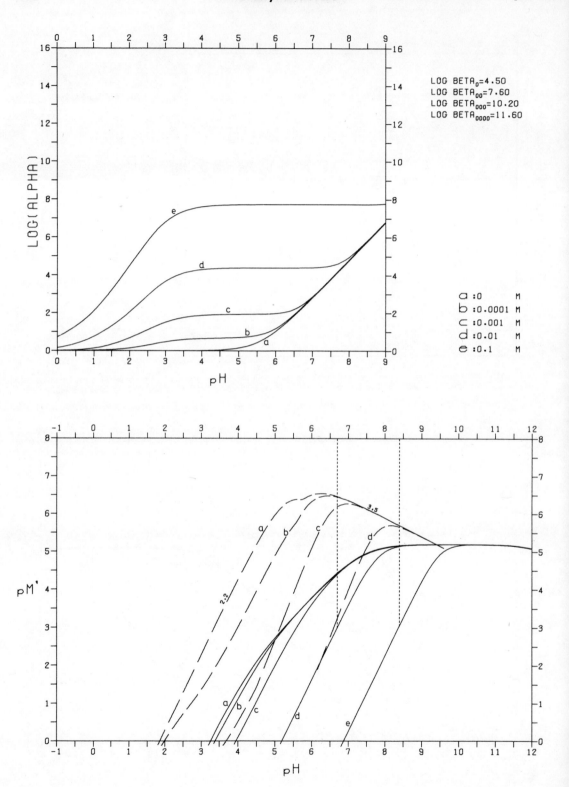

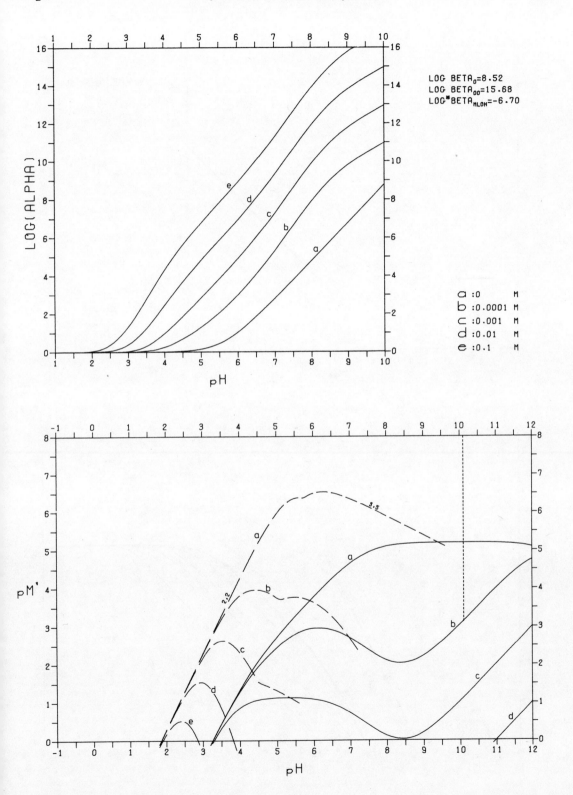

694 Uranyl-Iminodiacetate [Ch. 40

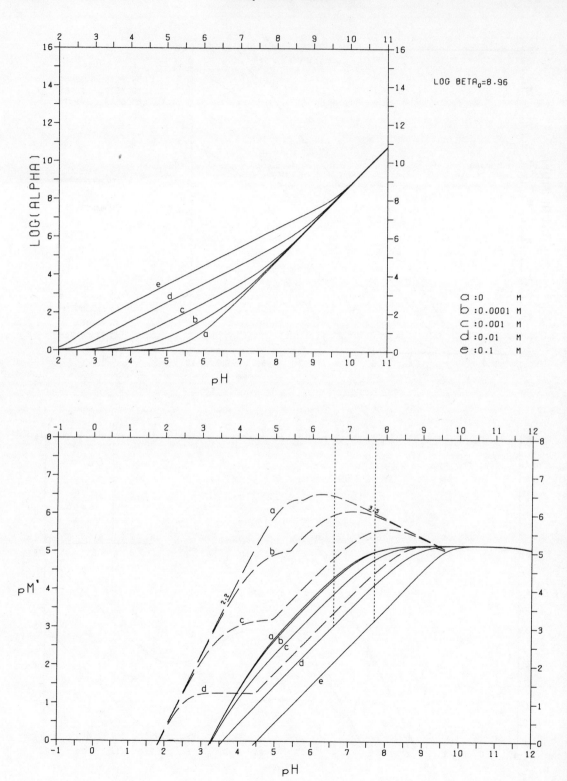

Uranyl-Salicylate

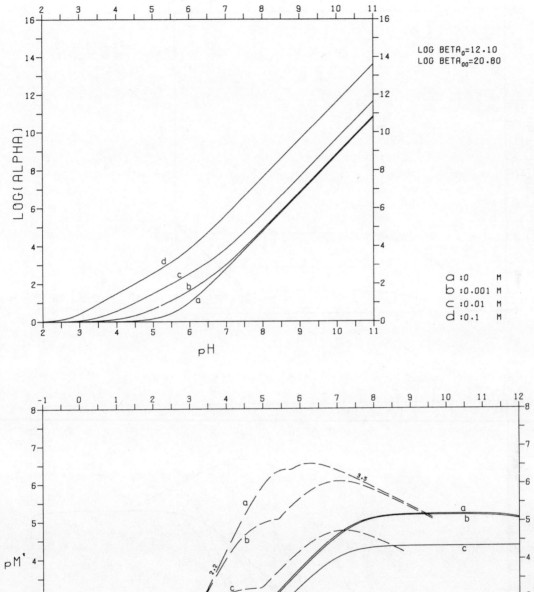

LOG BETA$_0$ = 12.10
LOG BETA$_{00}$ = 20.80

a : 0 M
b : 0.001 M
c : 0.01 M
d : 0.1 M

Uranyl-Sulphate

[Ch. 40

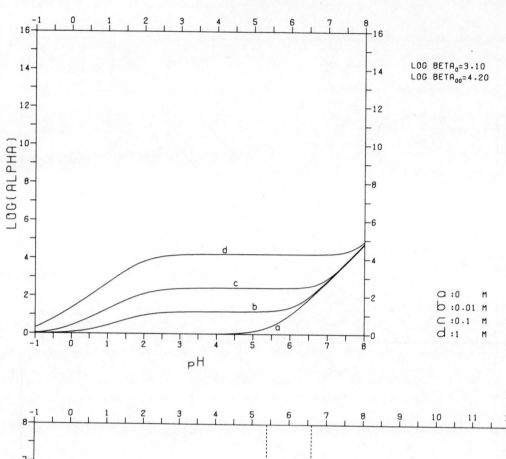

LOG BETA$_0$=3.10
LOG BETA$_{00}$=4.20

a : 0   M
b : 0.01   M
c : 0.1   M
d : 1   M

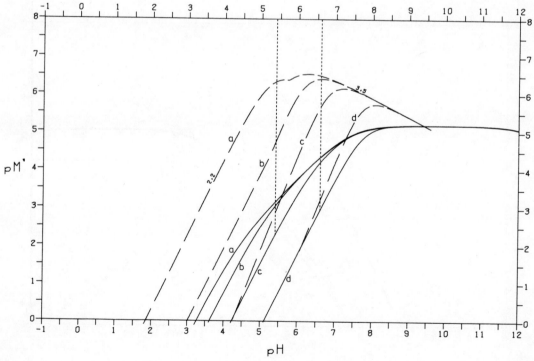

Uranyl-5-Sulphosalicyclate

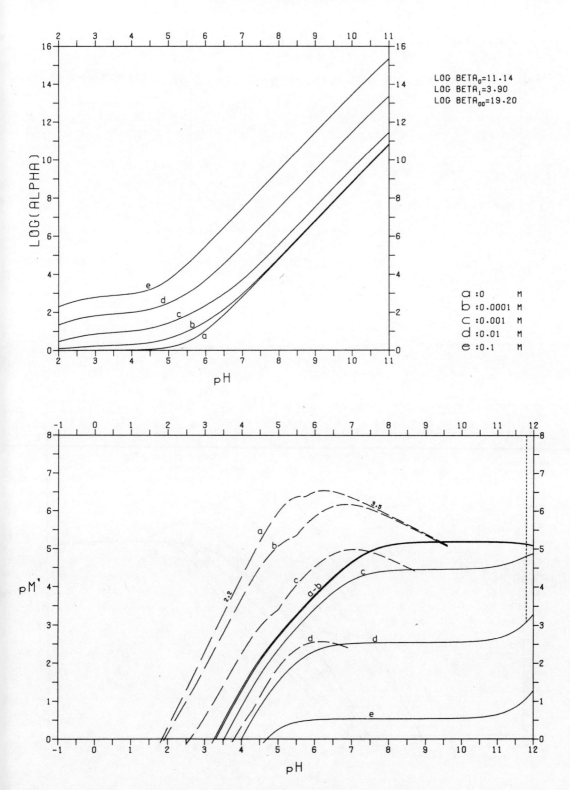

LOG BETA$_0$=11.14
LOG BETA$_1$=3.90
LOG BETA$_{00}$=19.20

a : 0 M
b : 0.0001 M
c : 0.001 M
d : 0.01 M
e : 0.1 M

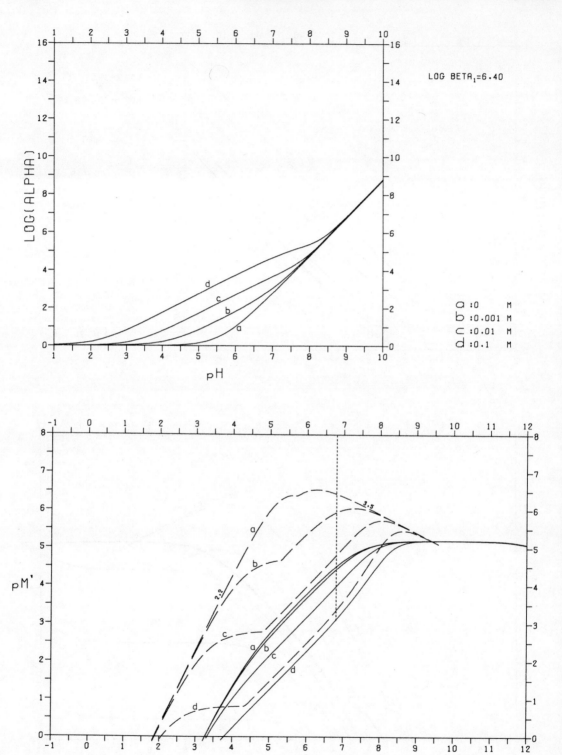

CHAPTER 41

Vanadium (III)   V

V^{3+} ions hydrolyse rapidly above pH 1 to form $V(OH)^{2+}$ and $V_2(OH)_2^{4+}$. $V(OH)_2^+$ and $V_2(OH)_3^{3+}$ have been suggested, in order to improve the agreement with experimental data, but the existence of $V_2(OH)_3^{3+}$ (introduced to explain the slow kinetics in weakly acidic medium) has not really been confirmed. The behaviour in neutral and basic media has not been very well established. The following data adopted from the literature agree with our own preliminary solubility measurements up to pH 8.

$$\log {}^*\beta_1 = -2.70 \qquad \log {}^*\beta_{22} = -3.8$$
$$\log {}^*\beta_2 = -6.50 \qquad \log {}^*\beta_{32} = -8.0$$
$$\log {}^*\beta_3 = -13.50 \qquad \log {}^*K_{s0} = 7.6$$

In the absence of complexing ligands, V^{3+} ions exist only in strongly reducing media; they are easily oxidized by air. V^{3+} ions form stronger complexes than VO^{2+} ions, so vanadium(III) complexes are more stable towards oxidation than V^{3+} ions.

Vanadium(III)

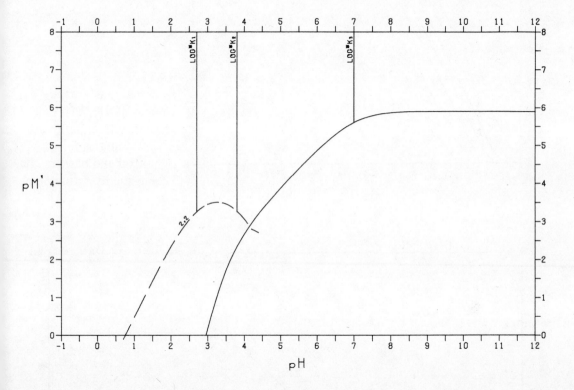

Vanadium(III)-EDTA [Ch. 41]

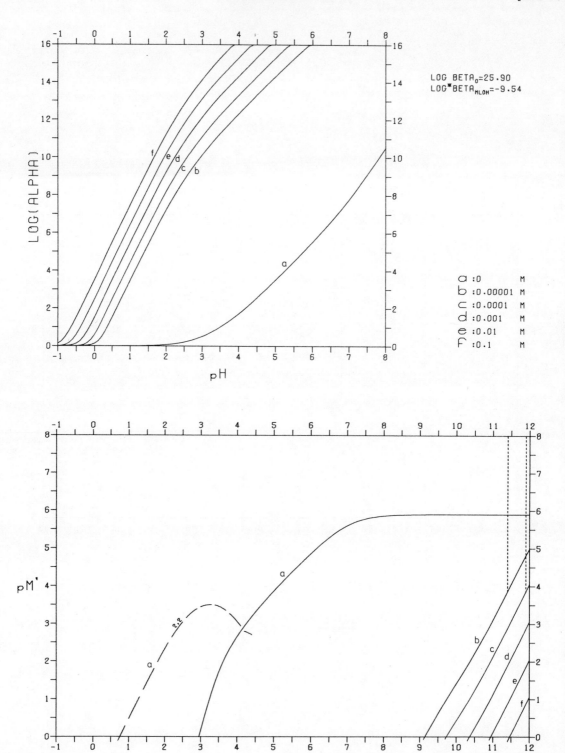

CHAPTER 42

Vanadyl, V(IV) VO^{2+}

The VO^{2+} ion is the least hydrolysed form of vanadium(IV) known with certainty. The most important hydrolysis products are $VO(OH)^+$ and $(VO)_2(OH)_2^{2+}$.

The solubility of the hydrated oxide has not been determined accurately; a reasonable value for the solubility product is log $*K_{s0}$ = 5.9, with log $*\beta_2$ = -12.0 There is evidence that the solubility increases in alkaline solution, but the data are only qualitative. This uncertainty decreases the reliability of the plots above pH 9.

The following values were used for constructing the plots.

$$\log *\beta_1 = -4.77 \qquad \log *\beta_{22} = -6.88$$
$$\log *\beta_2 = -12 \qquad \log *K_{s0} = 5.9$$

Quadrivalent vanadium is stable only in (moderately) reducing media; the equilibrium $VO(OH)_2(aq) \rightleftharpoons VO(OH)_2(s)$ is possible only at redox potentials between E = (400 - 59 pH) mV and E = (200 - 59 pH) mV, in the absence of complexing ligands. The range shifts to higher potentials when such ligands are present.

Vanadyl

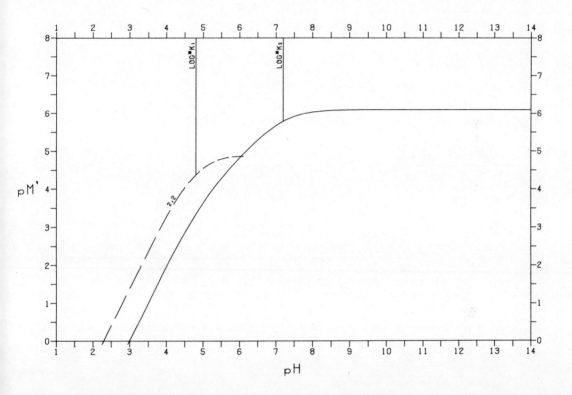

Vanadyl–Acetylacetone

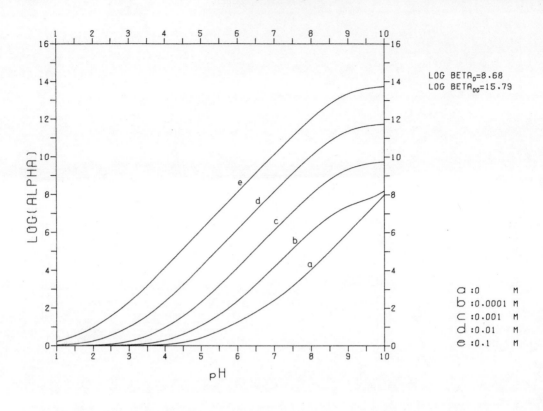

LOG BETA$_0$ = 8.68
LOG BETA$_{00}$ = 15.79

a : 0   M
b : 0.0001   M
c : 0.001   M
d : 0.01   M
e : 0.1   M

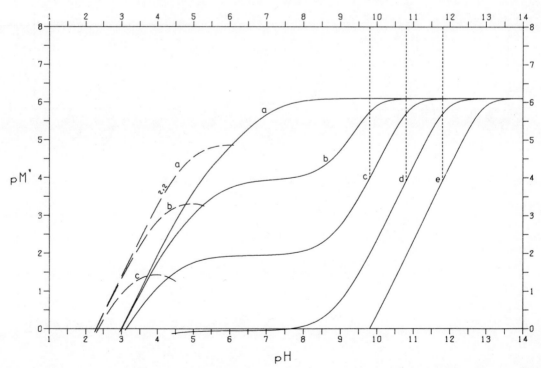

Vanadyl-Citrate

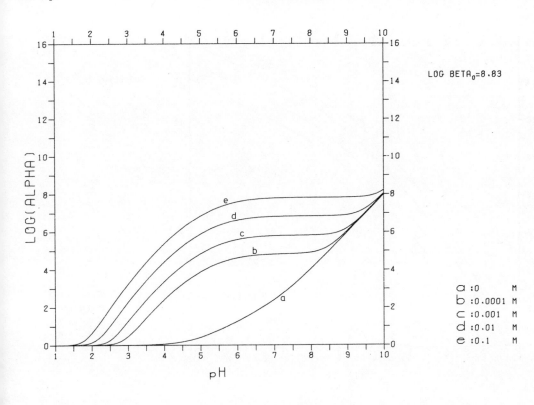

LOG BETA$_0$=8.83

a : 0 M
b : 0.0001 M
c : 0.001 M
d : 0.01 M
e : 0.1 M

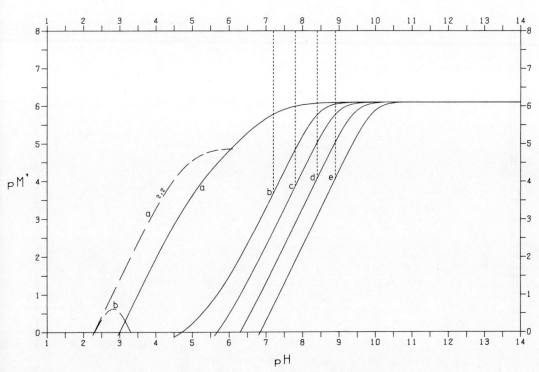

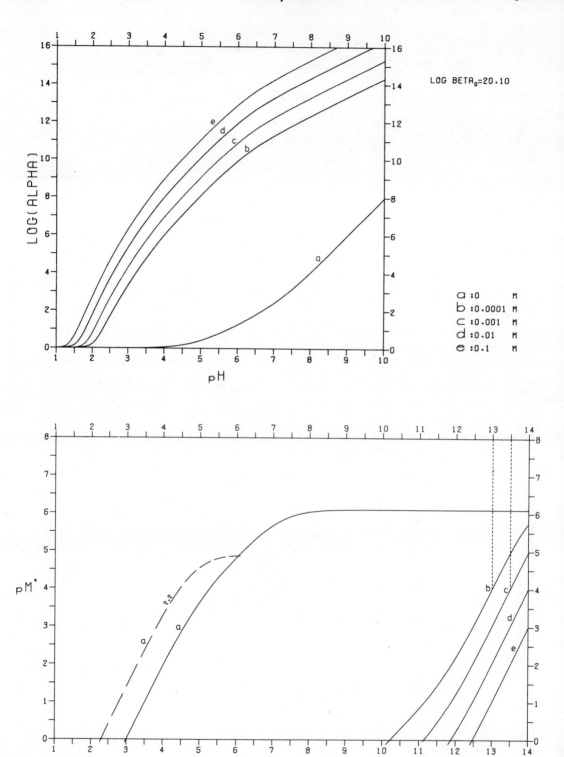

Vanadyl-EDTA

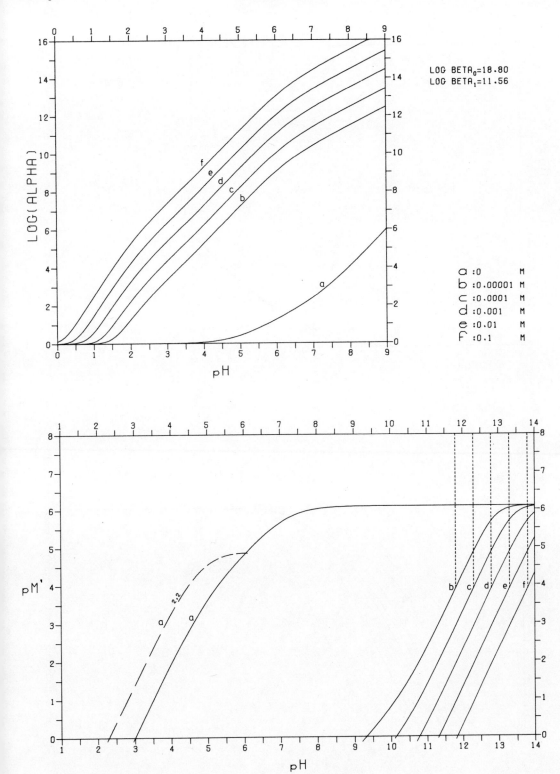

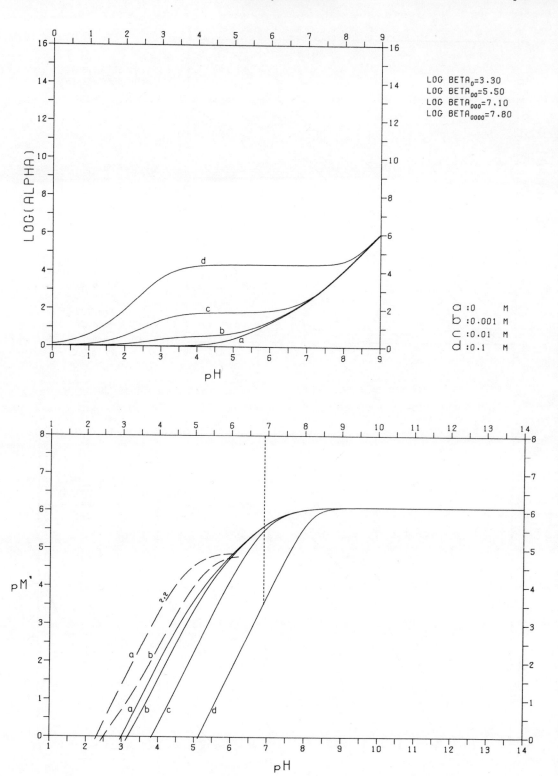

Vanadyl-(OH)-Quinolinesulphonate

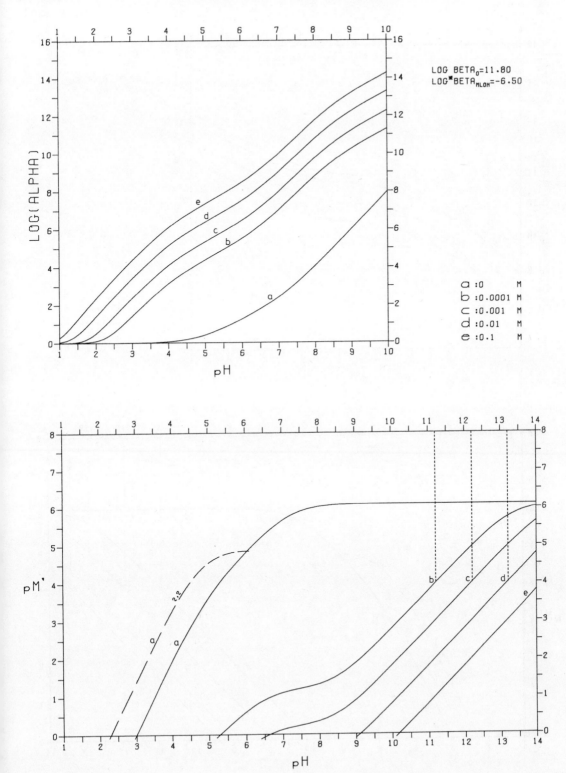

Vanadyl-Oxalate

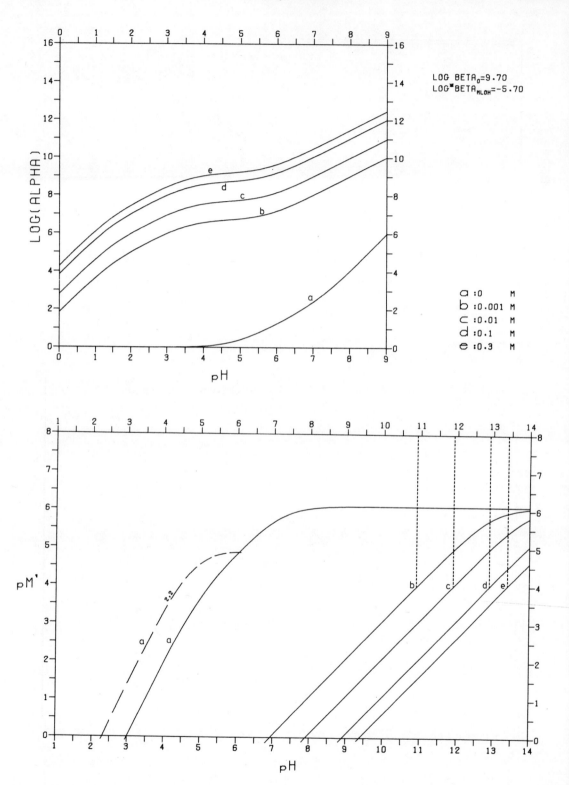

Vanadyl-1,10-Phenanthroline

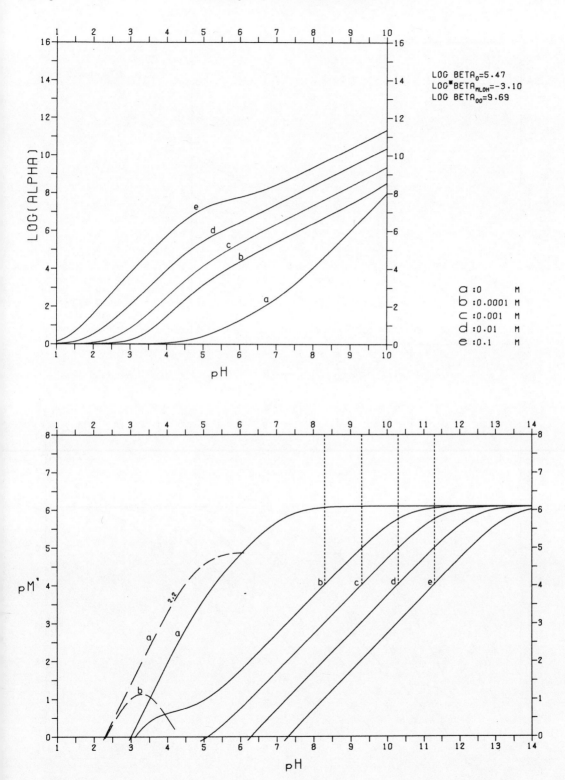

Vanadyl-Salicylate

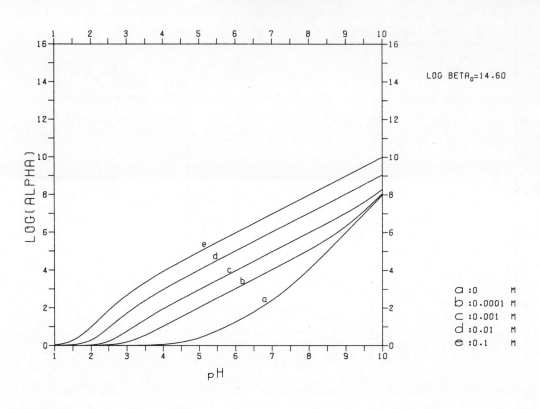

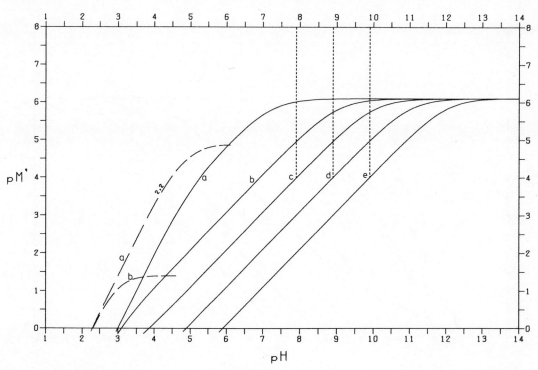

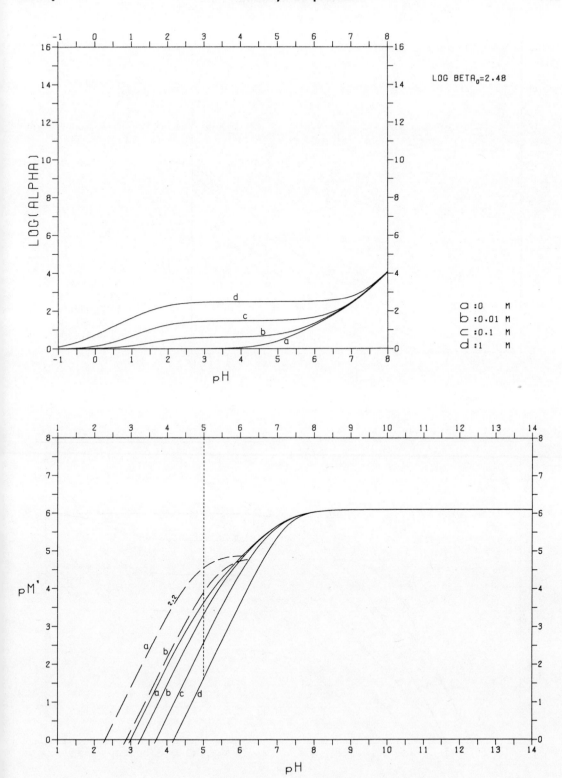

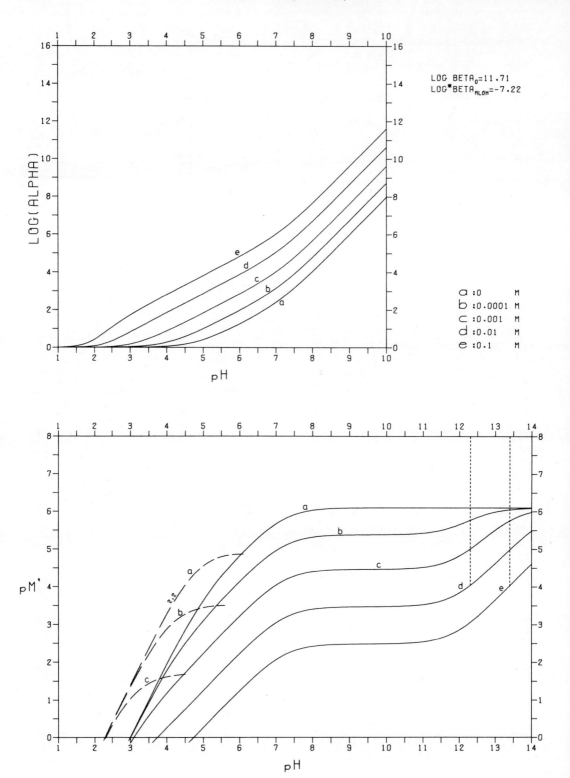

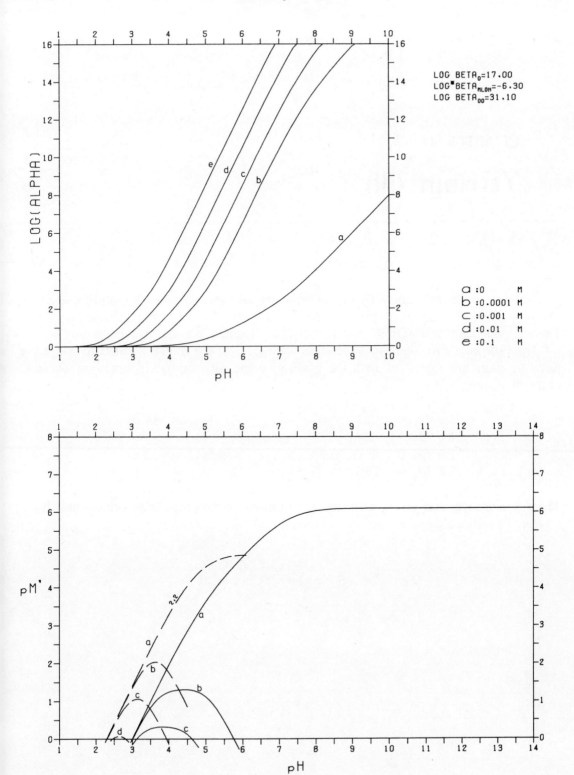

CHAPTER 43
Yttrium (III)   Y

Y^{3+} ions hydrolyse only slightly in aqueous solution before precipitation begins at pH values above 6, to form mononuclear and polynuclear species in small amounts. There is evidence that $Y_2(OH)_2^{4+}$ and $Y_3(OH)_5^{4+}$ are formed in about equal amounts.

The following data seem to be reliable estimates for the stability constants, since the resulting plots are consistent with the qualitative and quantitative information available in the literature.

$$\log {}^*\beta_1 = -\,8.1 \qquad \log {}^*\beta_{22} = -14.0$$
$$\log {}^*\beta_2 = -16.4$$
$$\log {}^*\beta_3 = -25.0 \qquad \log {}^*\beta_{53} = -32.6$$
$$\log {}^*\beta_4 = -37.0 \qquad \log {}^*K_{s0} = 19.0$$

There is evidence that in air-saturated solutions carbon dioxide is absorbed, and that a mixture of carbonate and hydroxide is precipitated.

Yttrium(III)

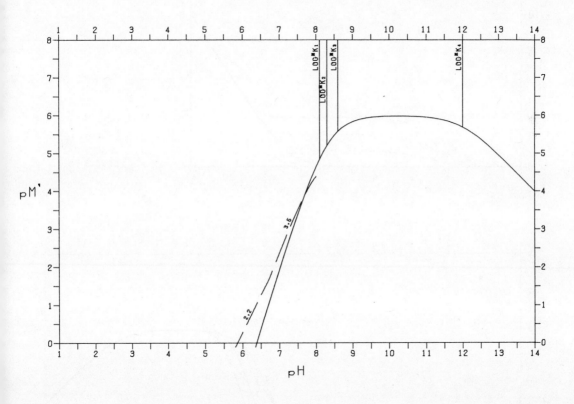

Yttrium(III)-Acetate

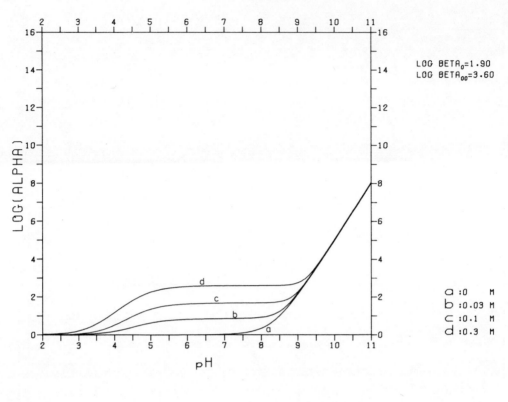

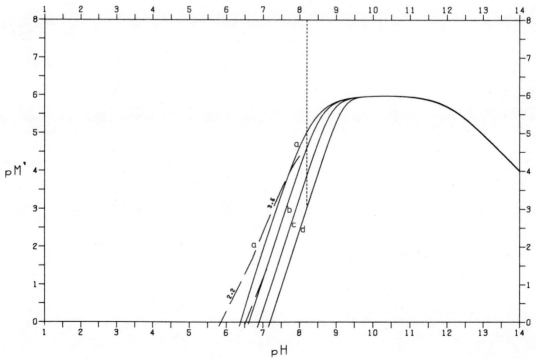

Yttrium(III)-Acetylacetone

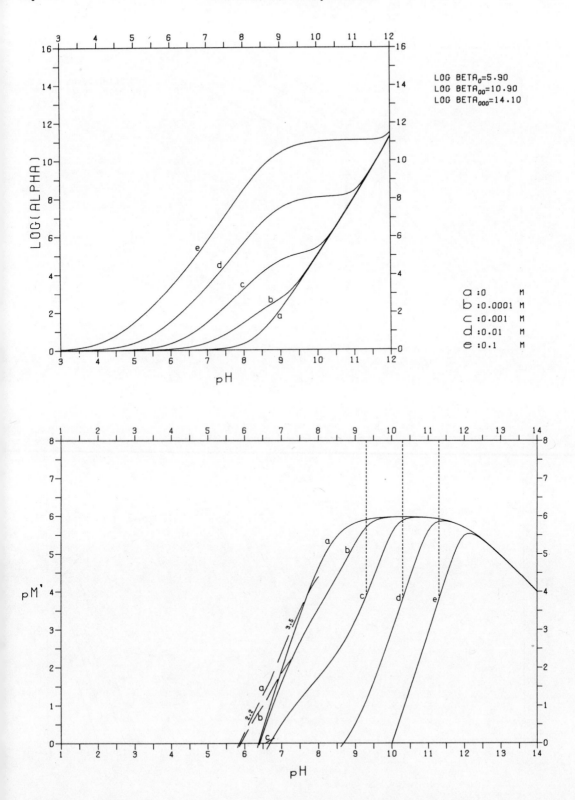

Yttrium(III)--DCTA

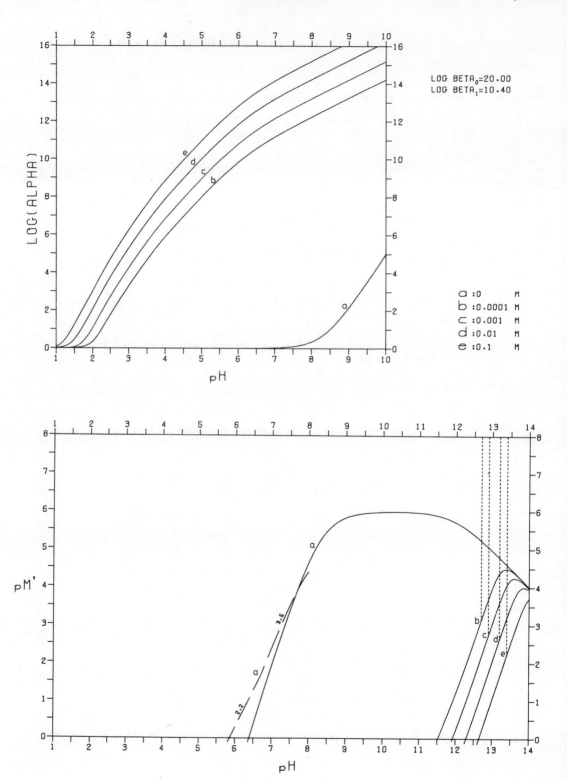

Yttrium(III)-DTPA

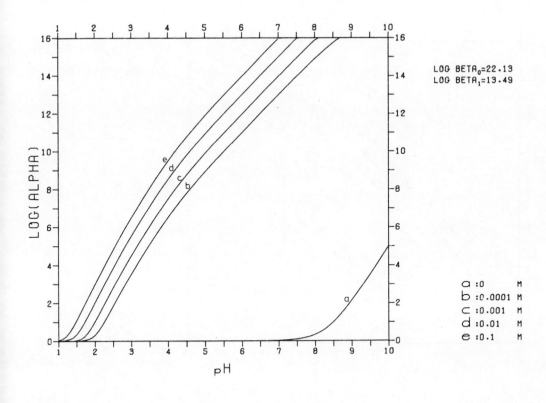

LOG BETA₀=22.13
LOG BETA₁=13.49

a : 0   M
b : 0.0001 M
c : 0.001 M
d : 0.01   M
e : 0.1   M

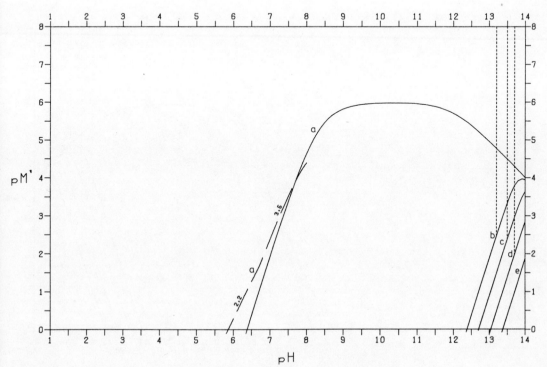

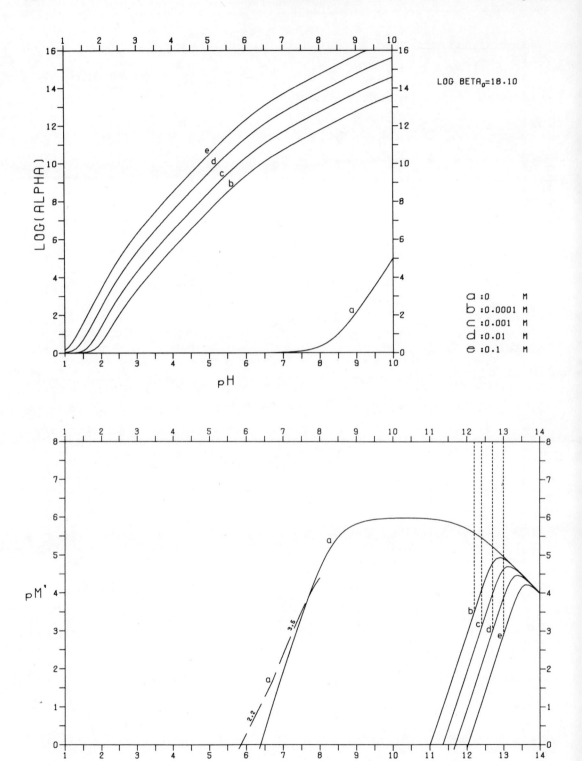

Yttrium(III)-Iminodiacetate

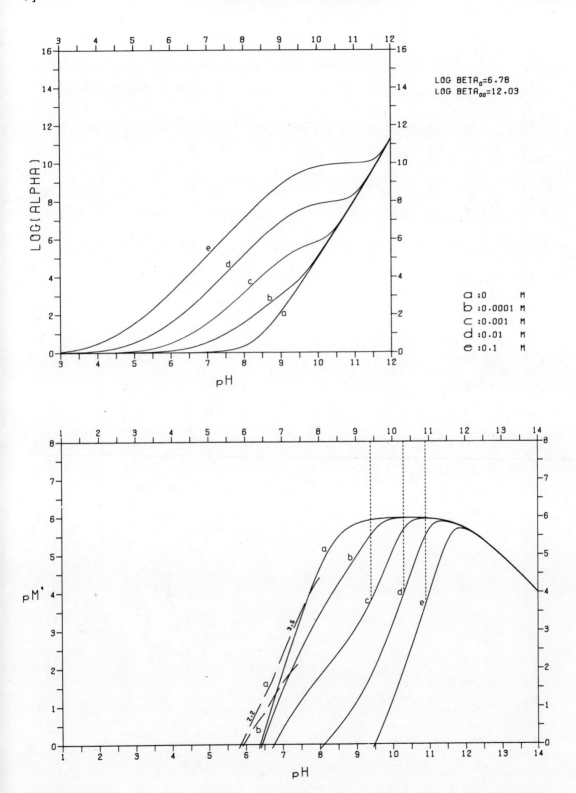

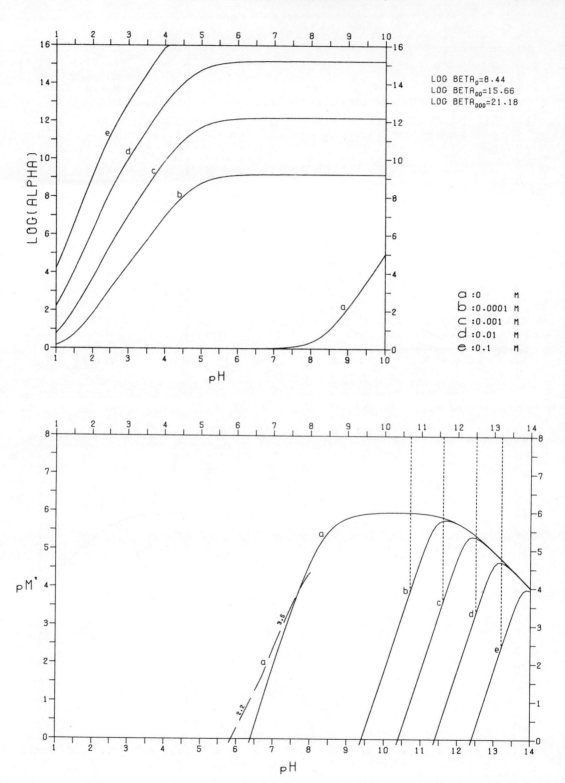

Yttrium(III)-Sulphate

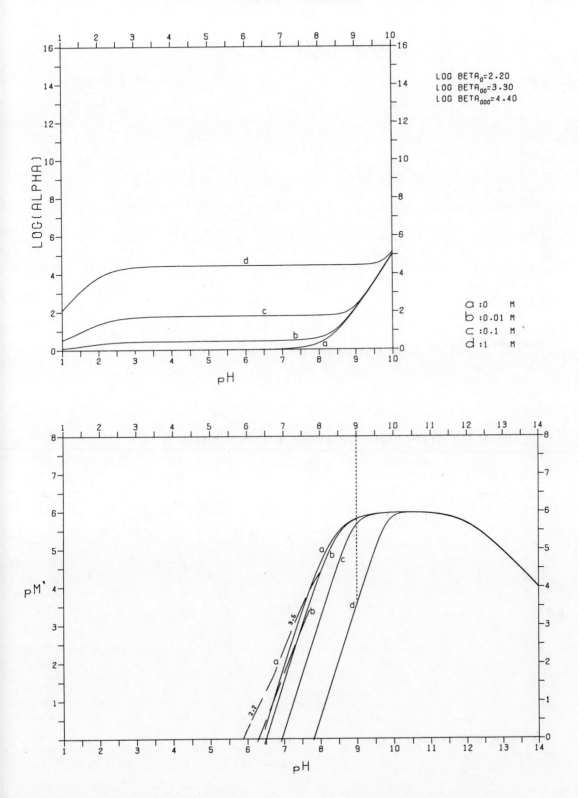

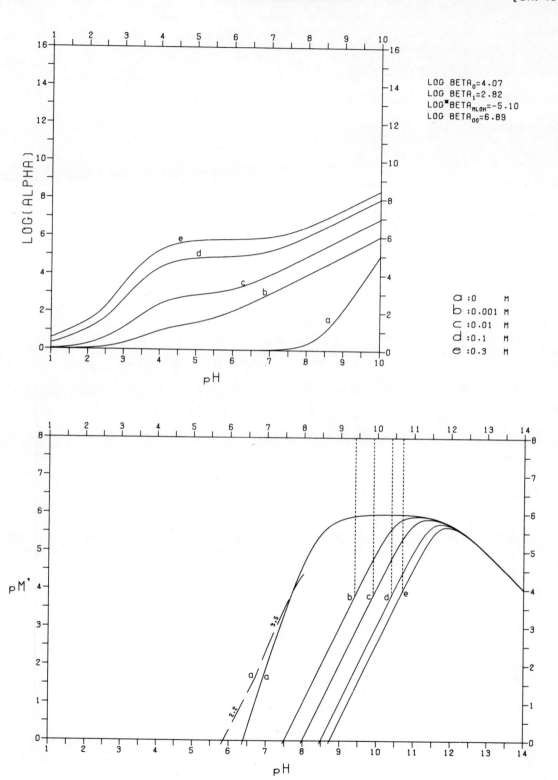

CHAPTER 44

Ytterbium (III) — Yb

Little information is available about the hydrolysis of Yb^{3+} ions, and there is no evidence for polycomplexation. The data used for the construction of the plots are estimates made by comparison with other rare earth elements; they agree with preliminary atomic-absorption experiments.

$$\log {}^*\beta_1 = -7.7 \qquad \log {}^*K_{s0} = 17.1$$
$$\log {}^*\beta_2 = -15.75$$
$$\log {}^*\beta_3 = -24.1$$
$$\log {}^*\beta_4 = -34.1$$

The value of the solubility product of $Yb_2(CO_3)_3$ means that in air-saturated solutions the solubility of the carbonate is equal to that of the hydroxide. This means that the pM′–pH plot for the carbonate system is the same as the one for the hydroxide system.

Ytterbium(III)

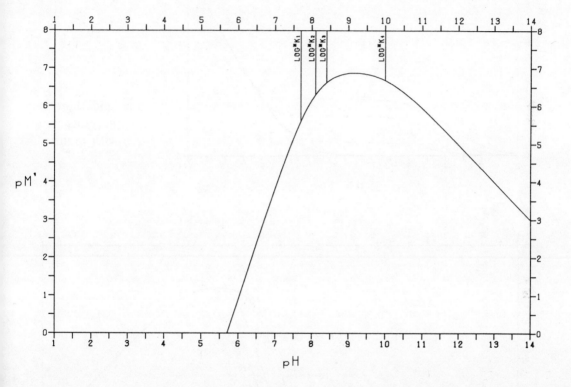

Ytterbium(III)-Acetate

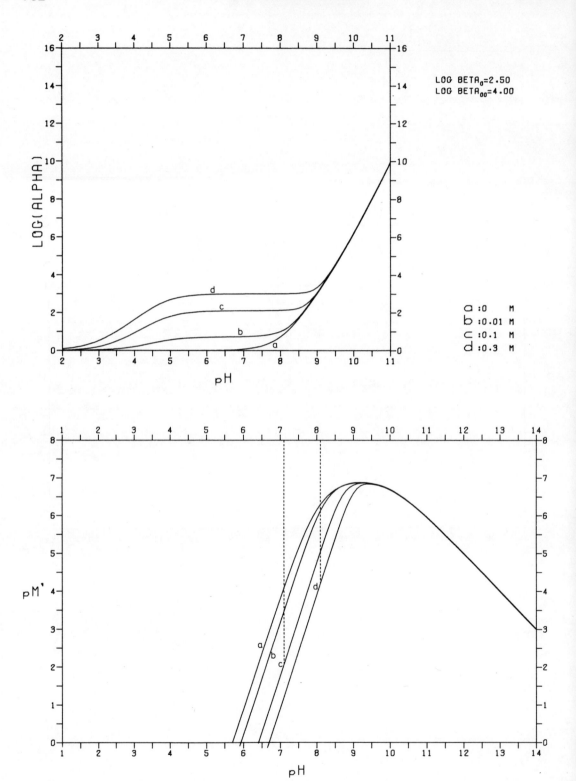

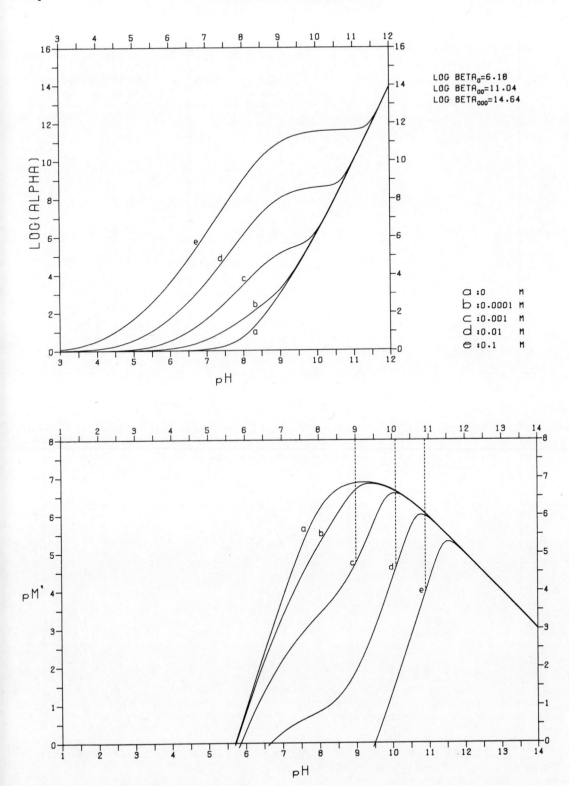

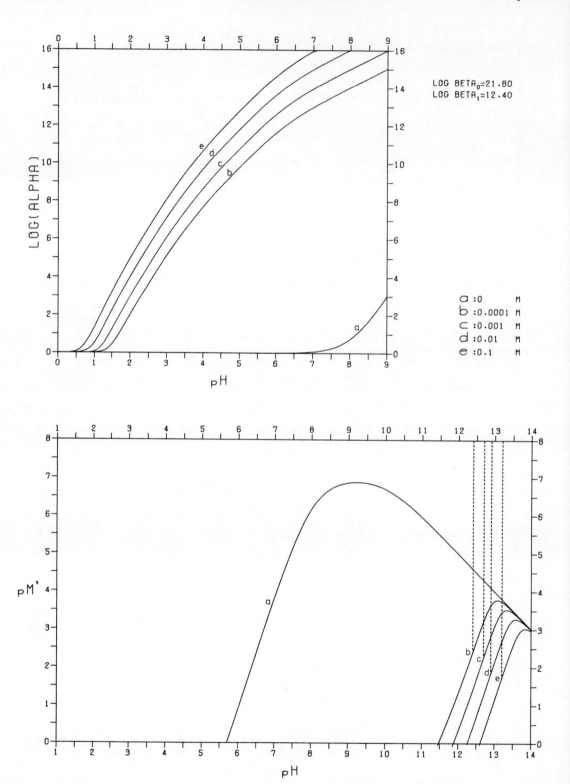

Ytterbium(III)-DTPA

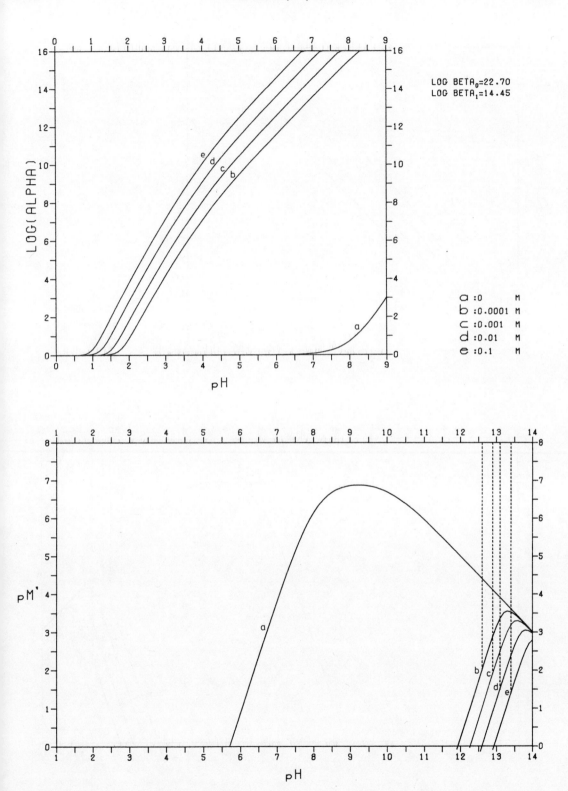

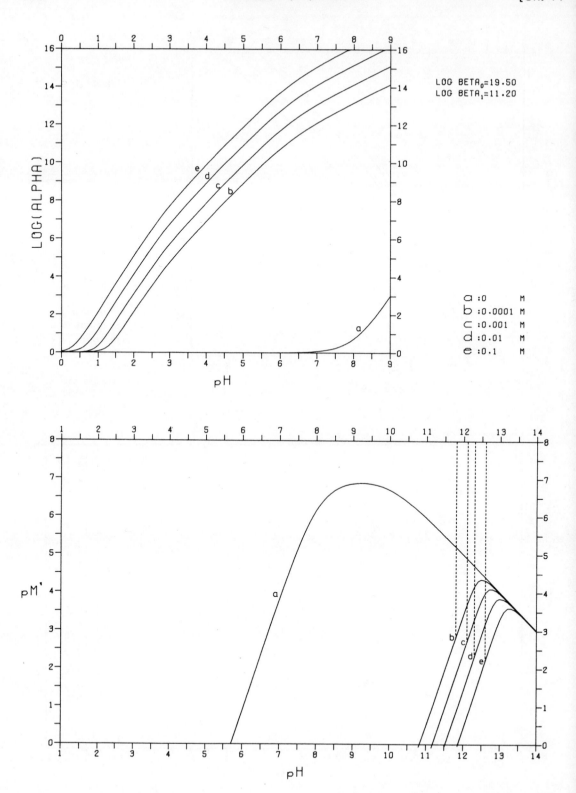

Ytterbium(III)-Iminodiacetate

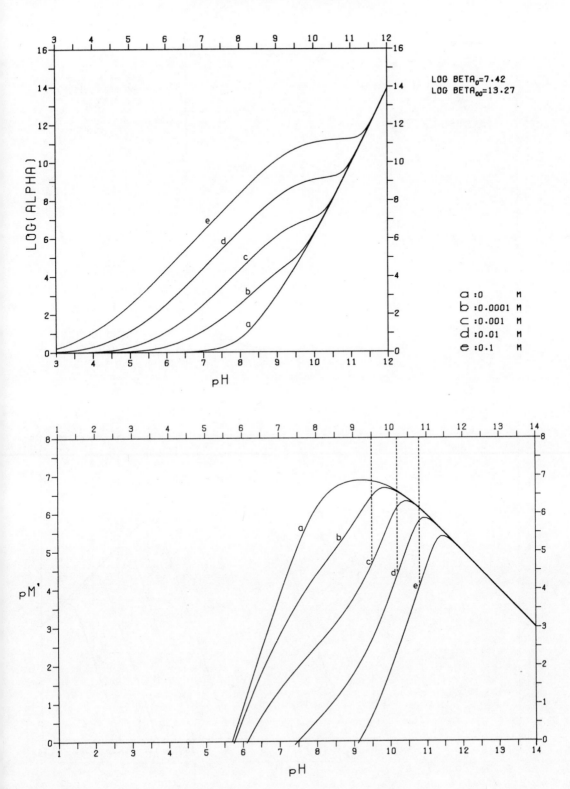

Ytterbium(III)-Pyridine-2,6-dicarboxylate [Ch. 44

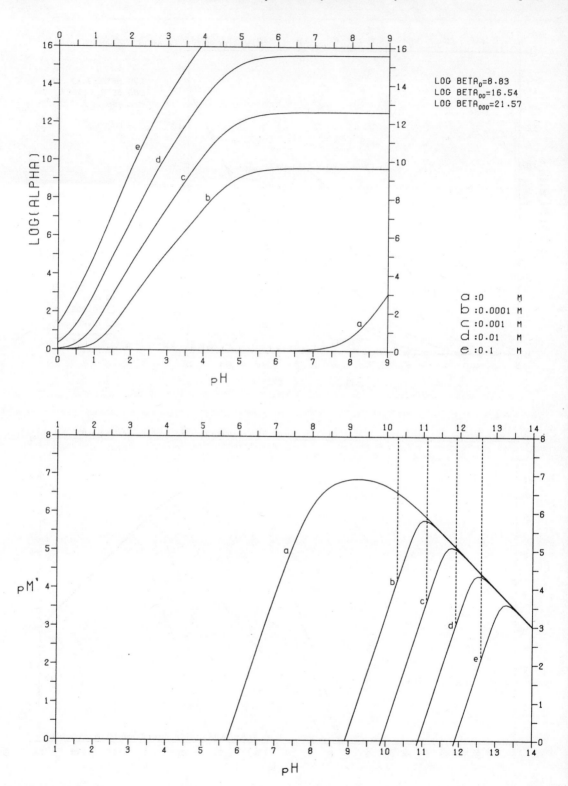

Ytterbium(III)-5-Sulphosalicylate

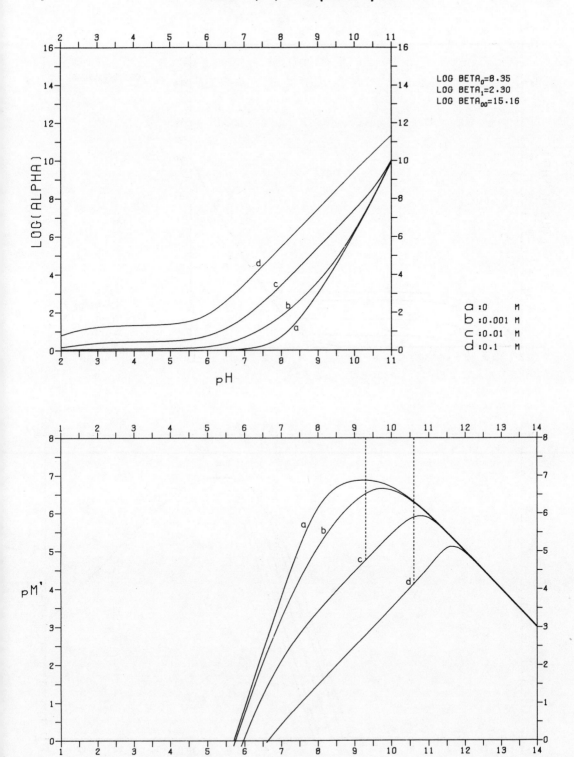

Ytterbium(III)-Tartrate

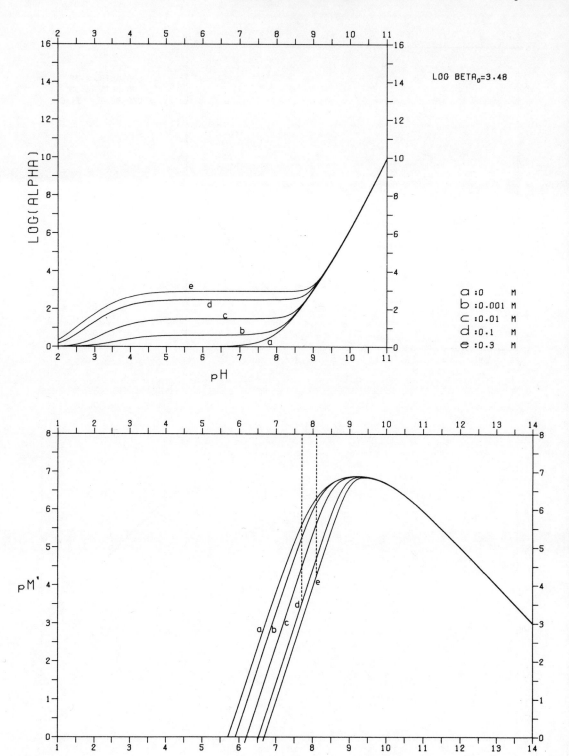

CHAPTER 4

Zinc(II)

Zinc is a bluish-grey, lustrous metal, widely used in galvanising, and in alloys such as brass. Zinc(II) is the only oxidation state in aqueous solution.

The stable oxidation state of zinc is Zn(II). Zn(OH)₂ and ZnO are amphoteric.

The stability of Zn(II) in aqueous solution arises from its filled d-shell configuration.

CHAPTER 45
Zinc (II)   Zn

Zn^{2+} ions hydrolyse only to a small extent before precipitation begins. It is generally agreed that the monomer $Zn(OH)^+$ and the dimers $Zn_2(OH)^{3+}$ and $Zn_2(OH)_6^{2-}$ are formed on both sides of the precipitation region.

The stability constants of $Zn(OH)_2(aq)$, $Zn(OH)_3^-$ and $Zn(OH)_4^{2-}$ have been determined from solubility measurements.

The solubility products of the various well-defined $Zn(OH)_2$ solid phases have been determined accurately; they differ only by a few tenths of a logarithmic unit. The active amorphous phase formed by zinc solutions on addition of a slight excess of base is ten times more soluble; it changes to a more stable form on aging for a few hours.

The following constants, which include the solubility product of the amorphous solid, were used for construction of the plots.

$$\log {}^*\beta_1 = -9.17 \qquad \log {}^*\beta_{12} = -8.9$$
$$\log {}^*\beta_2 = -17.1 \qquad \log {}^*\beta_{62} = -57.5$$
$$\log {}^*\beta_3 = -28.4$$
$$\log {}^*\beta_4 = -40.7 \qquad \log {}^*K_{s0} = 12.1$$

The difference between the solubilities of zinc carbonate and zinc hydroxide is insufficient to justify separate plots.

Zinc(II)

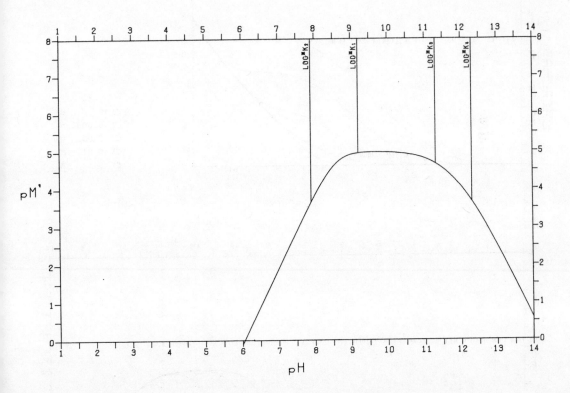

Zinc(II)-Acetylacetone

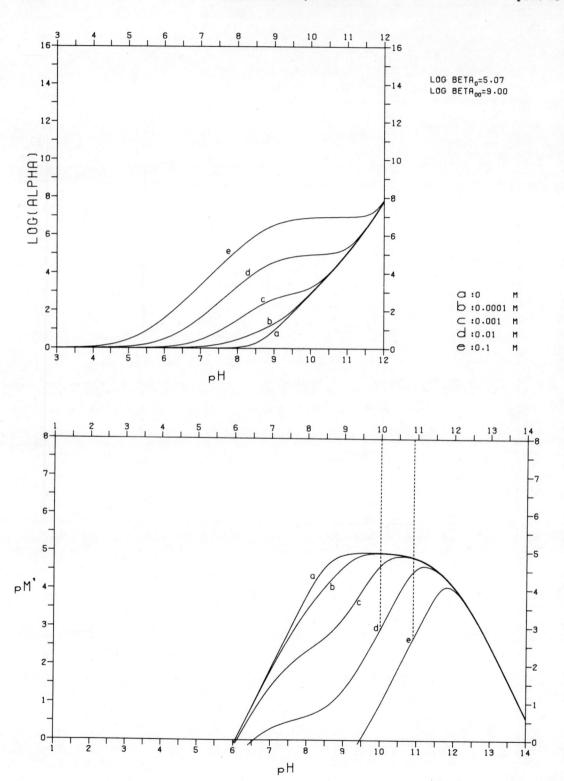

LOG BETA$_0$=5.07
LOG BETA$_{00}$=9.00

a : 0 M
b : 0.0001 M
c : 0.001 M
d : 0.01 M
e : 0.1 M

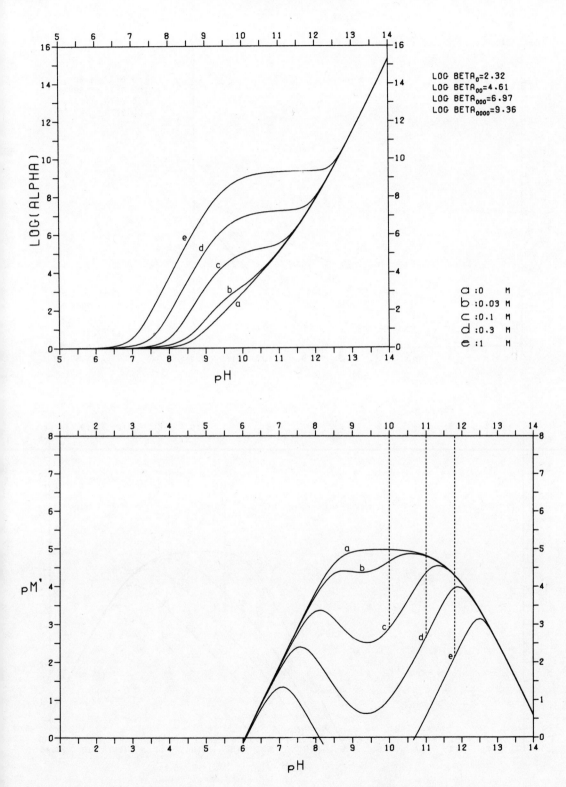

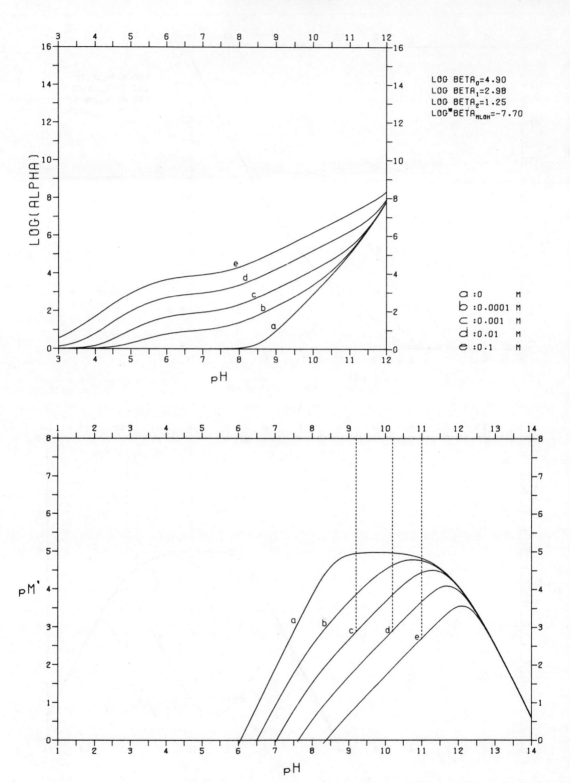

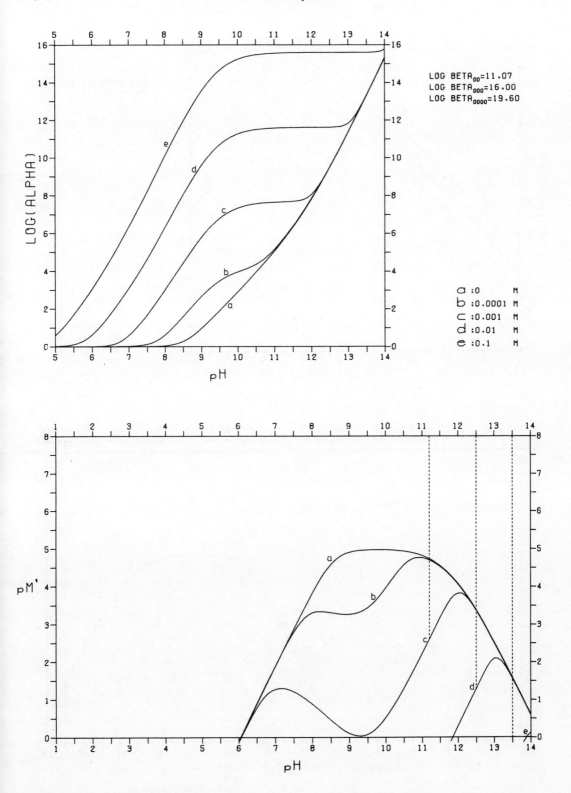

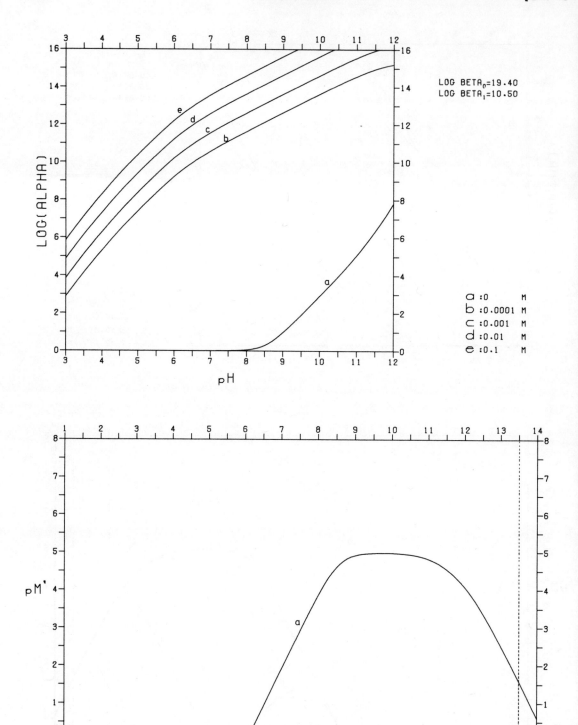

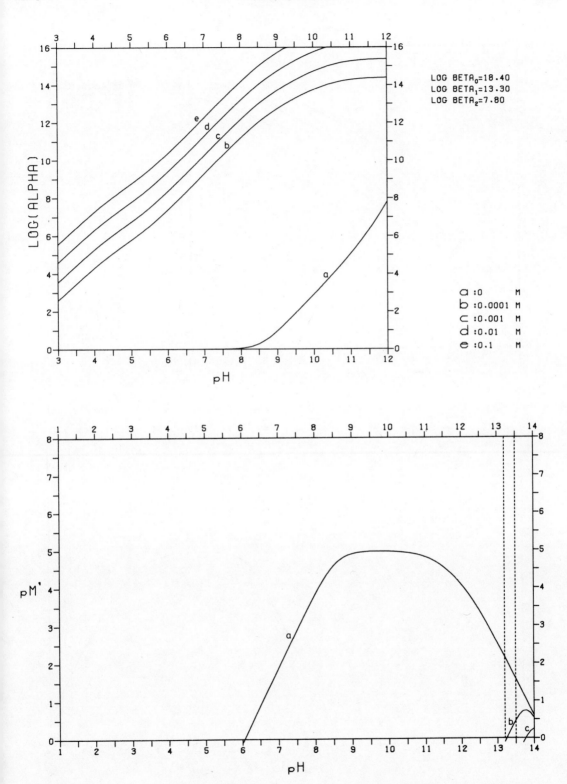

Zinc(II)-EDTA

[Ch. 45

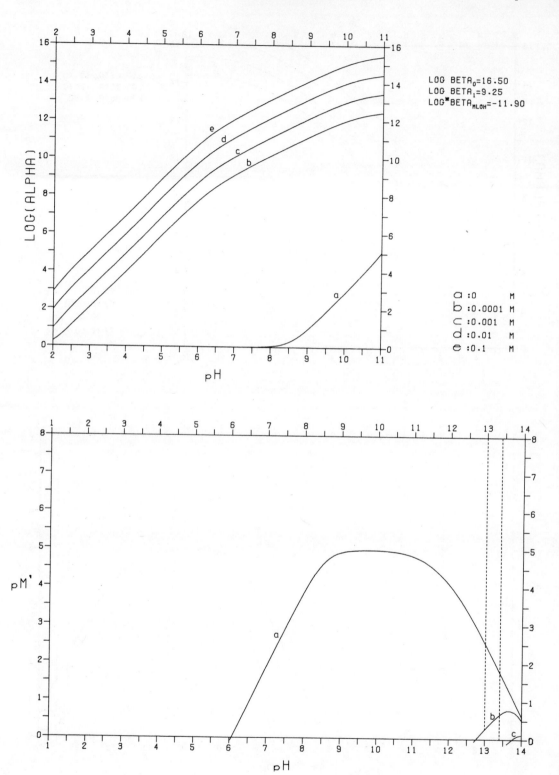

Zinc(II)-EGTA

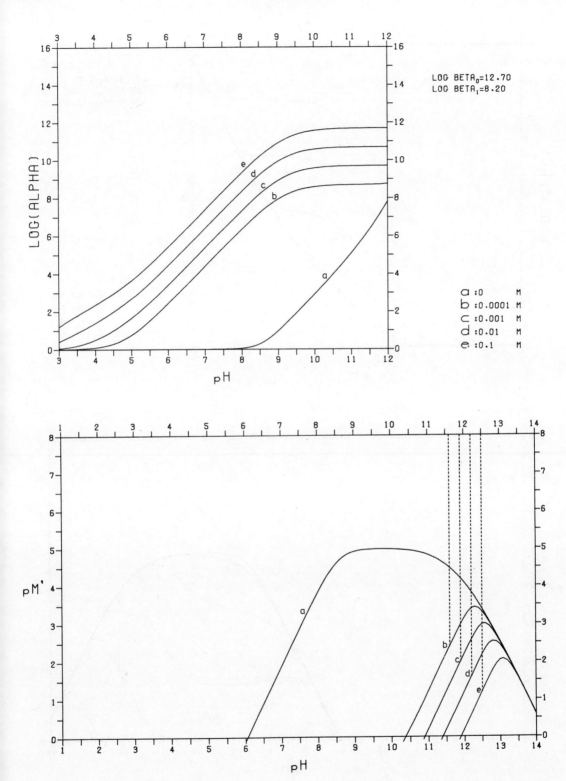

Zinc(II)-Ethylenediamine

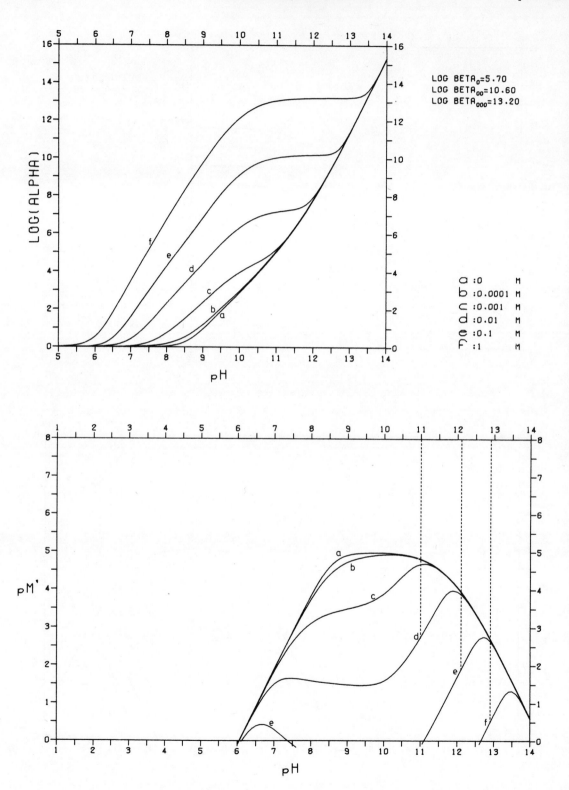

LOG BETA$_0$=5.70
LOG BETA$_{00}$=10.60
LOG BETA$_{000}$=13.20

a : 0 M
b : 0.0001 M
c : 0.001 M
d : 0.01 M
e : 0.1 M
f : 1 M

Zinc(II)-Glycine

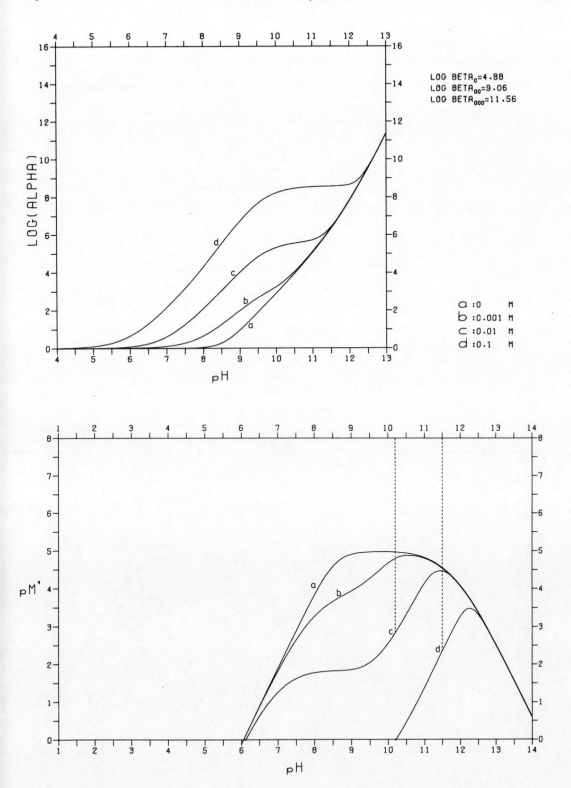

Zinc(II)-(OH)-Quinolinesulphonate

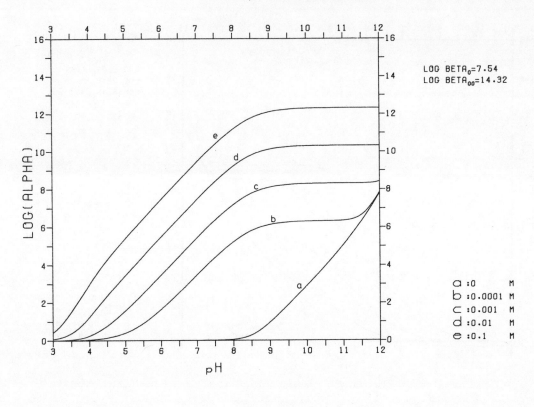

LOG BETA₀ = 7.54
LOG BETA₀₀ = 14.32

a : 0   M
b : 0.0001 M
c : 0.001   M
d : 0.01   M
e : 0.1   M

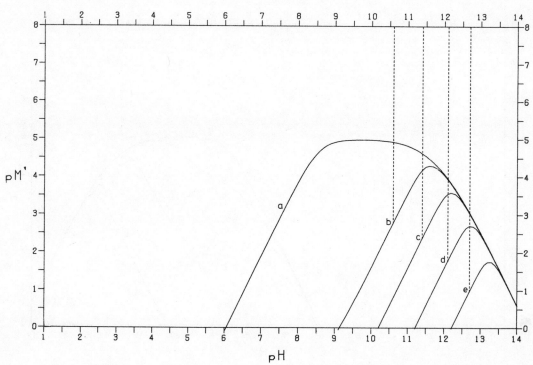

Zinc(II)-Iminodiacetate

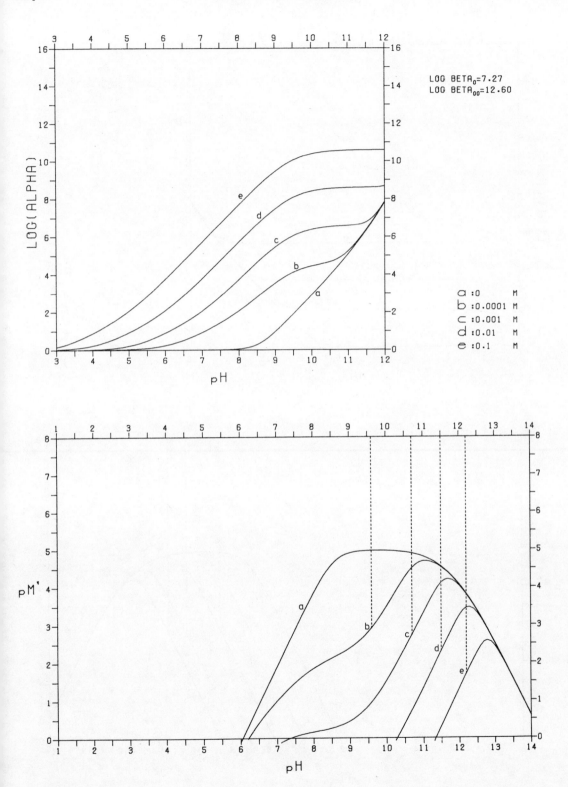

Zinc(II)-Oxalate [Ch. 45

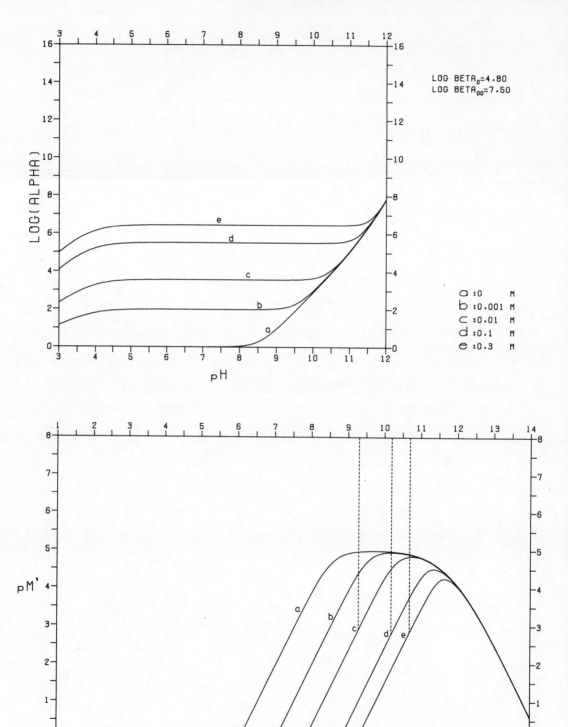

Zinc(II)-Pyridine-2,6-dicarboxylate

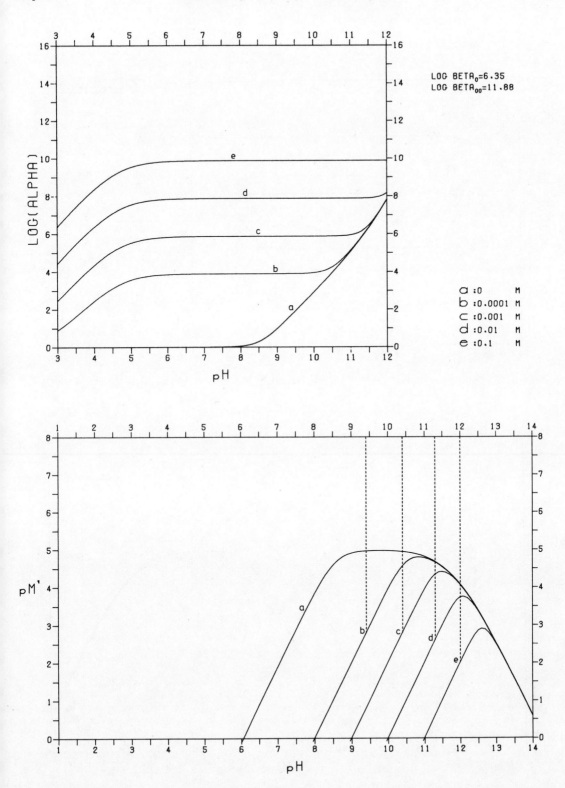

LOG BETA$_0$=6.35
LOG BETA$_{00}$=11.88

a : 0 M
b : 0.0001 M
c : 0.001 M
d : 0.01 M
e : 0.1 M

Zinc(II)-1,10-Phenanthroline

LOG BETA$_0$=6.40
LOG BETA$_{00}$=12.20
LOG BETA$_{000}$=17.10

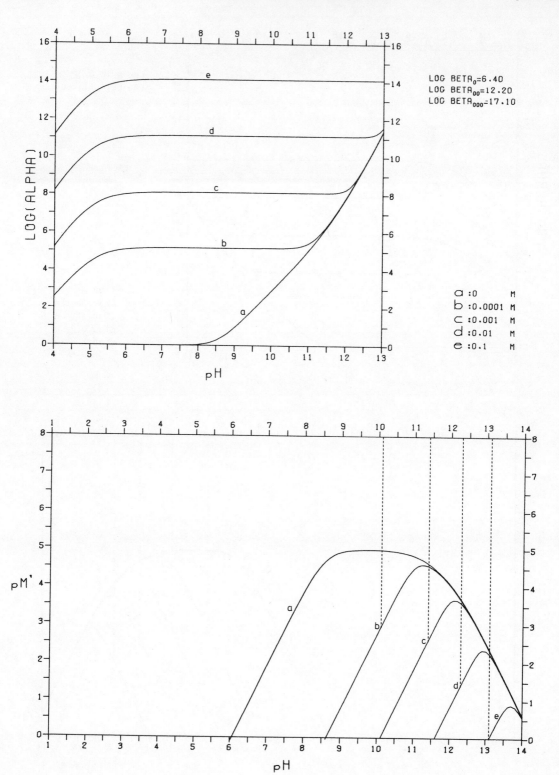

Zinc(II)-Sulphate

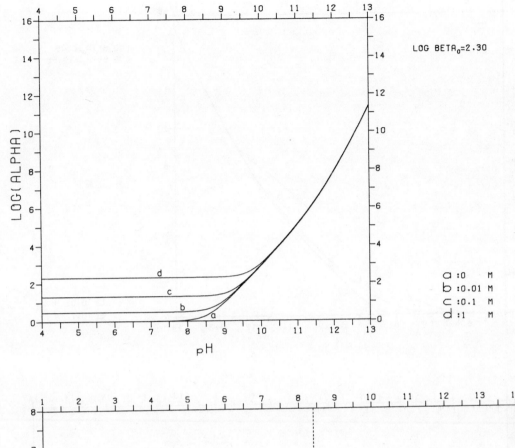

LOG BETA$_0$=2.30

a : 0 M
b : 0.01 M
c : 0.1 M
d : 1 M

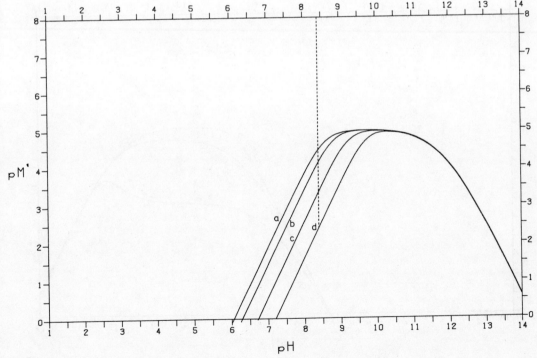

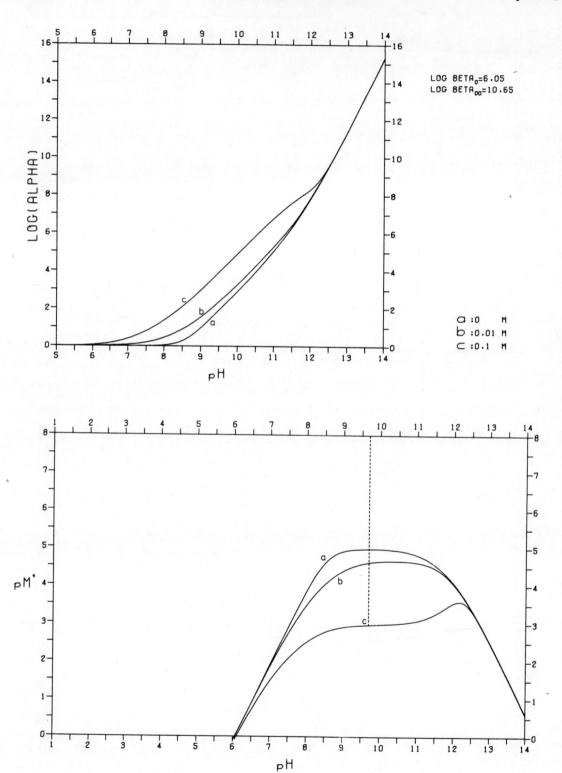

Zinc(II)-Tartrate

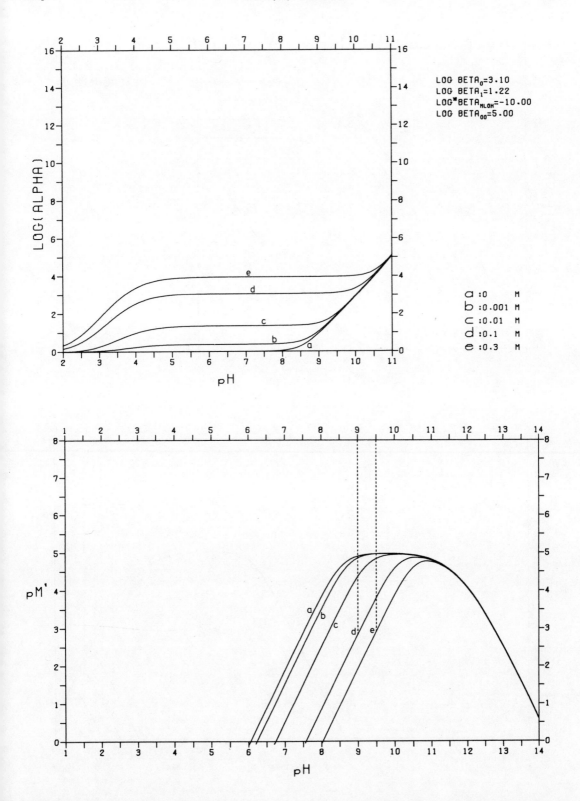

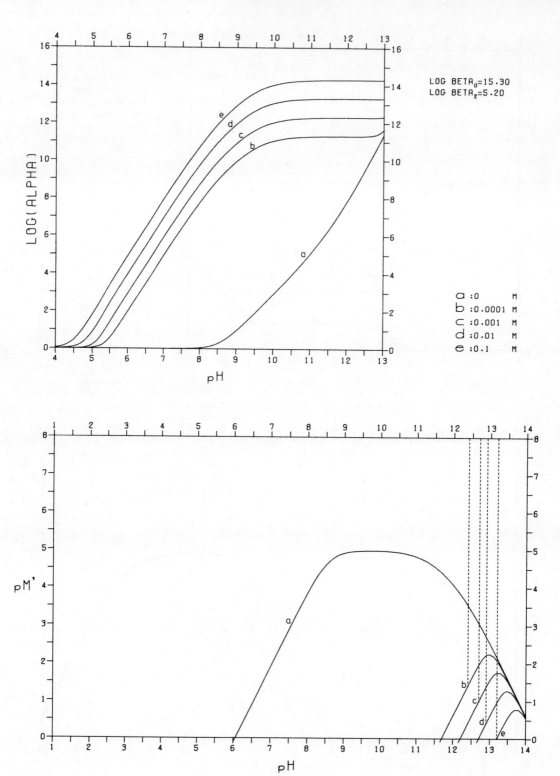

Zinc(II)-Tiron

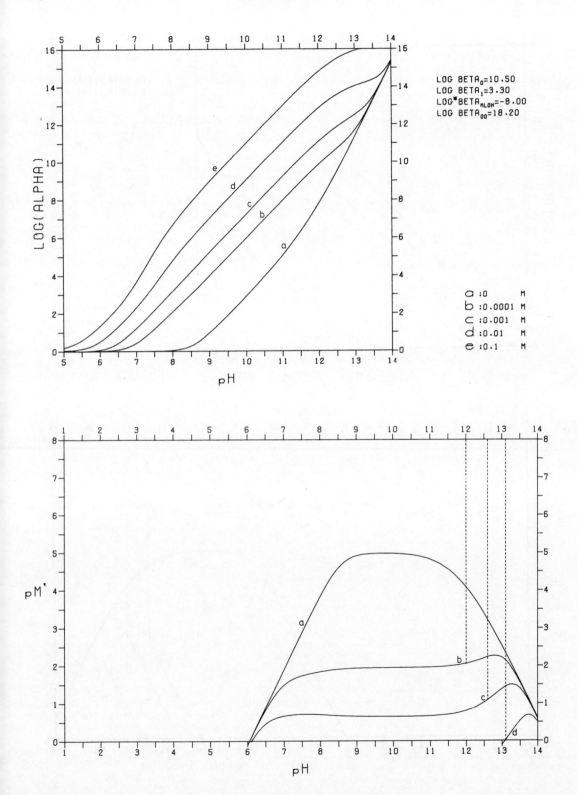

Zinc(II)-Trien

[Ch. 45

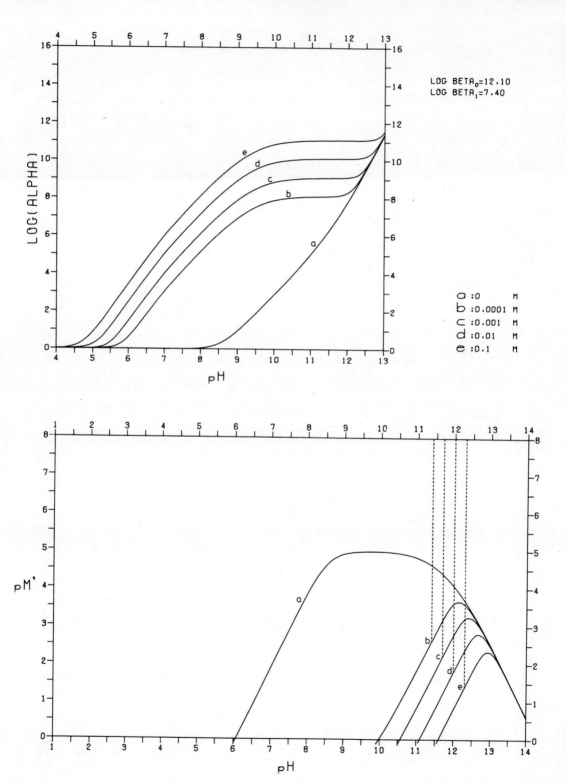

LOG BETA$_0$=12.10
LOG BETA$_1$=7.40

a : 0 M
b : 0.0001 M
c : 0.001 M
d : 0.01 M
e : 0.1 M

CHAPTER 46

Zirconium (IV)   Zr

The aqueous chemistry of Zr^{4+} ions is similar to that of Hf^{4+}, but Zr^{4+} has a greater tendency to form polycomplexes and more information is available. The stability constant of $Zr(OH)^{3+}$ has been determined with reasonable accuracy; the formation constants of the species from $Zr(OH)_2^{2+}$ to $Zr(OH)_5^-$ are rough estimates based on measured solubilities of ill-defined forms of the oxides. It is not possible to decide with certainty whether the trimeric species is $Zr_3(OH)_4^{8+}$ or $Zr_3(OH)_5^{7+}$, or whether both are formed, but there is strong evidence that $Zr_4(OH)_8^{8+}$ is formed when other anions are present.

The composition of the freshly precipitated hydrated active ZrO_2 depends on the anions present in solution, the temperature, the ionic strength and the degree of aging. The values for $*K_{s0}$ and the other constants $*\beta_i$ are estimated for ionic strength $\cong 2$, because Zr solutions are usually strongly acidic.

The data used for the plots are:

$\log *\beta_1 = -0.6$ $\log *\beta_{43} = 5.1$

$\log *\beta_2 = -2.1$ $\log *\beta_{53} = 5.5$

$\log *\beta_3 = -6.5$ $\log *\beta_{84} = 8.0$

$\log *\beta_4 = -11.2$

$\log *\beta_5 = -17$ $\log *K_{s0} = -0.4$

Zirconium(IV)

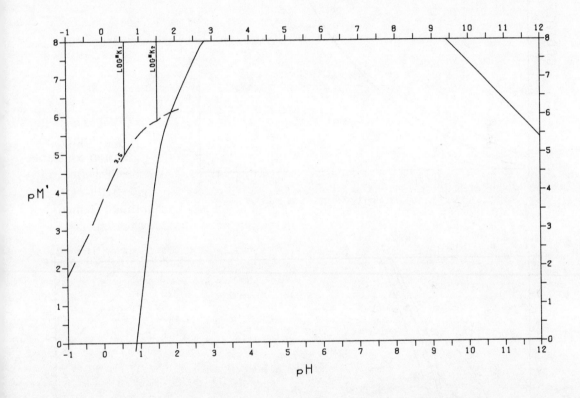

Zirconium(IV)-Acetylacetone

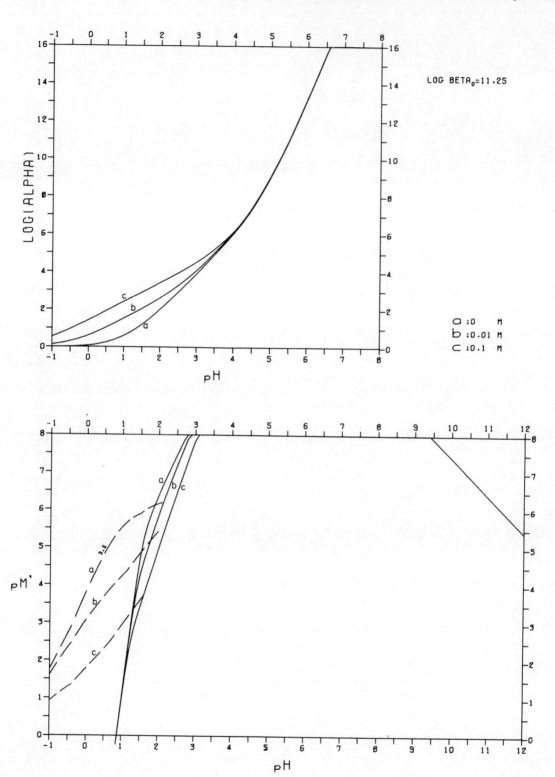

LOG BETA$_0$=11.25

a : 0   M
b : 0.01 M
c : 0.1 M

Zirconium(IV)-Citrate

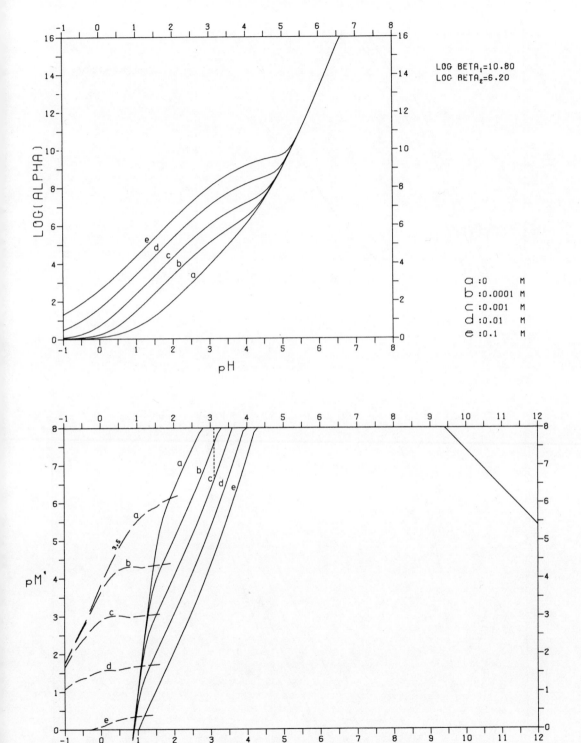

LOG BETA₁=10.80
LOG BETA₂=6.20

a : 0 M
b : 0.0001 M
c : 0.001 M
d : 0.01 M
e : 0.1 M

Zirconium(IV)–DCTA [Ch. 46

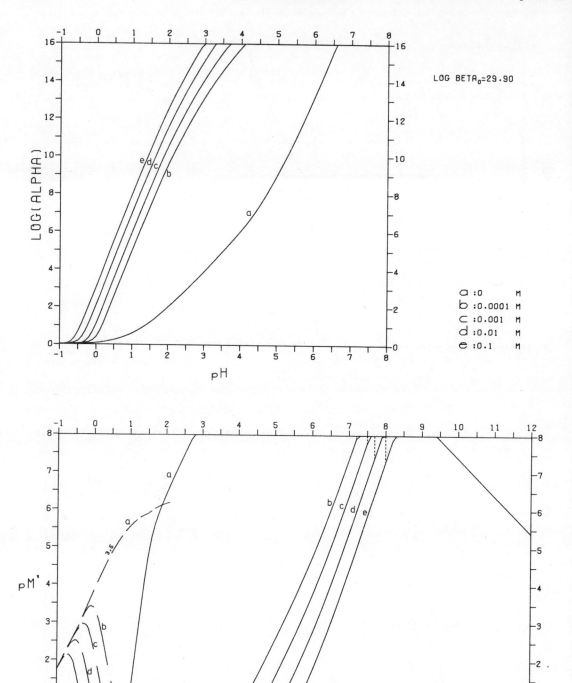

Zirconium(IV)-DTPA

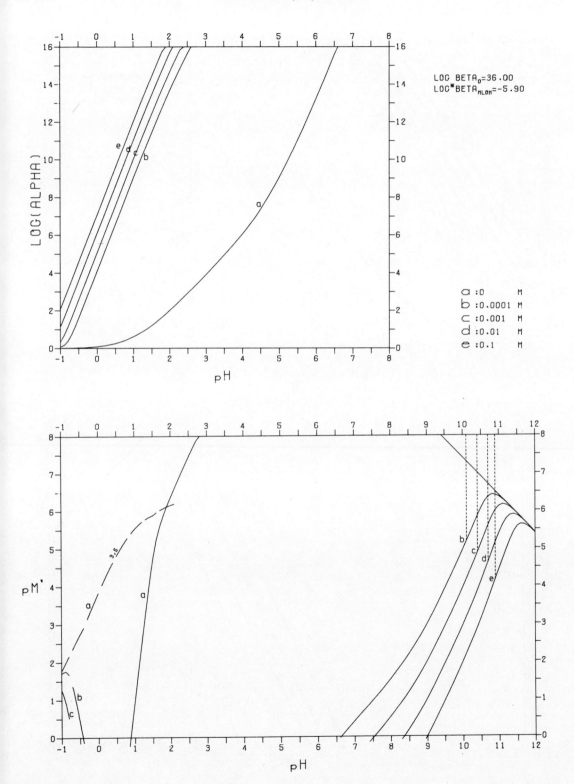

Zirconium(IV)–EDTA [Ch. 46

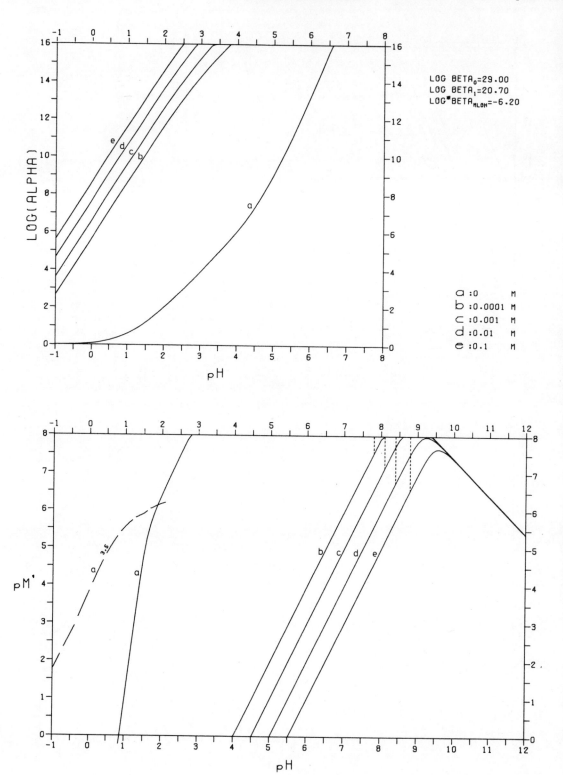

Zirconium(IV)-Fluoride

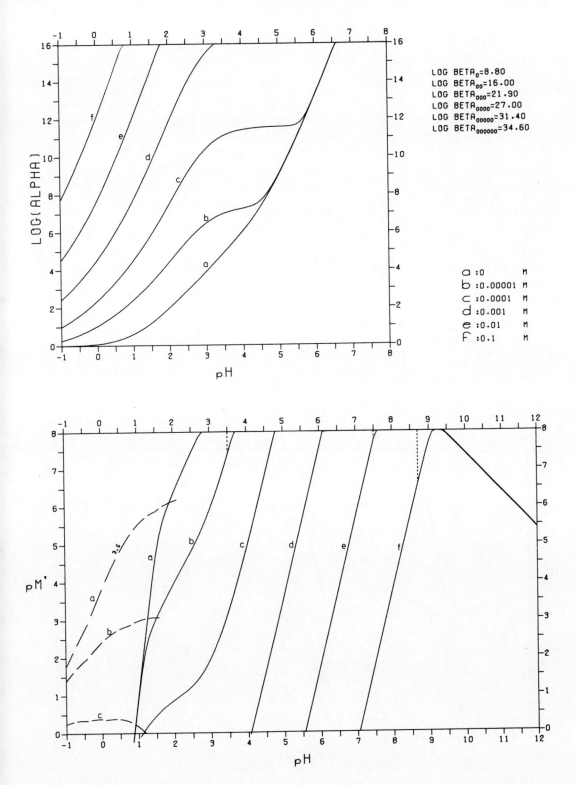

Zirconium(IV)-Oxalate

[Ch. 46

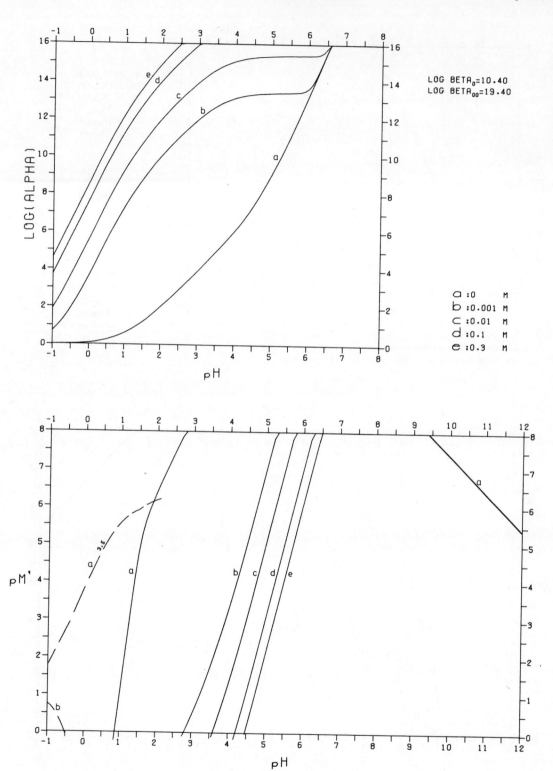

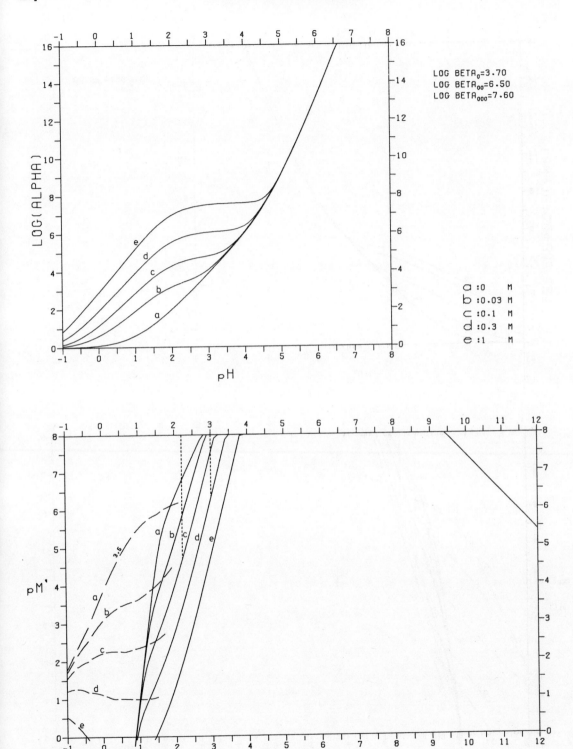

Zirconium(IV)-Tartrate [Ch. 46]

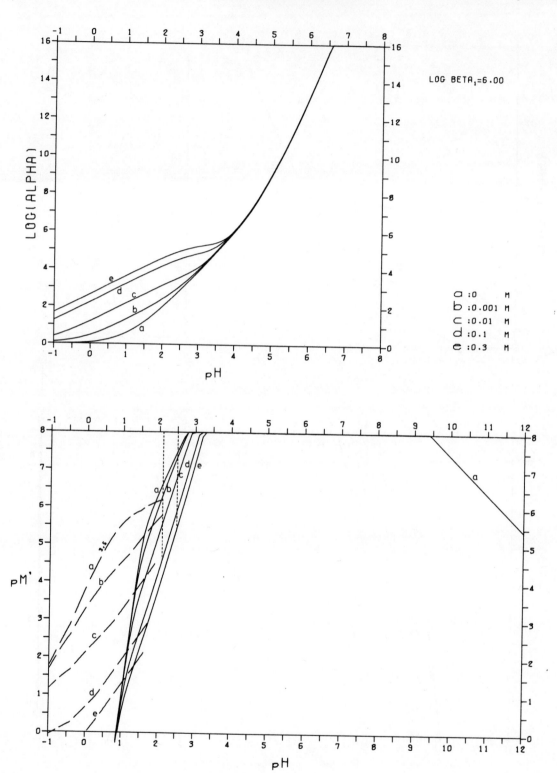

Zirconium(IV)-Tiron

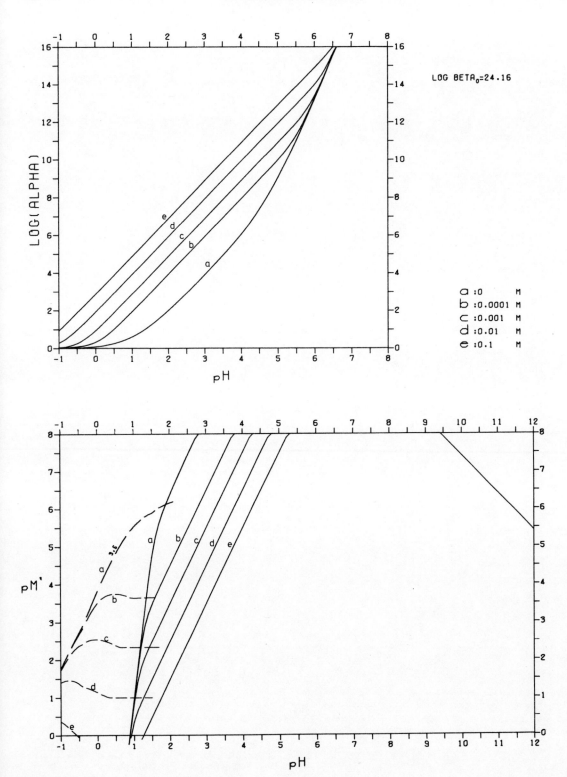

Appendix

Stability Constants for Proton-Ligand Complexes

The constants selected refer to 25°C or in some cases to 20°C, and ionic strength 0.1.

Ligand	$\log K_1$	$\log K_2$	$\log K_3$	$\log K_4$	$\log K_5$	$\log K_6$
acetate	4.70					
acetylacetone	8.82					
ammonia	9.4					
chloroacetate	2.84					
citrate	5.8	4.48	2.96			
cyanide	9.36					
DCTA	11.78	6.20	3.60	2.51	1.7	
DTPA	10.55	8.59	4.30	2.66	1.82	
EDTA	10.34	6.24	2.75	2.07		
EGTA	9.47	8.85	2.66	2.0		
ethylenediamine	10.11	7.30				
fluoride	2.90					
formate	3.75					
glycine	9.52	2.35				
hydroxylamine	6.19					
8-hydroxyquinoline-5-sulphonate	8.42	3.93	1.3			
iminodiacetate	9.44	2.60	1.76			
oxalate	3.64	1.22				
pyridine-2, 6-dicarboxylate	4.68	2.10				
1, 10-phenanthroline	4.93	1.9				
salicylate	13.6	3.0				
sulphate	1.8	−1.0				
sulphosalicylate	11.87	2.47				
tartrate	3.96	2.80				
tetren	9.85	9.28	8.19	4.80	3.06	
tiron	12.6	7.66				
trien	9.8	9.1	6.7	3.3		

Tabular Index

Chapter	element	leading plot	acetate	acetylacetone	ammonia	bromide	chloroacetate	chloride	citrate	cyanide	DCTA	DTPA	EDTA	EGTA	ethylenediamine	fluoride	formate	glycine	hydroxylamine	8-OH-quinolinesulphonate	iminodiacetate	oxalate	pyridine-2,6-dicarboxylate	phenanthroline	salicylate	sulphate	sulphosalicylate	tartrate	tetren	tiron	trien
2	Ag	41	W	—	42	P	W	—	P	P	—	43	44	45	W	46d	W	—	47	48	—	—	—	—	49	—	W	—	W	50	—	51
3	Al	53	—	54	—	—	—	—	55	—	56	57	58	W	—	59	60	—	—	—	—	—	61	62	63	—	64	65	66	67	68	—
4	*Ba	71	W	—	—	—	—	W	72	—	73	74	75	76	P	P	W	W	—	—	—	—	77	—	—	W	P	W	—	78	—
5	Be	81	—	82	—	—	—	W	83	—	84	—	85	—	86	—	—	—	—	—	—	87	—	—	88	89	90	91	—	92	—
6	Bi	95	—	—	96p	—	97p	U	—	—	98	99	100	—	101	—	—	—	—	—	—	—	—	—	—	—	—	102	—	—	—
7	*Ca	105	W	—	—	—	W	—	106	—	107	108	109	110	P	—	P	W	111	—	—	P	112	W	—	P	—	W	—	113	—
8	Cd	115	W	116	117	118	W	119	120	121	122	123	124	125	126	W	W	W	127	128	129	130	131	132	W	133	U	W	134	135	136
	*Cd	137	W	138	139	140	W	141	142	143	144	145	146	147	148	W	W	149	—	150	151	152	153	154	W	155	U	U	156	157	158
9	Ce	161	W	162	—	—	W	—	P	—	163	164	165	U	—	P	W	W	166	167	—	P	168	—	W	W	169	170d	—	W	—
10	Co	173	W	174	175	W	W	176	—	—	177	178	179	D	180	W	—	181	—	182	183	184	185	186	W	187	188	189	190	191	192d
	*Co	193	W	194	195	W	W	196	U	U	197	198	199	D	200	W	—	201	—	202	203	204	205	206	W	207	208	209	210	211	212d
11	Cr	215	216	—	U	—	W	217	—	U	—	—	218	—	219	—	—	220	—	—	—	227	—	—	W	—	—	—	—	—	—
12	Cu	223	W	224	225	W	W	226d	227u	228	229	230	D	231	W	232	233	234	235	236	237	238	239	W	240	—	241d	242	243	244
	*Cu	245																													
13	Dy	247	248	249	—	—	W	P	P	—	250	251	252	U	P	—	W	—	253	—	—	P	254	—	—	—	W	255	256	—	—
14	Er	259	260	261	—	W	W	P	P	—	262	263	264	U	P	—	—	—	—	265	266	267p	268	—	W	—	W	269	270	—	—
15	Eu	273	274	275	—	W	W	P	P	—	276	277d	278	U	P	—	W	—	—	—	—	279	280p	281	—	W	—	W	282	283	—	—
16	Fe(II)	285	W	286	W	—	W	W	287	288	289	290	291	D	292	P	W	293	—	294	—	295	296	297	298	W	299	300	301	—	302d
	*Fe(II)	303	W	304	W	—	W	W	305	306	307	308	309	D	310	W	—	311	—	312	—	313	314	315	316	W	317	318	319	—	320d
17	Fe(III)	323	W	324	—	W	W	W	325	326p	327d	328	329	—	—	330	—	331	—	332	—	333	334	335	336d	337	338	339	340d	341	342
18	Ga	345	—	346	—	—	—	—	—	—	347	348	349	—	—	—	W	—	—	—	—	—	351	—	352	353	354	355	356u	—	Wu
19	Gd	359	360	361	—	—	W	P	P	—	362	363	364	U	P	—	W	—	—	365	366	367p	368	—	W	369	W	370	—	—	—
20	Hf	373	—	U	—	—	—	W	374u	—	—	375	376	—	—	377	—	—	—	—	—	378	—	—	—	379	—	380	—	381	—

#	Elem																																
21	Hg	383	384	–	385	386p	–	387p	388	389p	390	391	392	393	394 W	–	395	–	396u W	397 P	–	W	–	U	398	–	399
22	In	401	402	403	–	404	–	–	405	406	–	407	408	409	–	–	–	410 411	–	–	–	412	–	413	–	414	–	415	–	–
23	La	417 W	418	–	–	–	–	419	420	421	–	–	–	–	–	–	–	–	–	422	423	424	–	–	P	–	U W	425d	–	–
	*La	426 W	427	–	–	–	–	428	429	430	–	–	–	–	–	–	–	–	–	–	431	432 P	433	–	–	U W	434d	–	–
24	Mg	437 W	438 W	–	439	–	440	441	442	443 W	P W	444	–	445	446	447	448 W	–	–	W	–	449	–	W	–	450
25	Mn	453 W	454	455 W	–	456	457	458d	459	460d	461	–	W	462 W	–	463	–	464	465	466	467	468	469	470	471	472	473
	*Mn	474 W	475	476 W	–	477	478	479d	480	481d	482	–	W	483 W	–	484	–	485	486	487	488	489	490	491	492	493	494
26	Nd	497	498	499	–	–	–	–	500	501	502	U	–	P	W	–	503	504	P	505	–	–	–	–	W	W	506	507	–	–
27	Ni	509 W	510	511 W	–	–	W	512	513	–	W	514	515d	516	517d	518 W	–	519	520	521	522	523	524	525	526	527	528	529	530	531	532d
28	Pb	535	536 U	–	537 W	538p	539u	–	540	541d	542	543d	544	P W	545	546	547	548 W	–	549	550	–	–	–	P	–	551	–	552	553	554
	*Pb	555	556 U	–	557 W	558p	589u	–	560	561d	562	563d	564	P W	565	566	567	568 W	–	569	570	–	–	–	P	–	571	572	573	574
29	Pd	577	–	578	579	580p	–	581	–	582	–	583	–	584	–	585
30	Sc	587	–	588	–	–	–	Wp 589	590	591	592 W	U	–	–	–	593p	U	–	594	595	596	–	597	598
31	Sm	601	602	603	–	–	–	–	604	605d	606	U	–	–	–	–	–	607	608	–	609	–	–	W	610	611
32	Sn(II)	613	–	–	–	614	–	615	–	616	617	618	619	–	620	–	–	–	–	–	621
33	Sn(IV)	623	–	–	–	–	–	–	–	624	–	625
34	*Sr	627 W	–	–	–	–	–	–	628	–	629	630	631	632 W	–	P W	–	–	633 P	–	634	–	W	–	535
35	Tb	637	638	639	–	–	–	–	–	640	641	642 U	–	W	–	–	–	643	–	644	–	Wu	645	646
36	Th	649	650	651	–	–	652 W	653	654d	655	656	–	P	–	–	657	–	P	–	–	–	W	658	659	D	–	D
37	Ti(OH)$_2^{2+}$	661	–	–	W	–	–	–	–	662	–	663	–	–	–	–	664	U	–	665 Wu	666	667	–	668
38	Tl	671	672	673	–	W	–	674	–	675	676	677	–	–	–	–	–	–	–	–	U	–	678	–	W	679 W	680
39	U(IV)	683	–	Wu	–	W	–	684	–	685d	–	686d	–	–	–	–	–	–	–	–	–	–	–	–	–	–	687
40	UO$_2^{2+}$	689	690	691	–	–	–	W	W	–	–	–	D	D	–	P	–	692 W	–	–	–	P	–	693	694	P	–	695	696	697	UD	–	698d
41	V(III)	701	–	–	D	W	–	–	–	–	702
42	VO^{2+}	705	–	706	–	–	–	W	707d	–	708	–	709	–	W	710	–	–	–	–	–	–	711d	–	712	–	713	–	714	715	716d	–	717d
43	Y	719	720	721	–	–	W	P	–	722	723	724	U	–	W	–	–	–	725	726	–	727	–	729d
44	Yb	731	732	733	–	–	W	P	–	734	735	736	U	–	P	–	–	–	737	738	–	–	W	739	740
45	Zn	743 W	744	745 W	–	W	–	746	W	747	748	749d	750	751d	752 W	753	754 W	755	756	757	758	–	W	759	760	761p	762	763	764
46	Zr	767	–	768	–	–	–	W	769	–	770	771	772d	–	773	–	774	–	–	–	–	–	775	U	776u	–	777

Bold numerals denote chapter numbers
Other numerals denote page numbers

P : precipitate formation W : weak complexes
D . binuclear complexes p : precipitate at higher concentrations
U, u: unreliable data in literature d : binuclear complexes at higher concentration
  — : no data found
* . refers to diagrams with carbonate precipitation under air-saturated conditions instead of hydroxide precipitation